10th Anglo-French Physical Acoustics Conference

(AFPAC 2011)

Journal of Physics: Conference Series Volume 353

Villa Clythia, Frejus, France
19-21 January 2011

ISBN: 978-1-62276-770-0
ISSN: 1742-6588

Printed from e-media with permission by:

Curran Associates, Inc.
57 Morehouse Lane
Red Hook, NY 12571

Some format issues inherent in the e-media version may also appear in this print version.

Printed by Curran Associates, Inc. (2013)

Institute of Physics
Dirac House, Temple Back
Bristol BS1 6BE UK

Phone: 44 1 17 929 7481
Fax: 44 1 17 920 0979

techtracking@iop.org

Additional copies of this publication are available from:

Curran Associates, Inc.
57 Morehouse Lane
Red Hook, NY 12571 USA
Phone: 845-758-0400
Fax: 845-758-2634
Email: curran@proceedings.com
Web: www.proceedings.com

TABLE OF CONTENTS

Journal of Physics: Conference Series **353** (2012) 011001 doi:10.1088/1742-6596/353/1/011001

Foreword

The Anglo-French Physical Acoustics Conference (AFPAC) had its 10th annual meeting in Villa Clythia, Fréjus, France, from 19–21 January 2011. This series of meetings is a collaboration between the Physical Acoustics Group (PAG) of the Institute of Physics and the Groupe d'Acoustique Physique, Sous-marine et UltraSonore (GAPSUS) of the Société Française d'Acoustique.

The conference has its loyal supporters whom we wish to thank. It is their loyalty that has made this conference a success. AFPAC alternates between the UK and France and its format has been designed to ensure that it remains a friendly meeting of very high scientific quality, offering a broad spectrum of subjects, welcoming young researchers and PhD students and giving them the opportunity to give their first presentations in an 'international' conference, but with limited pressure.

For the third consecutive year AFPAC is followed by the publication of its proceedings in the form of 18 peer-reviewed papers, which cover the most recent research developments in the field of Physical Acoustics in the UK and France.

Alain Lhémery, CEA, France
Nader Saffari, UCL, United Kingdom

All papers published in this volume of *Journal of Physics: Conference Series* have been peer reviewed through processes administered by the proceedings Editors. Reviews were conducted by expert referees to the professional and scientific standards expected of a proceedings journal published by IOP Publishing.

Design and fabrication of nanoscale ultrasonic transducers

R Smith[1], A Arca[1], X Chen[1], L Marques[2], M Clark[1], J Aylott[2] and M Somekh[1]
[1]Electrical Systems and Optics Research Division, Electrical Systems and Optics Research Division, University of Nottingham, UK.
[2]School of Pharmacy, University of Nottingham, UK.

Richard.j.smith@nottingham.ac.uk

Abstract. The development of nanometre sized ultrasonic transducers is important in both biological and industrial applications. The small size can be important in its own right or necessary in order to generate acoustic waves with nanometric wavelengths. Potential applications of nanotransducers range from embedded sensors through to sub optical wavelength acoustic imaging. In this paper we discuss the design and fabrication of nanoscale ultrasonic transducers. The transducers rely on optical and mechanical resonances, they can be used to generate and detect high frequency ultrasound in a sample. The mechanical and optical performance of the devices have been extensively modelled using both analytical techniques and finite element modelling. This allows the fine tuning of the design parameters to ensure optimised performance for the experimental configuration. The devices can be fabricated in a number of ways, we present one method for building these types of devices, a 'top down' approach where plate structures are built up and patterned using standard photolithographic techniques. This method produces nanoscale devices in one dimension only (the others being a few microns) but produces excellent devices for testing in situ and for comparison to the models as they are easy to handle and measure. Approaches for reducing the other dimensions to the nanoscale will also be considered.

1. Introduction

Nanometre sized ultrasonic transducers are important in both biological and industrial applications. The small size is necessary in order to generate nanometric wavelengths or access small structures such as cells. There has been much interest in the vibration of micro[1] and nano[2] objects excited by femtosecond laser pulses and the presented transducers build on this work. The transducers we introduce in this paper act as simultaneous optical and ultrasonic resonators designed so that they couple from the optical to the ultrasonic and vice versa. We have realised these in planar structures using optical resonators and are developing spherical transducers using optically and ultrasonically resonant nanospheres (see figure 1a). There are a number of different approaches to producing these small scale structures and in this paper we shall focus on one method for building planar devices. At this time the devices are nanoscale in one dimension only and we will discuss methods to reduce the size of the devices to be nanoscale in all dimensions. The devices have been modeled so that the parameters governing their behaviour could be tuned to produce a device that is effective from a mechanical and optical point of view. We have produced devices to be measured in a pump probe picosecond ultrasound system[3], where a pump pulse is absorbed by the sample and generates an acoustic wave packet. The probe beam monitors the sample at different time delays to measure the interaction of the propagating acoustic wave with the sample.

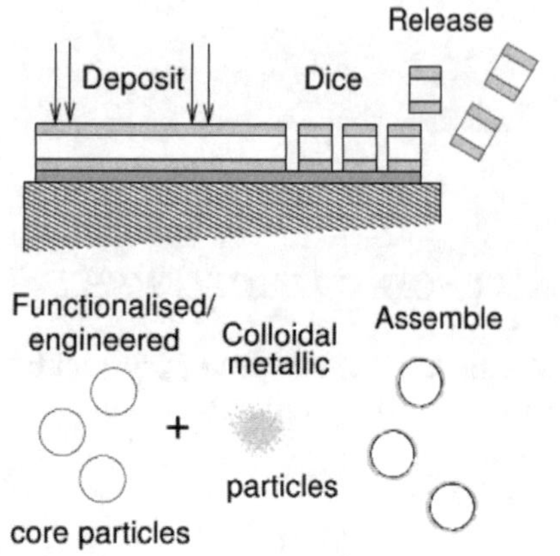

Figure 1. Two approaches to producing transducers.

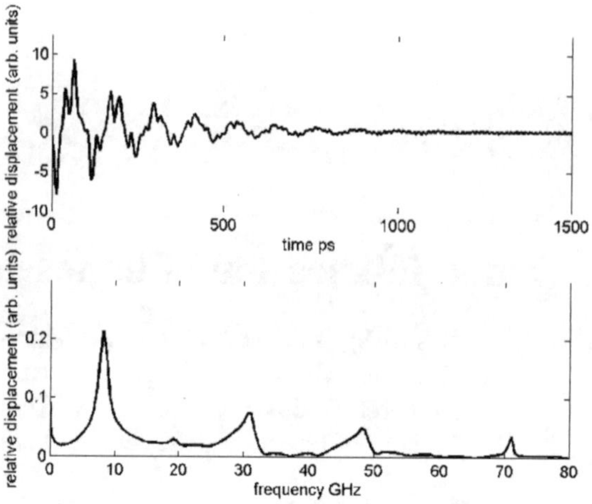

Figure 2. FE simulated ultrasound signal generated on nanostructure device .

These devices rely on optical resonances and therefore require the presence of partially reflecting mirrors sandwiching an optically transparent filling. The incident light reflects and scatters off the partially reflecting boundary layers and interfere. The sensitivity of the device depends on the energy ratio of the reflected beams and the device size. If the size changes (for example due to the presence of an acoustic field) then the phase of the reflected components change causing a variation in the measured intensity at the detector. We have approached the creation of these transducers from a 'top down' direction, where 'top down' means building the structures from the top, a layer at a time. The produced resonating layer structure can then be patterned via photolithography or diced up using laser machining or focused ion beam techniques. Once diced the transducers can then be liberated from the substrate into solution.

2. Mechanical operation

The transducers generate ultrasound thermoelastically from the pump beam light absorbed in the metal layers. Because the absorption of the laser light in these layers is not uniform we use an EM FE model to predict this. The model calculates where in the structure the laser pulse is absorbed, knowing where and over what time scale the energy is absorbed the model can work out the change in temperature throughout the structure and from that the mechanical motion that arises. For the gold:ITO:gold structure we find that most of the light is absorbed in the top gold layer, which is as expected as gold is very absorbing at the pump laser wavelength of 400nm.

The result from the mechanical model for a structure of gold:ITO:gold:glass substrate (40nm:160nm:40nm:inf) is shown in figure 2 below, the figure shows the difference in displacement between the two gold layers. We observe a decaying oscillating signal composed of 3 or 4 main frequency components. The frequencies that are present are related to the round trip times of the acoustic waves through the structure.

The frequencies produced by the transducer vary with the thickness of the layers used. Figure 3 and 4 show the changes to the dominant frequency component as the layer thickness varies. In figure 3 the ITO layer thickness is kept fixed at 160nm and the gold layer is varied. We see a shift in the frequency from 14 – 9.5 GHz. Figure 4 shows the case where the gold layer is kept fixed at 20nm and the ITO layer is varied. Again, we see the frequency shift by a significant amount 13 – 9 GHz. This allows some tuning of the device to operate at a specific frequency by carefully choosing the layer thicknesses.

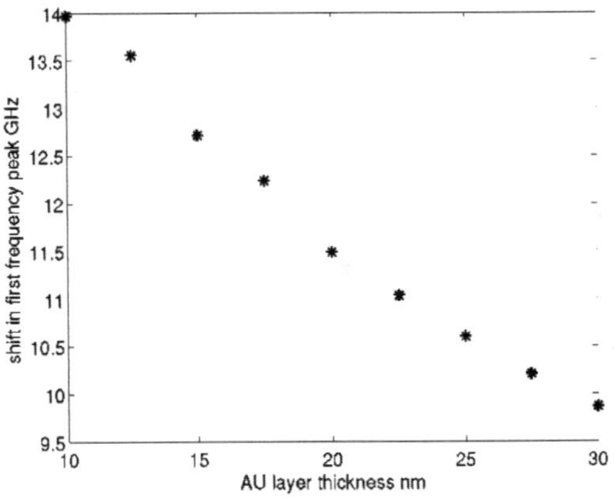

Figure 3. Effect of changing AU thickness

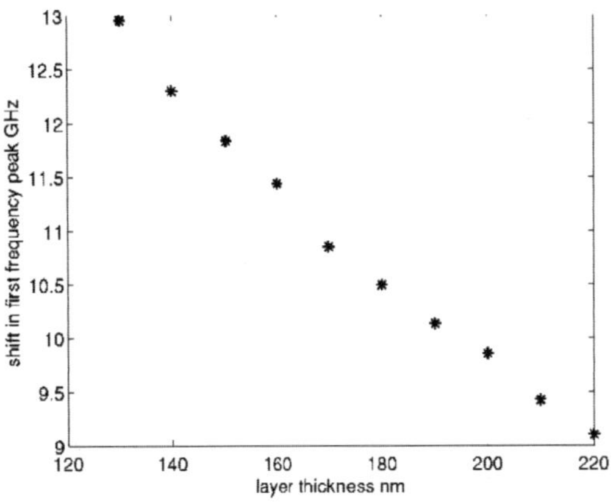

Figure 4. Effect of changing ITO thickness

The fact that the devices are so sensitive to the layer thicknesses means that this can be used as transduction method. Figure 5 shows the effect of adding a very thin layer of material to the top gold layer of a device. In this case the added thin layer was ITO as this is optically transparent to the pump beam and would therefore not affect the location of the generation of the sound waves. As the thickness of the loading layer is increased we see that the resonance peak shifts by a small but measureable amount.

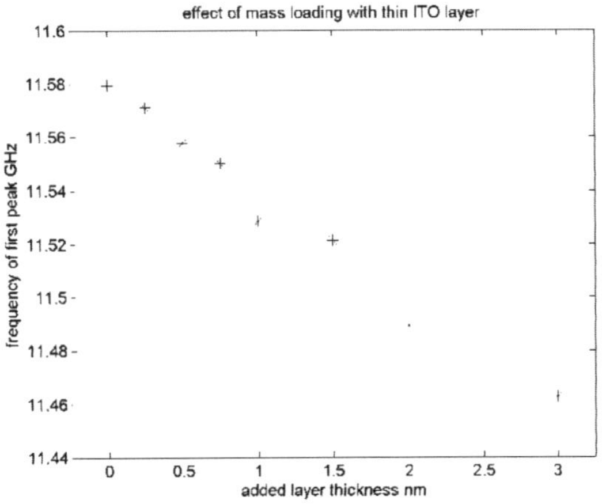

Fig 5. Effect of mass loading on the main frequency peak

3. Optical operation

The devices operate in a manner similar to that of a Fabry-Pérot interferometer. The layer structure is designed such that the top metal layer is a partial reflector allowing the probe beam to penetrate through the structure and reflect off the bottom layer of the transducer. The two reflected beams then interfere with a strength that is dependent on the metal layer separation. The acoustic field will change the separation of these partial mirrors, and so the optical path difference between the beams will change creating a difference in the strength of interference. The individual thicknesses of the films are designed to give the optimum change in the reflectivity for a given change in mirror separation/radius, while maintaining desirable acoustic properties.

When the transducers are bigger than the probe beam spotsize, they can be modelled analytically under the infinite width assumption using Fresnel coefficients. The result of this analytical optical model is shown in figure 6. The plotted quantity is the sensitivity of the device (dI/dt) i.e. the change in the reflectivity with respect to the mirror separation. From figure 6 we can see that for a gold:ITO:gold sandwich the best sensitivity is achieved with a 40nm gold layer and 160nm ITO layer, although there are broad range of device configurations that have good performance.

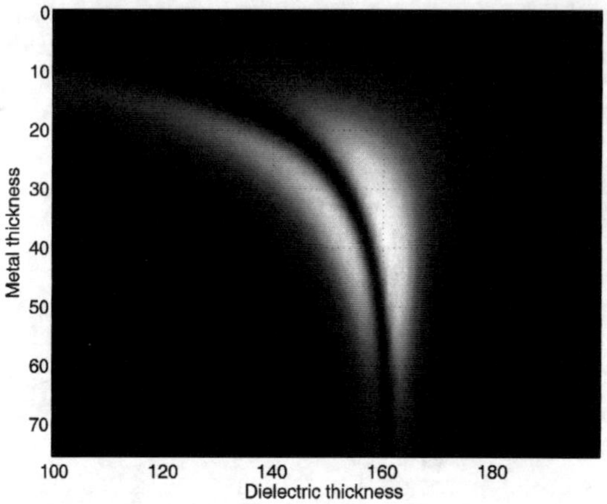

Figure 6. Sensitivity of the gold:ITO:gold transducer from the analytical model

To model the devices as they get smaller a full 3D FEM model is required. In this case the input field is incident on the device in the positive *x* direction, with the device layers extending in the x direction, the transverse size of the device can be varied (y,z direction) as shown in figure 8. We then calculate the reflected and transmitted far field spectra. This is achieved using a near field to far field transformation.

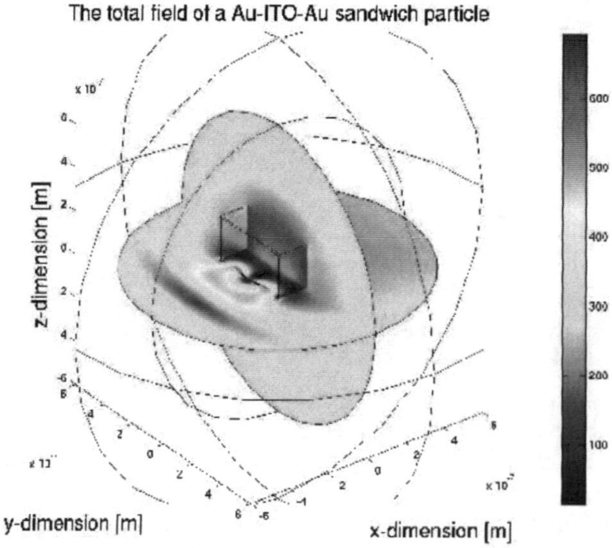

Figure 7. 3D FEM model of gold:ito:gold transducer (40:160:40)

To obtain the sensitivity we calculate the scattering cross section(C) as a function of separation (S) and then differentiate this to get dC/dS which is our "sensitivity" measure (ie how much the signal changes when the transducer is squeezed by the ultrasound). This is then repeated for different device layer thicknesses and transverse sizes. We find that as the devices get smaller and smaller (i.e. the transverse size gets smaller so there is less of the device interacting with the input field) then the sensitivity peak shifts to longer probe wavelengths. If the wavelength is fixed then the sensitivity will reduce as the device size is reduced. To keep optimal performance the device configuration must be changed. The change is required as the refractive index of the filling material (ITO in this case) starts to have reduced influence on the response, such that eventually the optimal point depends most strongly on the refractive index of the surrounding medium (air or water). As the refractive index of the medium is less than that of ITO the devices must be made thicker as the transverse size decreases. This can be seen in figure 9 which shows a plot of the sensitivity peaks of the devices shifting to thicker layer sizes (particle size) when the cross section of the devices reduce.

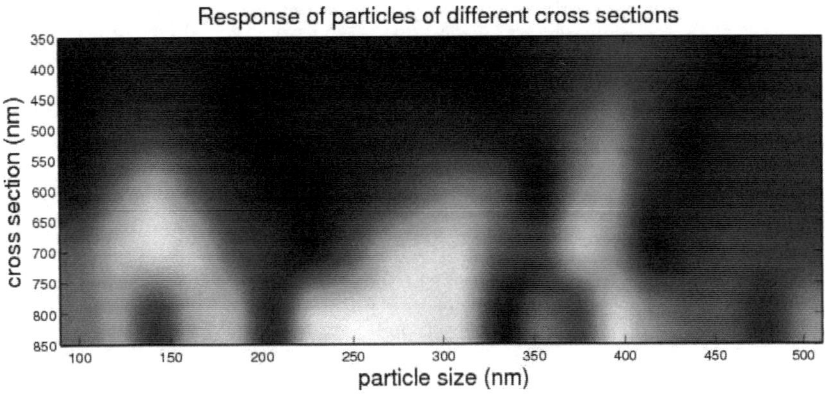

Figure 8. Effect of reducing the transverse size of the devices – the sensitivity peaks shift to taller device sizes

4. Manufacturing the devices

We have fabricated the devices using standard photolithography techniques to produce patterned substrates. Use of a sputter allows the gold and ITO layers to be deposited on the patterned substrate. This allows very fine control over the devices made. It allows easy monitoring of the devices and allows *in-situ* measurements to be made before release to fine tune the design and fabrication process.

However, the size and total number of devices made is limited, which in our case leads to a minimum lateral device size of ~ 5 microns and a few million devices per run. Using a better photolithography process or employing a different dicing technique (such as focused ION beam or E-beam lithography techniques) could produce transducers with lateral nanoscale dimensions.

5. The first prototype transducers

We have fabricated and tested a set of planar transducers. The experiment is based around an ASOPS [3] laser from Menlo Systems. This uses two femtosecond lasers that are jointly controlled with a set of high speed electronics to control the repetition rates of the two lasers. The electronics allow one laser to be locked at 100 MHz repetition rate and the other laser to be locked at 100 MHz + 10 KHz. This means that the lasers provide the equivalent of a 10 ns delay line sweep every 100 microseconds. The pump and probe beams are focused to the same location on the top of the transducer and probe response is recorded as the delay is swept.

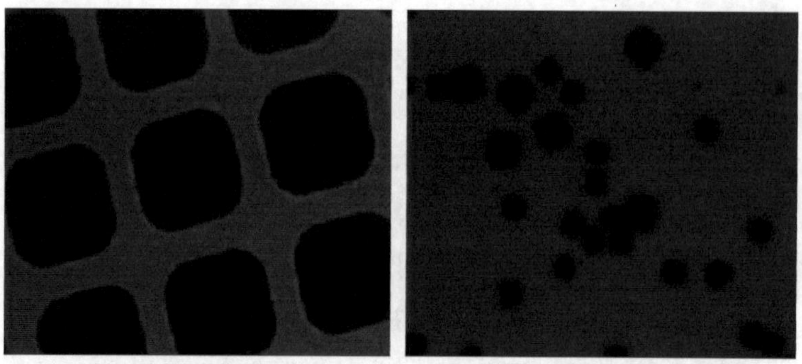

Figure 9. Optical image of transducers in situ and released and reattached on a glass substrate

Figure 9 shows an optical image of the 10 micron patches and a lower magnification image of a mix of 5, 10 and 20 micron patches reattached to a glass slide after being lifted off into solution. This shows that the transducers survive the lift off process and can be reattached successfully.

We measured a 10 micron transducer on top of a glass substrate that had a polystyrene buffer layer. The buffer layer has two purposes, firstly to provide acoustic isolation for in-situ measurements, thus increasing the length of time the signals exist, and secondly as a way to release the devices into solution. The recorded signal is shown in figure 10 after the coincidence peak and background have been removed.

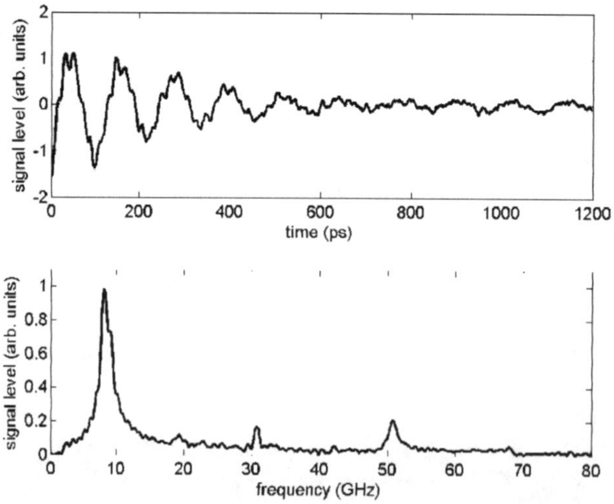

Figure 10. Experimental result from 10 micron patch transducer

The signal is very similar to the modeled result in figure 2. The differences are likely due to slight variations in actual layer thicknesses and difference in the actual mechanical properties of the layers. The oscillations in the experiment are longer lived than for the modeled result. This is likely due to the inclusion of the buffer layer and the reduction of acoustic energy lost into the substrate that it provides.

To demonstrate the influence of the surroundings on the response of the transducer we have measured a set of devices in both air and water (figure 11). The water layer acts as a mass loading layer and as such influences the mechanical resonance of the device causing the measured frequencies to shift.

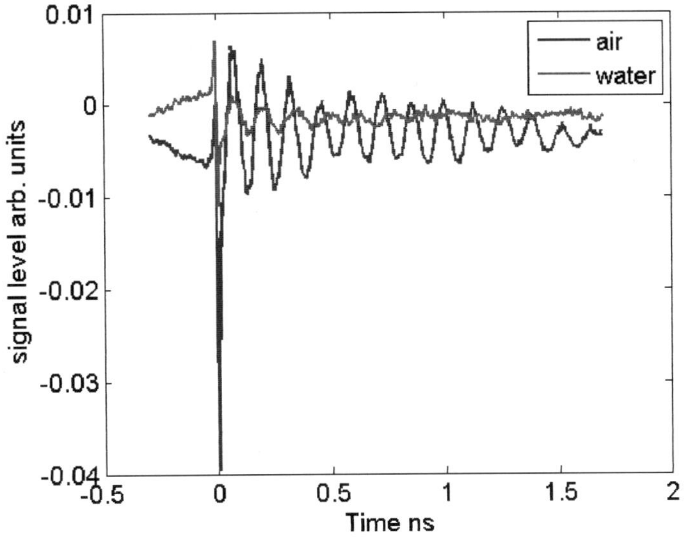

Figure 11. Measurement of 25 micron patch transducer (40:160:40) in air and water

Figure 11 shows that the signals levels are reduced in water significantly compared to the levels obtained in air. There are a number reasons to explain this; firstly for the water layer to be included an additional cover slip was added with standoffs to trap a water layer of ~300-400 microns above the transducers. This will have reduced the amount of pump and probe light reaching the transducers reducing the size of the generated waves and reducing the detection signal to noise ratio. Secondly water is very attenuating at high frequencies [4] and so we expect to see a reduction in signal level with much stronger damping and for the high frequency peaks.

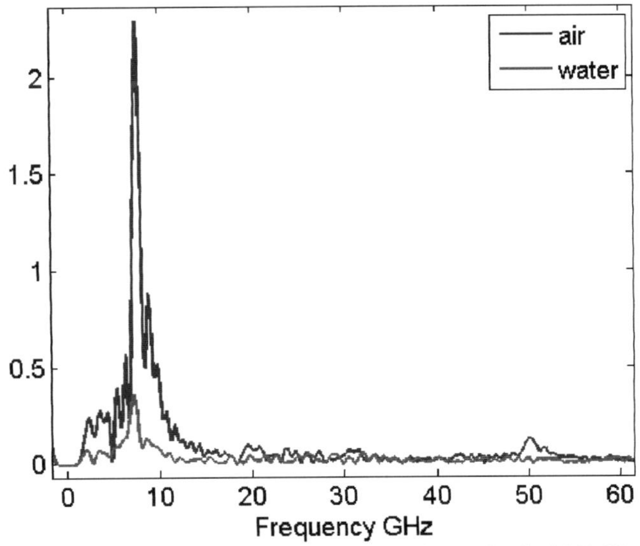

Figure 12. Frequencies measured showing shift in main peak of ~300MHz in air and water

Figure 12 shows the spectrum of the traces for the air and water experiments, we see that the main peak has shifted down by ~300MHz and the 50 GHz peak which was clear in the air case is no longer visible above the noise. More averaging was used in the water case to improve the chance of seeing these attenuated peaks but they have been damped to such a level that they are still not visible.

6. Future Directions

The devices that have been made and tested have only been nanoscale in one dimension. In the future we hope to reduce all dimensions to the nanoscale. This could be achieved by using other patterning techniques such as improved photolithography processes or ebeam lithography or focused ion beam techniques. We have also started to tackle this problem from another direction – that of self assembled nanoparticles. We have made a series of particles with thin gold coating shells around a transparent core. These transducers are essentially a spherical version of the transducers we have already tested. The gold shell acts as the partially reflecting layers and the clear soft inner silica core acts as the separation layer.

The advantage of using the self assembly approach is that it is easier to make small devices, but the difficulty arises in controlling the distribution of sizes so that most of the batch produced are within the correct size range for good operation at our laser wavelength.

We have manufactured devices with a core size of ~180nm and a 10 nm gold shell (Figure 13) but the distribution of sizes is still quite high. Testing of these devices is ongoing.

Figure 13. Self assembled nano particle transducers, the gold particles can be seen coating the silica core

As well as reducing the size we will need to improve our handling techniques for these small devices so that they can be reattached for measurements. In the future we hope to encapsulate and functionalise the devices so that they can be used for site specific measurements. This will increase their utility as transducers considerably as the ability of the transducers to only attach to the things we wish to measure will be very useful.

7. Conclusion

We have modelled, fabricated and tested an ultrasonic / optical transducers of approximately 200nm in height and with lateral dimensions of 10-5 microns in size. The transducers can be liberated from the base substrate and attached for non contact measurements. These transducers produce and detect ultrasound at frequencies in the 5-50GHz range. They have been remotely excited and probed using femtosecond lasers. We have the ability to manufacture even smaller devices using molecular self assembly.

8. References

[1] Dehoux T, Kelf T, Tomoda M, Matsuda O, Wright O B, Ueno K, Nishijima Y, Juodkazis S, Misawa H, Tournat V and Gusev V 2009 *Optics Letters* **34** 3740-42
[2] Guillet Y, Rossignol C, Audoin B, Calbris G, Ravaine S 2008 *Applied Physics Letters* **95** 061909
[3] Stoica V, Sheu Y, Reis D and Clarke R 2008 *Opt. Express* **16** 2322-35
[4] Hadimioglu B and Quate C F 1983 *Applied Physics Letters* **43** 1006-7

Optical MEMs transducers with enhanced efficiency and sensitivity

Xuesheng Chen, Theodosia Stratoudaki, Steve D Sharples and Matt Clark

Electrical Systems and Optics Research Division, University of Nottingham, University Park, Nottingham, NG7 2RD, UK

E-mail: xuesheng.chen@nottingham.ac.uk

Abstract. We introduce a highly sensitive laser ultrasonic detection MEMs transducer and an efficient laser ultrasound generation MEMs transducer. The detection transducer consists of a series of cantilevers with the same dimensions; any two cantilevers next to each other are separated by a solid with the same width as finger's. When ultrasound is incident upon this transducer, as long as there is a vibration component perpendicular to each finger's top surface and with a frequency the same as the finger's first resonance in the ultrasound, each finger will resonate upon the ultrasound. The moving fingers and the still solid separations form an optical phase grating, and therefore the ultrasound can be readout by a detection laser remotely. Because the ultrasound amplitude is amplified many times by the transducer's resonance before detection, the sensitivity of this transducer is much higher than that of traditional transducers. The generation transducer consists of a micro-disk seated upon a micro-stem. When a suitably focused laser pulse illuminated on the center of the disk, a certain order of flapping motion of the disk is mainly actuated, while other orders are just slightly, or not excited. This flapping motion couples a very narrow bandwidth of longitudinal wave propagating along the axis of the stem and into a sample. Because all absorbed optical energy is concentrated into this narrowband ultrasound, its amplitude is much higher than that of normal thermoelastic generation. It is possible to use these MEMs generation and detection transducers to form a simple but highly efficient laser ultrasound generation and detection system in the near future

1. Introduction

Laser based ultrasonic techniques have been paid great attention in recent decades. This is because laser ultrasonic inspection has advantages over traditional ultrasound generation and detection that are based on contact techniques. The ability to operate remotely from the sample surface makes the non-contact and couplant-free online generation and detection of ultrasound possible, allowing inspection of areas where access is difficult [1]. However, compared with conventional ultrasound techniques, the SNR of laser ultrasound is low due to the shot noise of optical detection. Raising the power of the laser pulse to produce ultrasound with a higher amplitude is one of the solutions; however this is constrained by the damage limit [2]. Great efforts have been made to overcome this drawback in the last two decades by either increasing the amplitude of the laser ultrasound or strengthening the detection sensitivity, or both [3, 4, 5].

In order to remotely generate and detect laser ultrasound, different complex and expensive systems have been developed, as was reviewed in [3], such as confocal Fabry-Perot interferometers, photo-emf detectors, electromagnetic acoustic transducers and knife-edge detectors [6]. This also leads to the development of hybrid ultrasonic systems [4]. For detection of ultrasound with high sensitivity, a cantilever optical MEMs detection transducer with a sensitivity a few hundreds times higher than that of the conventional methods has been also reported recently [5].

In order to increase the amplitude of laser-generated ultrasound, different techniques have also been successfully developed. One technique is to deposit a layer of material, such as graphite, with a higher optical absorption coefficient onto the surface of a sample [7]. This can effectively increase the absorbed optical energy. In [8], laser light was line-focused so that the optical power can be effectively increased while the power density of the laser pulse still remains within nondestructive regime. By combining the techniques in [7] and [8], generation transducers with patterned absorption layers have also been developed [9]; this can be applied to generate narrowband ultrasound with improved SNR. A different technique is to achieve higher amplitude by narrowing the bandwidth. This has been done by different methods; for instance different lasers were used to generate narrowband laser ultrasound [10, 11].

In addition to these, complex optical arrays [12, 13], special methods [14] and optical transducer systems [9] were also developed for the generation of narrowband ultrasound with higher amplitude/SNR.

In this paper, we present an optical MEMs detection transducer with high sensitivity and an optically powered MEMs generation transducer for the generation of narrowband ultrasound with high amplitude/SNR, providing a different solution to the optical shot noise problem for laser ultrasound generation and detection. The detection transducer is found to have a much higher sensitivity than that of traditional ones, and the generation transducer can produce narrow bandwidth ultrasound with a much higher amplitude than that of normal-thermoelastic generation at the same optical conditions.

2. The optical MEMs detection transducer

2.1. Analytical sensitivity of the detection transducer

The initial inspiration for the detection transducer came from experimental observations of anomalous ultrasonic responses of damaged coatings. Figure 1(left) shows a microphotograph and an acoustic scan of a ceramic sample with a damaged chrome coating. The coating was poorly adhered to the surface and peeled off using adhesive tape. At the edge of the damaged region there are delaminated regions of coating, some of which demonstrate very significant enhancements in an acoustic signal which is detected optically. These delaminated regions have, in response to the acoustic wave, a flap motion with a significantly higher amplitude than the surrounding bonded regions. The flap increases the amplitude of the vibration and because this occurs before optical readout it can increase the measured amplitude without increasing the noise. By combining this concept with a CHOT (cheap optical transducer) structure [9] we have produced an optically read ultrasonic transducer with high sensitivity as shown in figure 2. These transducers consist of several resonant 'fingers' that can be fabricated next to one or more solid spaces (usually the gaps between the fingers). This forms a zero-order interferometer where the normal 'object' beam is derived from the light reflected by the fingers and the normal 'reference' beam from the solid spaces adjacent to them. Normal construction uses several fingers which form an optical phase grating and generate discrete diffraction orders (if the pitch of the grating is less than the optical wavelength then these higher orders are evanescent)[9, 15]. The grating structure and presence of these orders make the optical detection simple as it is sufficient to place one of the diffracted orders (usually the zero-order) onto a photodetector to make the measurement. When an ultrasonic wave is incident on this device it causes the fingers to move, which changes the grating's optical properties and therefore modulates the optical power diffracted into the various orders. The amplitude of the ultrasound is mechanically amplified by the fingers' resonance from a few tens to several hundred times, depending on the material,

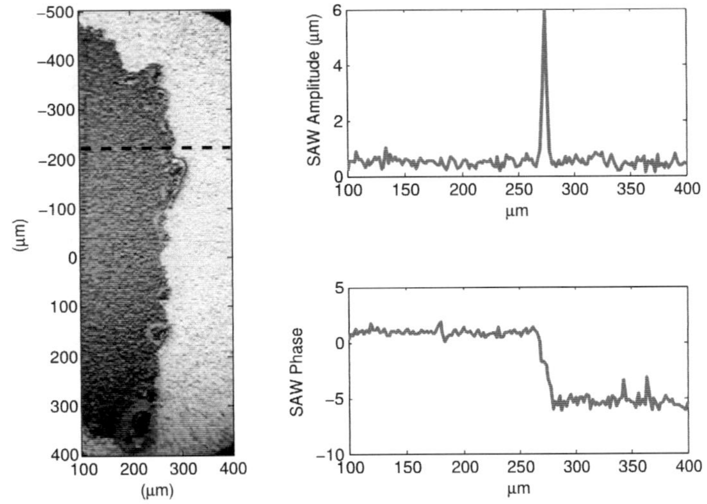

Figure 1. Evidence of amplitude enhancement by a loose structure. The image on the left shows a damaged coating (chrome on silicon nitride). At the edge of the damaged region there are delaminated areas of chrome. Scanning over these regions (dotted line) reveals significant amplitude enhancement at the delaminated region. The large amplitude (and phase change) suggests a resonant structure.

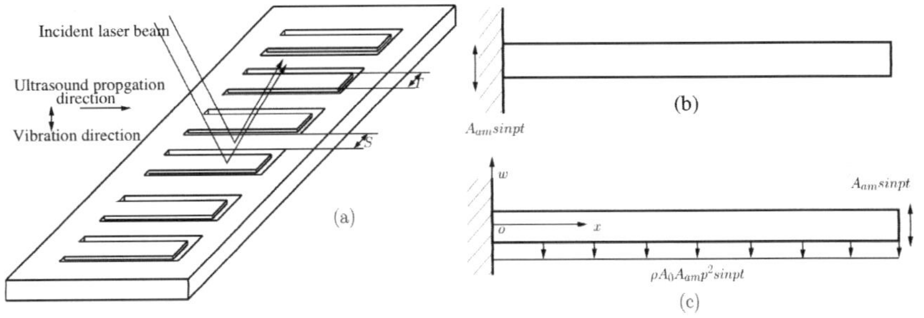

Figure 2. (a) Schematic design of a MEMs detection transducer. (b)and(c) Mechanical model of a finger of the MEMs transducer.

geometry and the working environment of the transducer. As this is done before detection, the sensitivity of the transducer to ultrasound is greatly enhanced.

Figure 2(right) shows the mechanical model of the finger of a detection transducer. From this model, analytically the sensitivity of the transducer is found out to be [5]:

$$\gamma_1 = \frac{|w_1(l,t)|_{max}}{A_{am}} \approx 1.566Q_1, \tag{1}$$

$$\gamma_2 = \frac{|w_2(l,t)|_{max}}{A_{am}} \approx 0.868Q_2, \tag{2}$$

$$\gamma_3 = \frac{|w_3(l,t)|_{max}}{A_{am}} \approx 0.508Q_3, \tag{3}$$

where γ_i (i=1,2,3) is the sensitivity of the transducer at i^{th} order of resonance, $w_i(l,t)_{max}$ is the

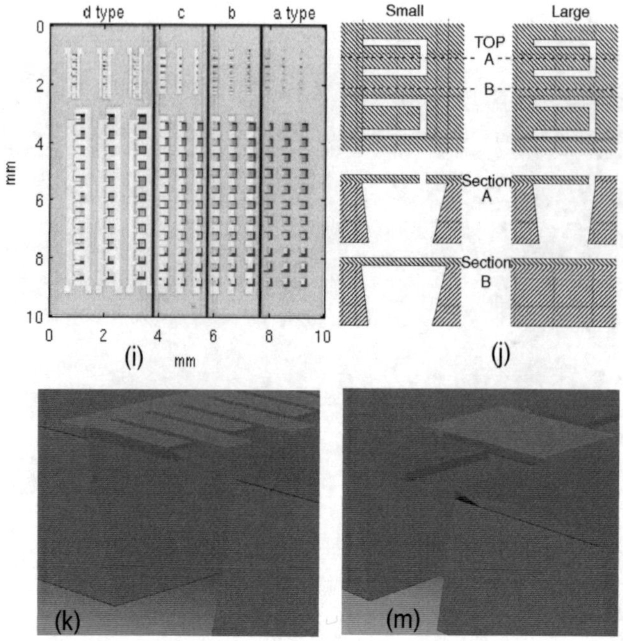

Figure 3. i) Top view of the designed device with 24 transducers on it. (j) Structure details of small and big transducers. Limited by the design rules of MEMSCAP, a big hole is opened up underneath small transducer fingers, leaving a residual membrane at one side and fingers being separated by bridges. For big transducer fingers, the large finger width allows the hole beneath the bridges and membrane to be removed, leaving solid separations between fingers. (k) and (m) Images of structural details of a small transducer (*c* type) and of a big transducer (*b* type).

maximum finger's tip amplitude, A_{am} is the amplitude of the ultrasound that is incident on, and Q_i is finger's quality factor at i^{th} order resonance. Analytically, this type of transducers have high sensitivity. For details of the analysis, reference [5] can be referred to.

2.2. Design and fabrication of the transducer

Several different types of transducers were designed with frequency of 1 MHz, 2 MHz and 3 MHz respectively. Figure 3(i) shows the top view of the designed 10×10mm device. 24 transducers with various different structure types were designed onto the chip in two rows; a row of small transducers (top) and a row of large transducers (bottom). In each row there are four different types of structures, referred to as a, b, c and d types from right to left: a type transducers consist of plain silicon fingers and separations. b type transducers are based on a type, plus a reference metal layer on the top surface of each separation. For c type transducers, both the fingers and the separations are covered by the metal layer. d type transducers have a thermal actuator on the top surface of each finger. In each of the four types of structure, there are three transducers with 1-, 3- and 5-MHz fingers respectively, arranged from left to right. We designed transducers with these four different structures in order to attempt different initial deflections of the fingers to satisfy the diffraction condition; these initial deflections can be different because of their different lengths and coatings [16, 17]. Finger lengths of transducers with the same resonance frequency are slightly different for different structure types due to the additional metal layer on the finger top surface; they were refined by finite element (FE) analysis.

The dimensions of each small transducer finger were approximately $(66 \sim 170) \times 50\mu m$ plus a

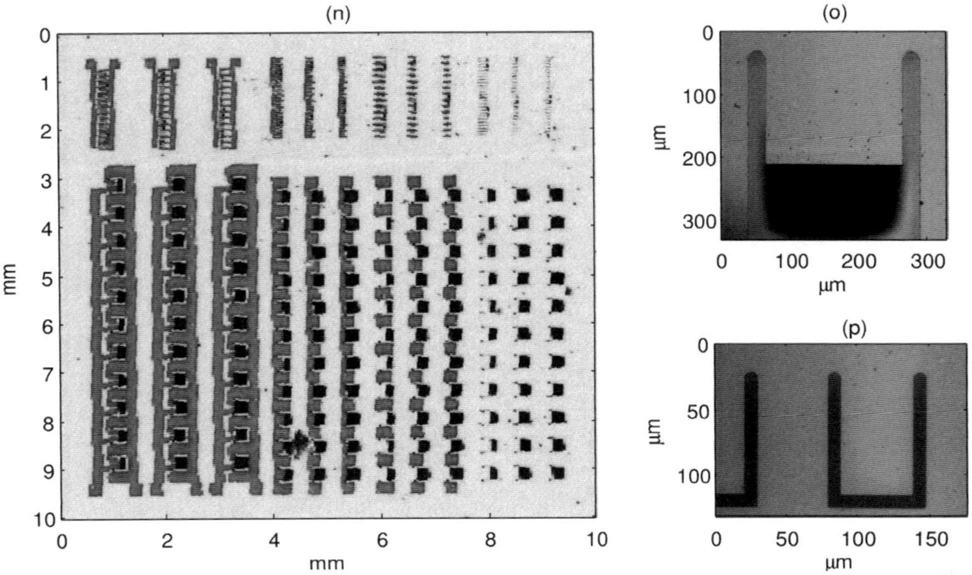

Figure 4. (n) Ultrasonic c-scanned image of the wafer with fabricated 24 detection transducers. (o) and (p) Images of fabricated transducers under a microscope to show local structure details.

$9\mu m$ air gap at each side of the finger. In order to satisfy the MEMSCAP design rules a large hole was opened up underneath the entire transducer (figure 3(j)∼(k)). This left the "solid" surfaces between the fingers as thin bridges across the hole. In addition there was a residual membrane on one side (figure 3(j)∼(k)). This has the disadvantage that resonances of the bridges and membrane exist and may be detected in addition to those of the fingers. However these could be exploited to extend or manipulate the frequency response of the transducer.

To overcome this limitation of the MEMSCAP process larger transducers were also fabricated. The dimensions of each big transducer finger are approximately $(66 \sim 170) \times 240\mu m$ plus a $30\mu m$ air gap at each side of the finger. The larger finger width allows the hole beneath the bridges and membrane to be removed leaving a simpler structure (figure 3(j) and (m)).

Transducers were fabricated using a typical MEMs process—a standard SOIMUMPs (Silicon on Insulator Micromachining Process) at MEMSCAP (Refer to http://www.memscap.com.) The fabrication material is (100) type of SOI (silicon on insulator) wafer with a crystal silicon design layer of thickness 25μm; this defines the finger thickness of the transducer. The primary flat direction of the wafer is (110) and the material properties of the wafer are defined in [18]. Figure 3((i)∼(m)) shows the designed transducers and their 3D structures details, and Figure 4(n) was an ultrasonically c-scanned image of a wafer with 24 detection transducers on it, and (o) and (p) of the figure are the images of fabricated transducers under a microscope to show local structural details of a transducer.

2.3. Experimental study of the MEMs detection transducer

The aim of the experiment was to confirm that an ultrasonic wave of a given frequency can be detected by using the appropriately designed transducer with high sensitivity, and to examine the transducer's dynamic behavior. To do this, an optical measurement system was configured, as shown in figure 5. Collimated light from a laser (optical wavelength 532 nm) was used to illuminate the MEMs transducer. An optional cylindrical lens can be inserted to focus the light to a line and ensure a greater fraction of the light lands across the fingers. The reflected light and

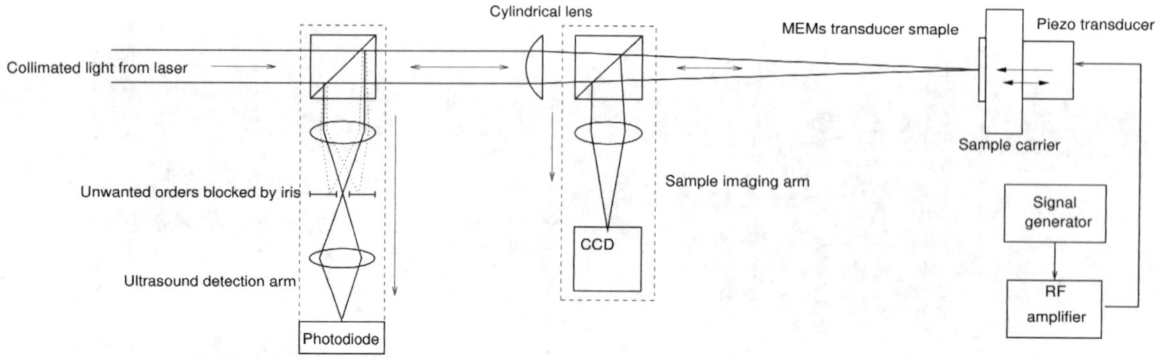

Figure 5. Schematic diagram of the experimental setup. Collimated light is used to illuminate the transducer (via an optional cylindrical lens which is used to increase the power directed to the fingers). The light is collected by the same optics and directed onto a CCD for imaging and alignment, and also into the detection arm where the diffracted orders are focused and separated. An iris is used to strip off the unwanted orders and the remaining 0-order is focused onto a photodiode for detection.

orders from the MEMs transducer travel back through the same optics and are diverted in turn to (1) a CCD which is used to image the sample and align the system, and to (2) a detection arm. In the detection arm the unwanted diffraction orders are removed by an iris. The remaining 0-order is then focused onto a photodiode for detection. The fabricated transducer was glued onto a carrier that was attached to a piezoelectric ultrasound generation transducer. We used a 5-MHz 'Olympus V609-SB 5.0/0.5 Videoscan' piezoelectric transducer with a bandwidth of about 110%. An RF signal generator was used to generate a signal to drive the piezoelectric transducer, which produces a required longitudinal wave to excite the sample.

2.3.1. Bandwidth measurements The dynamic response of an *a* type small transducer with 1-MHz fingers to the ultrasound generated by the 5-MHz generating transducer is shown in figure 6(a). Several peaks at different frequency bands were detected in the middle diagram, labeled A, B, C and D in the figure. Obviously, peak A includes the resonances of the 1-MHz fingers; this includes both the fingers' resonance with in-phase motion and the resonances of fingers with different phases. The trace corresponding to peak A is shown by figure 6(c); the SNR of the detected signal is so high that the signal can be observed without averaging. B, C and D in the figure are, by c-scan, shown to be different modes of bridges of the transducers instead of fingers'. Similar results for a large transducer with 1-MHz fingers were also obtained. Figure 6(b) shows the dynamic response for a *b* type large transducer with 1-MHz fingers, where the corresponding ultrasound was also generated by the 5-MHz piezoelectric transducer.

It should be emphasized that the ultrasound corresponding to the dynamic response in figure 6(a) was generated by a 5-MHz piezoelectric transducer (figure 5). In this case the amplitude of the generated ultrasound around 1 MHz should be smaller relative to that around 5 MHz due to the dynamic properties of the generating transducer. Despite this, much stronger ultrasonic signals around 1 MHz were detected, as shown by the frequency spectrum in figure 6(a and b); on the contrary, the detected ultrasonic signals around 5 MHz are rather weak. This shows that the fingers' resonance is a dominant factor in achieving high sensitivity for ultrasound detection; only signals of waves within the transducer's designed frequency band around its resonances are amplified by the transducer; sounds or noise out of this frequency band of the transducer

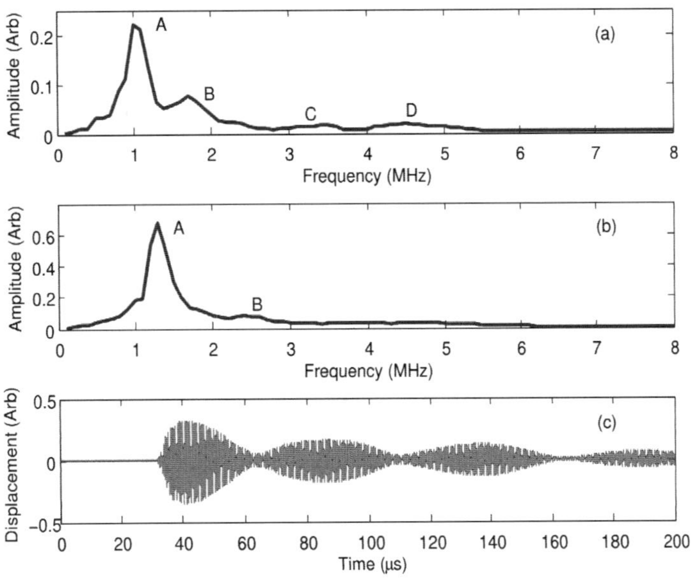

Figure 6. (a) Dynamic response of an *a* type small transducer with 1-MHz fingers. (b) Typical dynamic response of a *b* type large transducer with 1-MHz fingers. (c) Typical detected ultrasonic signal shown on an oscilloscope screen.

cannot cause the resonance of fingers, and therefore are not amplified by the MEMs fingers. This enables the transducer to be particularly noise resistant while its high sensitivity to ultrasound remains. The resonances of bridge separations only widen this frequency band of the transducer with a sensitivity lower than that of the fingers.

However, this dynamic property of the transducer also shows that the designed detection transducer has a narrow bandwidth. This may cause difficult in using of this type of transducer; for any ultrasound with a frequency out of the narrow bandwidth of the transducer, the sensitivity of the transducer will soon decrease to the same level as traditional transducers'.

2.4. Sensitivity measurements

In the experimental configuration shown by figure 5, we used a laser vibrometer to replace the optics in the system, so that the amplitudes at both fingertip and substrate around finger's root would be measurable for a given pulsed excitation on the piezoelectric generation transducer in the system. The scanned finger's modes are shown in figure 7 (b, c and d, top) respectively and the measured corresponding finger tip amplitudes are illustrated by figure 7 (b, c and d, middle). However, in this case, we found out that the finger's root noise level amplitude is nearly immeasurable, as shown by figure 7 (b, c and d, bottom).

A Tektronix AFG 3252 signal generator was used to drive the 5-MHz piezoelectric transducer with a continuous sinusoidal RF signal. The frequency of the sinusoidal signal was set to the measured finger resonance; this enables a good resonance of the finger. In this case, both the amplitudes of finger tip and substrate can be measured. However, it is still difficult to define the actual root amplitude; the root/substrate amplitude is found to vary across the substrate and in the vicinity of each root. Despite this, it is possible to establish a relative relationship between the root/substrate amplitude and the fingertip amplitude. An arbitrary but fixed point was chosen around a root to measure the relative root displacement. The measured results, processed by Fourier transformation, are shown by figure 8 for different fingers. They show

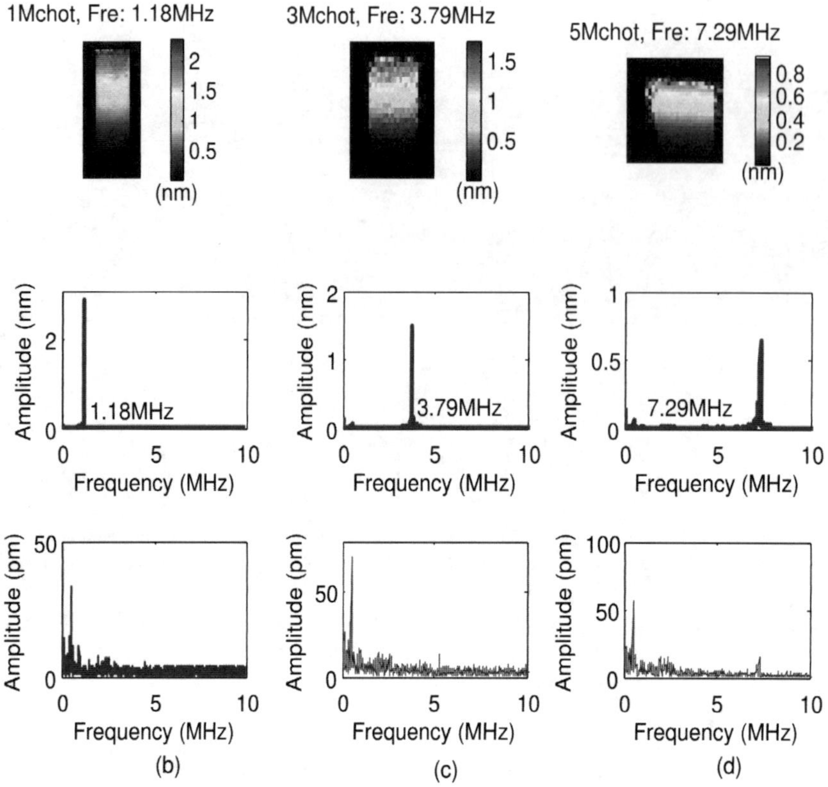

Figure 7. (b, c and d, top) C-scanned 1-, 3- and 5-MHz fingers' modes respectively, (b, c and d, middle) the corresponding fingers' tip amplitudes and (b, c and d, bottom) the corresponding substrate amplitudes at given frequency.

that the fingertip amplitude has a good linearity with the root/substrate amplitude. Slight nonlinearity for high fingertip amplitudes were also detected; this may be due to the contribution of air damping. This illustrates that the sensitivity/gain of each finger is nearly constant with the excitation amplitude at a given resonant frequency. This is reasonable, because the deflection in each case is far less than the linear range of small deflection defined by elasticity theory[19]; FE methods can be applied to analyze that the maximum strain along the axial direction of a finger—in the extreme test case with the highest tip amplitude—is only in the order of 10^{-5}; the vibration is indeed in the range that can be described as small deflections. In addition, the ratio between maximum tip displacement and finger length is at order of 10^{-4}. This shows the slope of each finger in vibration is relatively small and thus geometric nonlinearity caused by deflection can be ignored [20]. In this case from elasticity theory [19], the tip amplitude must be linear with the root excitation amplitude even at the resonance state. At a frequency away from a finger's resonance, this linear relationship should remain; the sensitivity/gain decreases as the frequency moves away from its resonance, however, it remains constant provided the frequency is fixed. In these cases the finger's tip amplitude is more sensitive to the frequency, as shown by figure 6.

The slopes of the curves in figure 8 broadly agree with the sensitivity/gain we defined in equation (1). However, we found out that although the linear relationship between input amplitude and output amplitude remains, the slope of each curve strongly depends on the root amplitude measuring location; for example with the 1-MHz finger, the slope varies from about

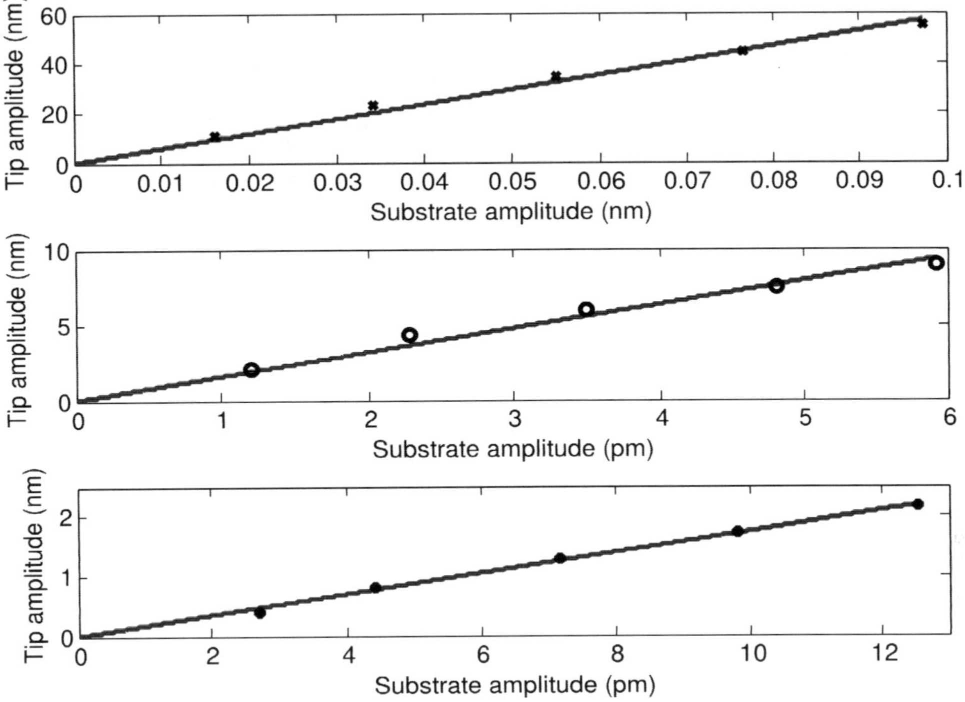

Figure 8. Sensitivity measurements for different fingers with 1 MHz (top), 2 MHz (middle) and 3 MHz (bottom) respectively, the slope of each curve is the corresponding measured sensitivity for different fingers.

700 to several thousand around the vicinity of the root. The variation in the substrate amplitude is thought to originate from the standing waves in the substrate, the influence of the vibrating finger and from waves, especially surface waves, being radiated from the fingers. Because the substrate amplitude was found to vary across the substrate (especially in the vicinity of the fingers) it was difficult to measure the driving amplitude and therefore difficult to measure the gain. As a result we can only put a lower limit on the gains as ≥ 700 for the 1-MHz transducer, ≥ 550 for the 3-MHz transducer and ≥ 60 for the 5-MHz transducer.

3. Optical MEMs generation transducer

3.1. Geometry and working principle

we present an optically powered MEMs transducer for the generation of narrowband ultrasound with high amplitude/SNR, providing a different solution to the optical shot noise problem for laser ultrasound generation and detection. It is a 3-D elastomechanical micro-structure, a typical MEMs device, consisting of a resonant microdisk, seated on a solid microstem, as shown by figure 9(a). This transducer generates longitudinal waves in a narrow bandwidth with higher amplitude than would be excited with the same optical power without the transducer present. It has high efficiency, compact size and a wide application range by using simple optics with a common pulsed laser source.

When a laser pulse is incident on a microdisk (figure 9(a)), it produces localized temperature differences and stresses which excite the vibrational modes of the entire structure due to its small size. A surface wave with a circular wavefront is actuated first at a laser pulse (see figure 9(b)). As this surface wave propagates towards the edge of the disk, it turns into an antisymmetric

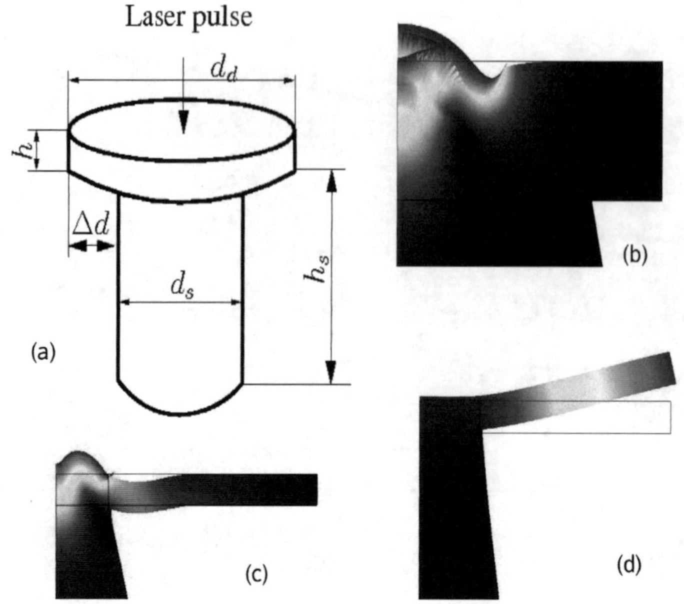

Figure 9. (a) Schematic design of the geometry of an optical transducer. It is a microdisk with a diameter d_d and a thickness h, seated on a microstem with a diameter d_s and a length h_s. As a laser pulse is incident on the center of the disk, a resonant flap motion of the disk is activated, coupling a longitudinal wave propagating along the axis of the stem. (b) A surface wave is actuated first at the disk center at a laser pulse. (c) The surface wave turns into an antisymmetric Lamb wave as it passes the flange root of the disk. (d) Reflections of the Lamb wave between the disk edge and root turn the Lamb wave into a flap motion of the disk.

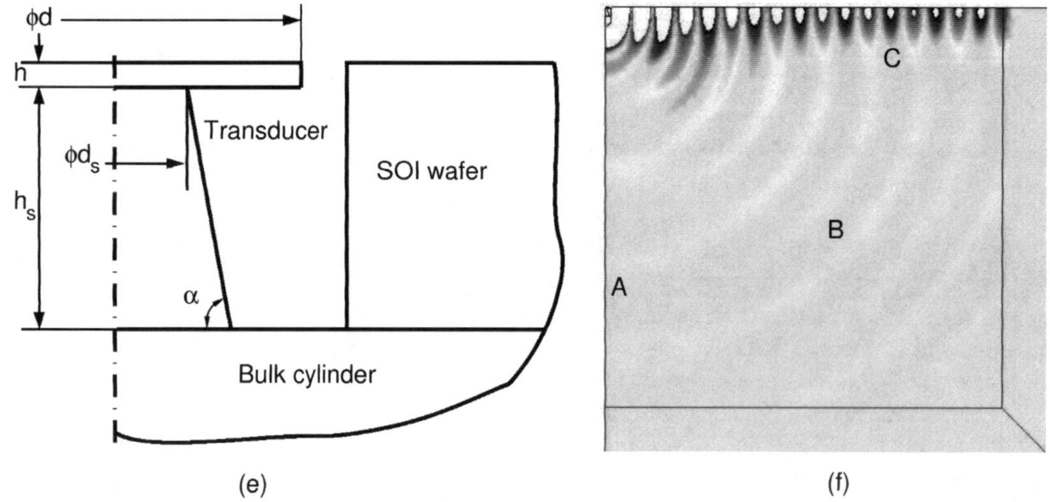

Figure 10. (e) Geometry of an optical generation transducer of an FE model. (f) The ultrasonic waves predicted by FE simulation from the transducer shown by (e) at top right corner of the figure, and by a laser pulse. C, B and A show the generated surface wave, shear wave and longitudinal wave respectively.

Lamb wave at the disk root (see figure 9(c)) because of the small thickness of the disk. After this, reflections of the Lamb wave between the disk edge and the disk root start because of the small step length Δd and soon turn the Lamb wave into a resonant flap motion of the disk (figure 9(d)).

The disk structure is a resonator and the resonance determines the frequency of the transducer. The modes activated by a laser pulse tend to be dominated by one of the natural resonances of the disk, referred to as the main frequency/mode of the transducer; this mode tunes the vibration of the disk. Usually it is the first resonance that is dominantly actuated if the disk is 2-D and axisymmetric. Depending on the parameter $\frac{\Delta d}{h}$, the material and the optical actuation, more modes—higher or lower than the main mode of the transducer—may be actuated at the same time, in particular for the case where the transducer is nonaxisymmetric. It is necessary to point out that although theoretically a transducer can be simplified to be 2-D axisymmetric, an actual transducer is usually nonaxisymmetric because of material anisotropy, fabrication errors or the optical alignment; the actual main frequency can be a higher order resonance. For an axisymmetric 2-D disk with a suitable $\frac{\Delta d}{h}$, and the laser pulse being incident on the disk center, the actuated modes are dominated by the first resonance of the disk, resulting in a nearly sinusoidal resonant flap motion of the disk. Because the disk is elastically supported by the solid stem, this resonant flap motion couples a surface wave traveling along the surface of the stem. A surface wave consists of a transverse wave and a longitudinal wave [21]. Due to the structural axisymmetry, the phase of the transverse wave is axisymmetric with respect to the cross section of the stem. Thus, the coupled transverse wave is canceled out because of the small diameter of the stem. However, the coupled longitudinal wave is in-phase around the stem cross section. Therefore, the coupling of the disk resonant flap vibration leaves an enhanced longitudinal wave, traveling along the axis of the stem, with a bandwidth the same as that of the disk resonant motion. Because the bandwidth is defined by the resonant motion of the disk, it should be narrower than normal laser generation of ultrasound reported in the literature [10, 11, 12, 13].

From this working principle, two conditions are important for a successful transducer design: only one mode is dominantly actuated and the phase of the mode should be the same around the circumference of the disk. The first condition ensures the generated wave is narrowband; the second enables most of the kinetic energy induced by the absorbed optical energy being tuned into this narrowband wave. Therefore, it is a key step to design a disk resonator, by which, for a given laser pulse, only one mode is dominantly actuated while other modes should be minimized. In a case where the disk of the transducer is not 2-D, or the transducer is not axisymmetric, the resonant mode may be nonaxisymmetric with a phase changeable around the disk circumference. In this case the coupled transverse wave cannot be completely canceled while the longitudinal wave may be partially canceled. This reduces the efficiency and decreases the SNR of the ultrasound system, or widens the corresponding bandwidth; as a consequence, the amplitude of the generated ultrasound is reduced. Figure 10 shows the FE simulated ultrasonic waves generated by a microdisk MEMS transducer.

3.2. Design and fabrication of the transducer

The purpose of the design process is to define the optimum geometry of a transducer for a required frequency to achieve the ultrasound with the highest amplitude. The geometries of the designed MEMs transducers are shown by Table 1 with frequencies of 1, 2 and 3 MHz respectively. The 1-MHz transducers were designed as being optimal and the 2- and 3-MHz transducers were designed for comparison. As a typical two-mask-level MEMs device, the transducer design is strongly influenced by the fabrication process. The transducers were fabricated by the standard SOIMUMPs process at MEMSCAP. SOI wafer was used once again as the basic material for the transducer design and fabrication. Therefore, the disk thickness

Table 1. Frequencies and dimensions of designed MEMs transducers

f (MHz)	$d_d(\mu m)$	$d_s(\mu m)$	$h(\mu m)$	$h_s(\mu m)$
1	412	100	25	401
	462	150		
	512	200		
2	310	100		
	360	150		
	410	200		
3	272	100		
	322	150		
	372	200		

is $h = 25\mu m$ for all generation transducers. Figure 11(a) shows the top view of the designed transducers. The 3-D structure for a 3-MHz transducer is shown by figure 11(b). As can be seen from the image the fabricated stem is not a solid cylinder as shown by figure 9(a) but a truncated cone due to the restrictions of the design rules of the fabrication process. In this case the stem diameter means the diameter of cross-section of the stem against the lower surface of the disk. In addition to this, four spokes are used to support the transducer; each spoke is only connected to the stem at the bottom location while the upper part of each spoke and the disk are completely free from each other.

Two deep reactive ion etching processes were used to fabricate the transducer. Details of the fabrication process can be referred to at the website of MEMSCAP. Figs. 11(c–f) are images of fabricated transducers under a microscope, where (c) shows the top view of two 1-MHz transducers with d_s=100 and $150\mu m$ and (e) shows the top view of two 3-MHz transducers with the same d_s. Figs. 11(d) and (f) demonstrate the corresponding back side images of the transducers, where the blurred part in each photo is the image of the disk for each transducer viewed from the back side of the wafer.

3.3. Experimental study of the MEMs generation transducers

In order to check if the designed and fabricated transducers can be applied to optically generate ultrasound with a high amplitude as expected, a system was configured, as shown in figure 11(g), to conduct experimental studies with these fabricated transducers. The optical transducer sample is attached to the surface of a metal block with a given thickness. The metal block is bonded to the surface of a piezoelectric transducer. As a laser pulse is incident on the disk of a transducer on the sample (figure 11(g, top)), the generated ultrasound travels through the metal block and can be sensed by the piezoelectric transducer (figure 11(g, bottom)). We used again the 5-MHz Videoscan piezoelectric transducer. Figure 12(h)$\sim$(k) illustrate the detected ultrasound by the test system with the piezoelectric detection transducer, and aluminum blocks with different thicknesses of 10 and 15mm respectively.

Figure 12(h) shows the generated longitudinal wave and the corresponding Fourier transformation by a 1-MHz MEMs optical transducer; it has a narrow bandwidth, centered at 0.89 MHz with a high amplitude at the frequency, and with a 15 mm aluminum metal block. Similarly figure 12(i) shows the generated strong signal around 1.98 MHz and the corresponding Fourier transformation for a 2-MHz transducer, and with a 15 mm aluminum metal block. For a 3-MHz transducer with a 10-mm aluminum block, the generated longitudinal wave is shown by figure 12(j).

In order to show the high efficiency of this optical transducer, a comparison between the

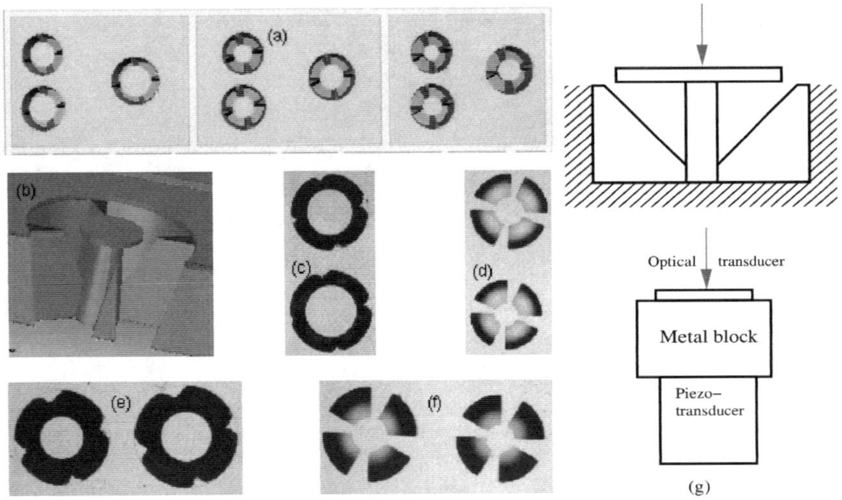

Figure 11. (a) Image of top view of the designed 1-MHz (left), 2-MHz (middle) and 3-MHz (right) transducers with stem diameters of 100, 150 and 200 μm respectively. (b) 3-D structure of a 3-MHz transducer. (c) and (d) Front and back images of the 1-MHz transducers with different stem diameters under a microscope. (e) and (f) Similar images of 3-MHz transducers. (g) Experimental configuration.

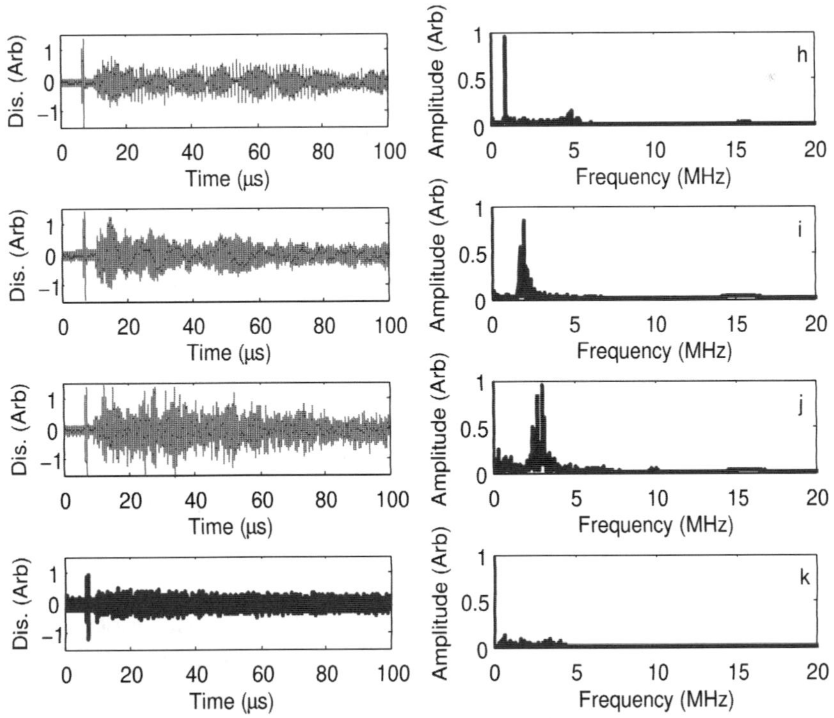

Figure 12. (h)∼(k) Ultrasound measured by test system (figure 11(g)) and produced by the fabricated MEMs generation transducers at laser pulses.

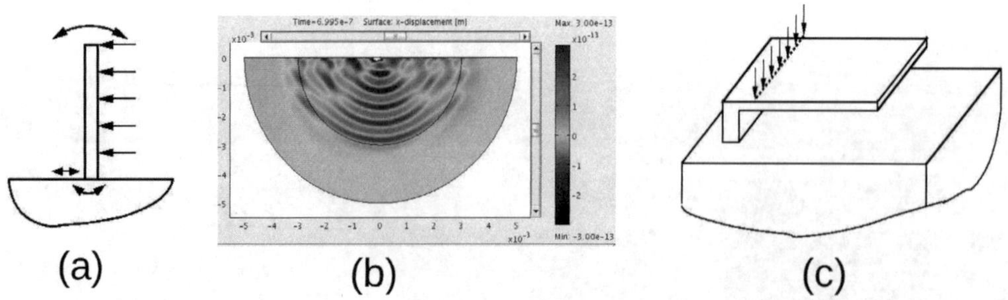

Figure 13. Possible future work on optical MEMs detection and generation transducers. (a) Working principle of a narrowband shear wave transducer. (b) FE predicted shear waves generated by a vertical cantilever shear wave transducer at a laser pulse. (c) manufacturable MEMs structure for an optical shear wave generation transducer.

transducer generation and the normal thermoelastic generation has been shown by figure 12(h) and figure 12(k). The diagram (k) shows the ultrasound by normal thermoelastic generation with the same optics using the substrate surface away from any structure on the transducer sample. It can be seen that amplitude of the broadband ultrasound by normal thermoelastic generation is only at the noise level of the narrowband ultrasound generated by the MEMs transducer; the amplitude of the generated narrowband ultrasound is at least 5 times higher than that of the normal thermoelastic generation (figures (h) and (k) are on the same scale). More important is that the SNR of the narrowband ultrasound is much higher than that of the broadband ultrasound generated. It is an interesting fact that the amplitude achieved by the MEMs transducer was difficult to reach by normal optical generation, even with the input optical power in the ablation regime. This is an advantage of this optical transducer over a simple absorption layer.

For the resonant flap motion of a disk, it is not necessary for the laser light to be well focused; a laser spot matching the disk radial dimension can increase the laser power while the power density still remains within thermoelastic regime. Thus, pulsed light from a laser with relatively higher power can be applied to this transducer. A hopeful improvement of the transducer is to combine with the technique presented in [7] with a highly absorbing layer deposited onto the disk surface of the MEMs transducer for a laser with less power. For high frequency narrow band ultrasound, a finer disk array transducer perhaps can be designed.

4. Future work

4.1. Optical MEMs shear wave transducer

Some possible work in the near future is shown in figures 13 and 14. The first work that can be done is to design an optical MEMs narrowband shear wave generation transducer, as shown by figure 13(a∼c). (a) shows the working principle of the MEMs structure. It is a vertically fabricated cantilever on a solid material surface. When a laser pulse is incident upon one side of the vertical finger, its first resonance would be activated. The connection between the finger and the bulk material underneath is elastic, so that the motion of the finger root can be decomposed into a rotation and a shear motion along the solid surface at the first resonant motion of the finger. The shear motion can produce a shear wave propagating into the material underneath, while rotation around the root can be equivalent to two vertical displacements, along the two-sides of the finger with opposite direction and with the same amplitude. Because these two motions are close to each other, they generate nearly nothing for any location far away from the root. This shows a vertical cantilever MEMs structure can be used as a shear wave generation

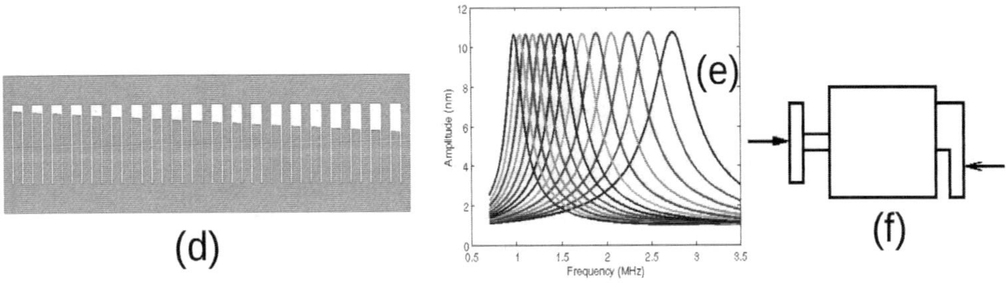

Figure 14. Possible future work on optical MEMs detection and generation transducers. (d) A MEMs structure with wide bandwidth for an optical detection transducer. (e) FE predicted bandwidth of the MEMs structure shown by (d). (f) A simple but efficient laser generation and detection system.

transducer. Figure 13(b) shows FE predicted shear wave generated by such a transducer at a laser pulse. However, this vertical cantilever structure is difficult to manufacture, we present an improved equivalent MEMs structure, as shown by figure13(c); it functions as a vertical cantilever and is possible to be manufactured by surface micromachining technology.

4.2. Improvement of detection transducer

Although the designed optical MEMs detection transducer has a much higher sensitivity to ultrasound than traditional ones [5], it has a narrow bandwidth; for any ultrasound with a frequency out of this narrow bandwidth, the sensitivity of the transducer will soon decrease to the same level as traditional ones. In order to increase the bandwidth of the transducer, a new MEMs structure is presented as shown by figure 14(d), with changeable finger lengths. Because the length of each finger is different, the first resonance of each finger is also different, thus ultrasound with different frequency will select different finger to resonate. Due to the actual damping of each finger, if one finger is in resonance, the fingers adjacent to it will also have relatively high amplitude provided the change rate of finger length is not too high. In actual design, it should make sure that for ultrasound with any frequency within the bandwidth, at least there should be more than three fingers in motion and have relatively high amplitudes to enable the transducer work[5]. Figure 14(e) shows FE predicted bandwidth of the new MEMs structure.

4.3. A possible simple but efficient optical laser generation and detection system

Our aim is to set up a simple but efficient optical laser generation and detection system, as shown in figure 14(f) as an example. Upon a metal block, at one side an optical generation transducer is attached, and on the opposite side, an optical detection transducer is amounted. We hope the optically generated ultrasound at a laser pulse, propagating through the metal block with given dimensions, can still be detected by the optical detection transducer on the other side. Thickness of the metal block will be an important assessment parameter to the system. The key factor to make sure the system works is the bandwidth of the detection transducer. It should be sufficiently wide to include the frequency of the generated ultrasound within it and to ensure more than three fingers of the detection transducer are activated at the same time.

5. Conclusions

- Microcantilever optical ultrasound detection transducers were successfully designed, fabricated and used to detect ultrasonic waves with high sensitivity. The sensitivity of the transducer is ∼1–3 orders of magnitude higher than conventional CHOTs or other optical detection techniques. The sensitivity is $\geq$ 700 for the designed 1-MHz transducer, $\geq$ 550 for the 3-MHz transducer and $\geq$ 60 for the 5-MHz transducer. The transducer is highly noise resistant, and geometrically compact. No wiring or power connections are required, lending the transducers themselves to remote non contact applications and to those in hostile environments. These transducers are also inexpensive even in prototype numbers; the cost of per transducer was less than $10. However, the transducer has a narrow bandwidth.

- Optical MEMs generation transducers were designed and fabricated. FE simulation and experimental study have shown that these transducers can be applied to optically generate narrowband ultrasound—surface wave, shear wave and longitudinal wave—with high efficiency, high amplitude/SNR and narrow bandwidth; compared with normal thermoelastic generation, ultrasound with at least 5 times higher amplitude can be achieved by a suitably designed optical transducer. Only a simple optical arrangement is required to use this transducer.
- It is possible to design an optical MEMs shear wave generation transducer, and to design an optical detection transducer with a suitable bandwidth in the near future. Our aim is to set up a simple but efficient optical laser generation and detection system.

References

[1] R E Green Jr 2004 *Ultrasonics* 42 9-16
[2] S J Davies, C Edwards, G S Taylor and S B Palmer 1993 *J. Phys. D: Appl. Phys.* 26 329-348
[3] A S Murfin, R A J Soden, D Hatrick and R J Dewhurst 2000 *Meas. Sci. Technol.* 11 1208-1219
[4] K C Baldwin, T P Berndt and M J Ehrlich 1999 *Ultrasonics* 37 329-334
[5] X, Chen, T Stratoudaki, S D Sharples and M Clark 2009 *IEEE Trans. Ultrason. Ferroelectr. Fre. Control* 56(12) 2722-2732
[6] D Karabacak and T Kouha 2006 *Appl. Phys. Lett.* 88 193122
[7] E Biagi, F Margheri and D Menichelli 2001 *IEEE Trans. Ultrason. Ferroelectr. Freq. Control* 48(6)
[8] I Arias and J D Achenbach 2003 *Int. J. Solids. Struct.* 40 6917-6935
[9] Stratoudaki T, Hernandez J A, Clark M. and Somekh M. G. 2007 *Meas. Sci. Technol.* 18(3) 843-851
[10] J W Wagner, J B Deaton and J B Spicer 1988 *Appl. Optics* 27(22) 4696-4700
[11] J B Deaton, Jr, A D W. McKie, J B Spicer, and J W Wagner 1990 *Appl. Phys. Lett.* 56(24) 2390-2392
[12] Yves H Berthelot and Jacek Jarzynski 1990 *J. Nondestruct. Eval.* 9(4) 271-277
[13] James W Wagner, Andrew D W McKie, James B Spicer and John B Deaton Jr 1990 *J. Nondestruct. Eval.* 9(4) 263-270
[14] Giannoula A and Cobbold RS 2008 *IEEE Trans. Ultrason. Ferroelectr. Freq. Control.* 55(2) 343-58
[15] Gregory R Osche 2002 *Optical detection theory for laser applications* (Hobpken, N.J.:Wiley-Interscience)
[16] P LemoineU, J F Zhao, A Bell, P Maguire and J McLaughlin 2001 *Diam. Relat. Mater.* 10 94-98
[17] Jeung-hyun Jeong and Dongil Kwon 2002 *J. Appl. Phys.* 91(10) 6333-6340
[18] Marc J Madou 2002 *Fundamentals of Microfabrication, the Science of Miniaturization, second edition* (CRC press, Boca Raton, London, New York, Washton DC)
[19] W. Weaver, Jr, S P Timoshenko and D H Young 1990 *Vibration problems in engineering* (John Wiley & Sons, New York)
[20] Tarsicio Beléendez, Cristian Neipp andAugusto Beléendez 2002 *Eur. J. Phys.* 23 371-379
[21] Peter J Shull 2002 *Nondestructive Evaluation, theory, techniques and applications* (CRC Press)

Orientation imaging using spatially resolved acoustic spectroscopy

R Smith, S Sharples, W Li , M Clark and M Somekh
University of Nottingham, Electrical Systems and Optics Research Division, University Park, Nottingham, NG7 2RD, UK

Richard.j.smith@nottingham.ac.uk

Abstract. The term 'orientation imaging microscopy' describes generic techniques for imaging the orientation of crystalline structures of heterogeneous media. Typically, scanning electron microscope techniques are used, such as electron backscatter diffraction or transmission electron microscopy. We have developed an acoustic technique that can perform equivalent measurements. The key to enabling a practical acoustic method is spatially resolved acoustic microscopy (SRAS), a robust laser ultrasonic technique for quantitatively determining the surface acoustic wave velocity at a high spatial resolution. Here we present quantitative determination of crystallographic orientation of large nickel grains by comparing the measured SAW velocity in multiple propagation directions, with a data base of calculated velocity surfaces for all orientations. As well as discussing the method of determining the crystallographic orientation details are also presented of the recent advances in the capabilities of the latest generation of instrumentation, including significant increase in the data acquisitions rate and the reduction in size and complexity of the SRAS instrument itself.

1. Introduction - Grain orientation and SRAS

Measuring the crystallographic orientation of grains of different materials is generally performed with electron microscope techniques, such as transmission electron microscopy [1] and electron backscatter diffraction microscopy [2]. These techniques provide the crystallographic orientation with exceptional spatial resolution across the sample of interest. The main drawbacks to these electron based techniques are the requirements on samples size and preparation. The samples have to be small in size to fit inside the electron microscope so typical samples sizes are of the order of 10x10mm. The electron microscopy techniques are very sensitive to the surface finish and so special care must be given to ensuring the finish of the sample is sufficient, this typically requires polishing the sample to a mirror like finish. We have developed an acoustic technique that can perform equivalent measurements, on samples of any size and varying surface finish, although with a lower spatial resolution than that obtained by electron techniques (10's microns).

Although the SRAS technique has been described previously [3, 4], it is worth repeating the salient details here as the new instrument uses the fixed grating method. The SRAS technique itself will be described, and a description of the new instrumentation will follow in the next section. We will then describe the process from going from velocity contrast maps to true orientation, before finally providing examples of experimental measurements on single crystal nickel.

2. Spatially Resolved Acoustic Spectroscopy for velocity imaging

The spatially resolved acoustic spectroscopy technique is analogous to optical spectroscopy techniques where analysis of the optical spectrum and the wavelengths that are most readily transmitted or absorbed tells us some properties of the sample being examined. In the acoustic sense we monitor the spectrum of sound waves emitted from a grating source. The technique can be performed two ways, the first is to use a fixed frequency source and sweep the grating spacing or the grating period can be fixed and the frequency is swept. The previous versions of the SRAS instrument used the first approach, the new instrument introduced here uses the second approach and this will be used to explain how the technique works.

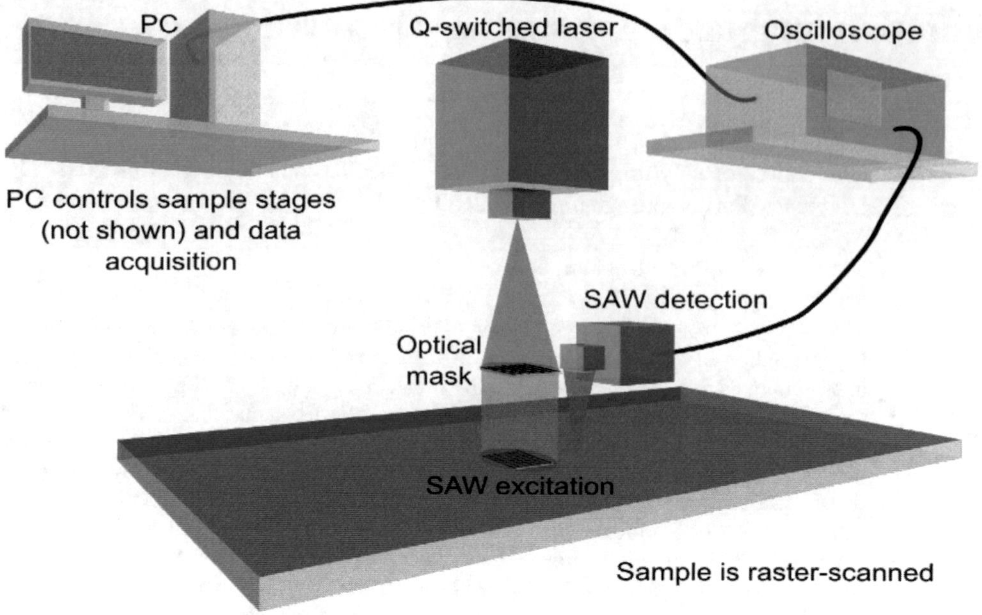

Figure 1. Cartoon of a SRAS system.

To measure the acoustic spectrum at each point on the sample a fixed grating is illuminated with a short laser pulse (typically ~1ns). As the laser pulse contains a broad range of frequencies, only the frequencies which match the grating spacing and acoustic velocity of that sample point will be generated. We measure the waves that have been generated close to the excitation region (see figure 1). We can then calculate the frequencies generated and with the knowledge of the grating spacing we can obtain the acoustic velocity using $v = f\ \lambda$. The advantage this approach has over the fixed frequency laser approach is that all the information we require to calculate the velocity is obtained in a single measurement, where as previously the grating spacing needed to be changed several times per measurement point leading to a lower data acquisition rate. To build up a velocity map the sample is raster scanned and the measurement is repeated for each point.

Using this approach the data rate is greatly increased, single shot measurements on well prepared samples can be carried out at 6000 points/second. The overall data rate obtained including data read out / download and flyback of the stages is around 1200 points / second. An example image of TiLG685 obtained with the new system is shown in figure 2. This 3.2 megapixel image was obtained in ~50 minutes. The new instrument is also considerably smaller than previous instruments. The main optics of the system fit in a head of around 350x200x150 mm with an umbilical to the electronics rack. When fully complete this new instrument will also have the ability to detect waves on optically rough surfaces and propagate the sound in any chosen direction without the need to rotate the sample itself.

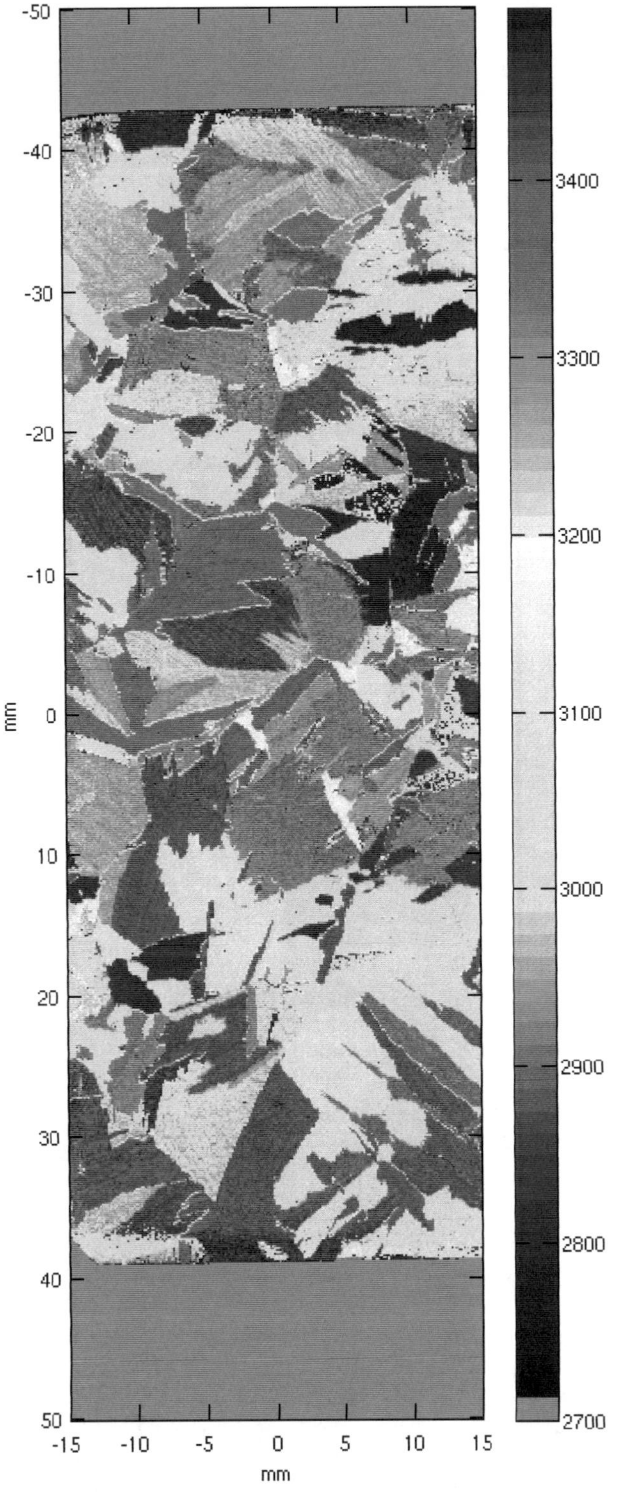

Figure 2. 3.2 Mega pixel velocity map of TiLG685 sample, waves propagating up/down direction.

3. From contrast to orientation measurements

To provide orientation measurements we need to convert from velocity contrast maps to true orientation information. This process will now be described before presenting experimental results on large nickel grains.

The acoustic velocity of a known material in a known orientation can be calculated analytically. We use the method outlined by Farnell [5]. This method is used to calculate SAW and pseudo-SAW velocities on different crystallographic planes by using iterative search procedures and the known materials' elastic constants. The method requires the iteration through different SAW velocities in a specified range to find out which velocities satisfy the boundary conditions. In the case of pseudo-SAW a similar method is used, with the additional assumption that the wave is attenuated along the propagation direction. Using this method we can calculate the velocity surface for any crystallographic orientation and a selection of surfaces are shown in figure 3.

We plot the residuals of the boundary conditions and the minimas of this curve suggest the possible propagation of different wave modes e.g. surface waves, pseudo surface waves and leaky modes. The detector is sensitive only to the out of plane motion of the waves, so we calculate the out of plane motion for all of the modes found during the search and choose the dominant mode. This then gives us an indication of the expected velocity surface for that crystallographic orientation that will be measured with our experiment.

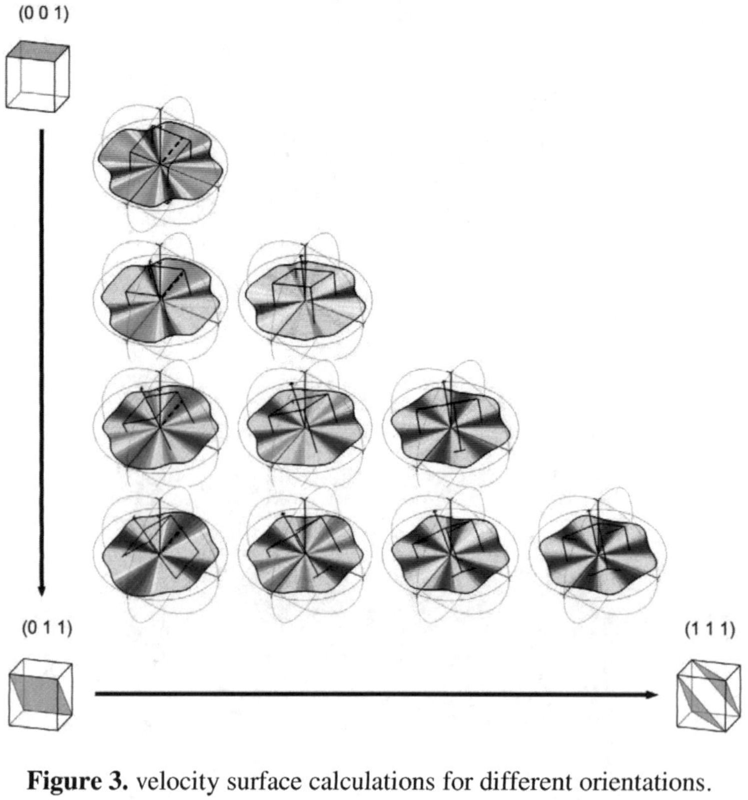

Figure 3. velocity surface calculations for different orientations.

The Inverse problem is difficult to solve, so to provide orientation measurements we use the measured velocity surface and fit it to the model data by searching a database of all known orientation angles to get the orientation of the measured point. As mentioned above the velocity surface maps we have calculated have been modified to take into account the workings of the detector so we used these adjusted maps for the fitting process. The fitting is achieved using cross correlation; we record the strength of the correlation of the experimental data with each plane and rotation in the calculated velocity values database. Once all planes and rotation angles have been compared we look for the largest correlation value as this corresponds to the best fit orientation.

This process is shown in figure 4 below. It shows the correlation strength for different planes and rotational angles and we can see various peaks in the image showing partial matches. The figure 5 shows the modelled velocities for the chosen plane superimposed on the experimental data. In this case we have only overlaid the modelled values for every 5 degrees of rotation for clarity. We see the fit is extremely good and we can be confident that the crystallographic plane has been recovered by the fitting algorithm.

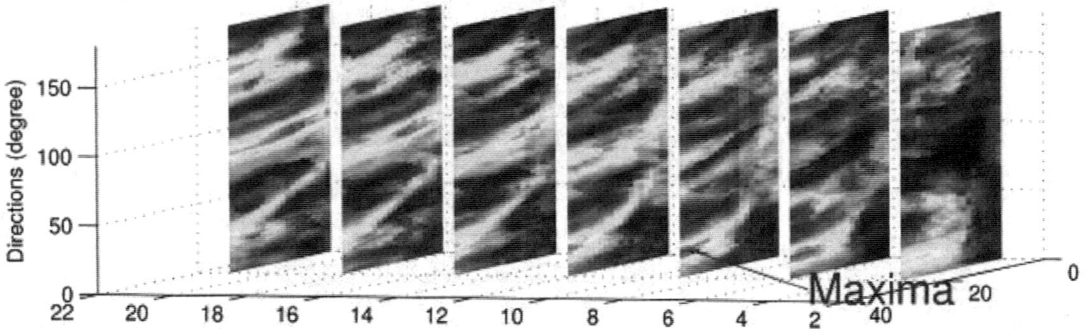

Figure 4. The correlation factors obtained during the fitting process for different rotations and plane orientations.

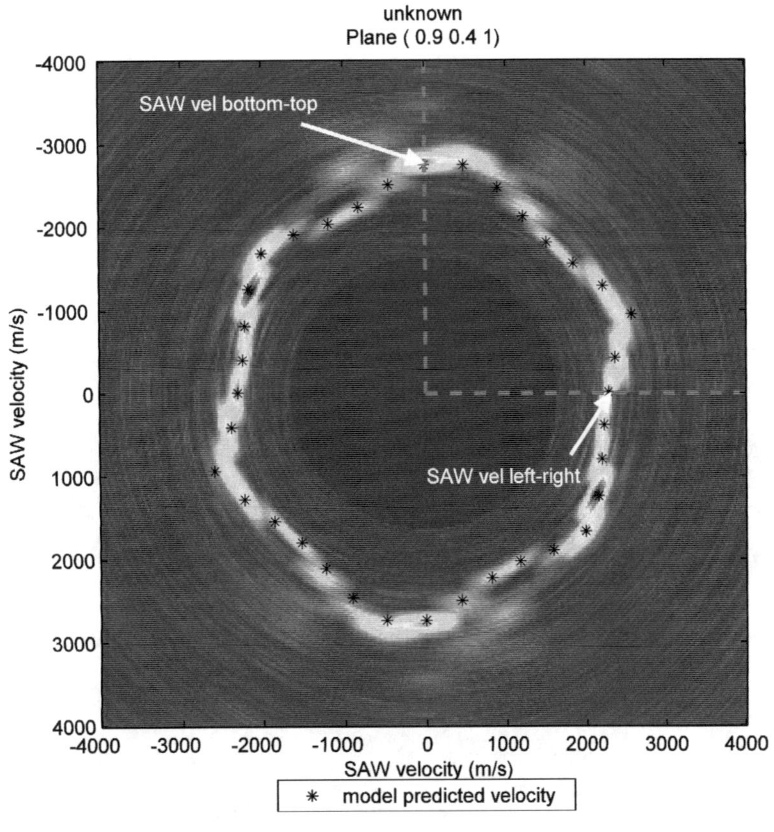

Figure 5. Unknown plane overlaid with best fit chosen velocity values.

4. Experimental Results – Single crystal nickel.

A velocity map was produced for a nominally single crystal nickel sample. This map (see figure 6 below) shows two regions of differing velocity showing that this sample is not a single crystal but is made of two grains with differing orientations. A velocity surface was then taken for each grain by measuring the surface wave velocity around 360 degrees of rotation with 5 degree increments.

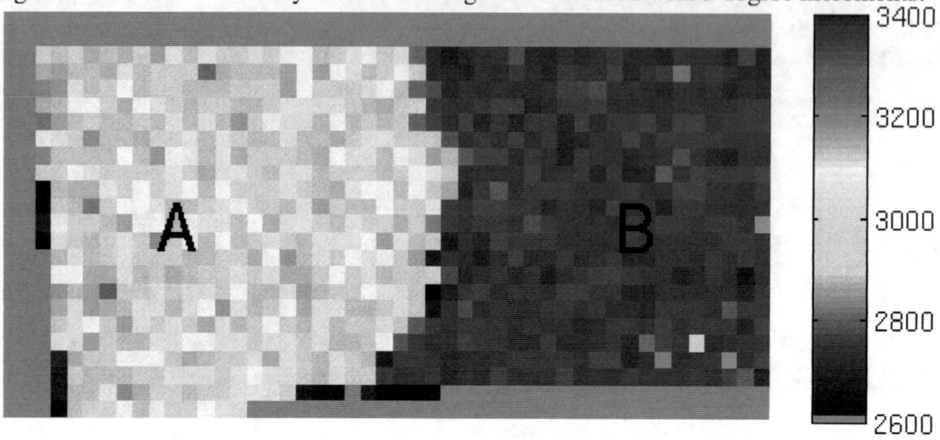

Figure 6. Velocity map of single crystal nickel sample.

The velocity surface is shown in figure 7 for the two regions. The white stars are the modelled velocity values for the output orientation obtained via the fitting process. We see excellent agreement between the experimental velocity surface and the chosen best fit plane. The chosen plane and angle are shown in the inserts of figure 7. Figure 8 shows the velocity map with a cube representing the orientation of the grain for comparison. The height of the sides of the cube show the slight change in the plane orientation and the top face shows the large rotational change.

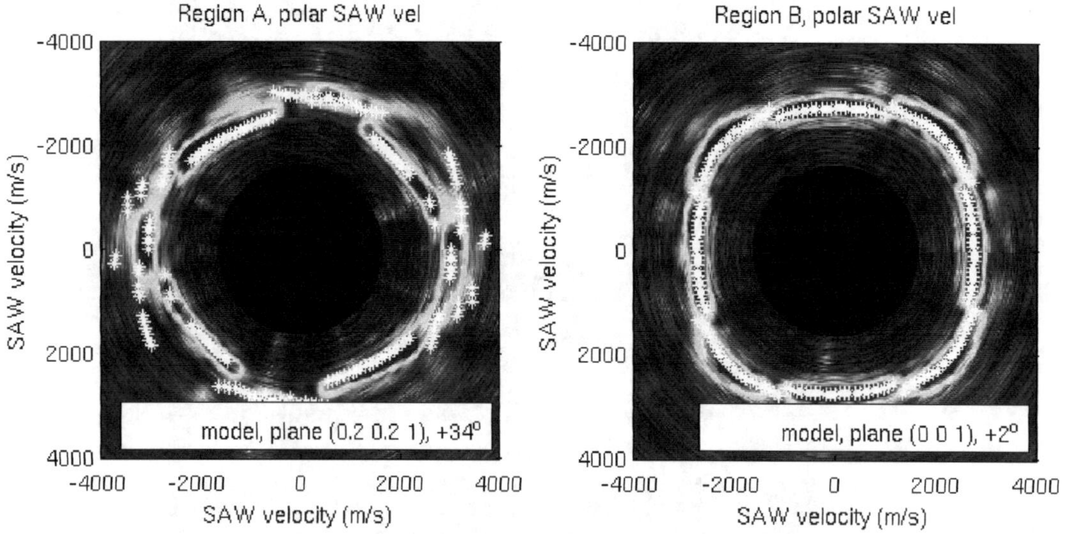

Figure 7. Velocity surface for region A and B with modelled grain orientation superimposed.

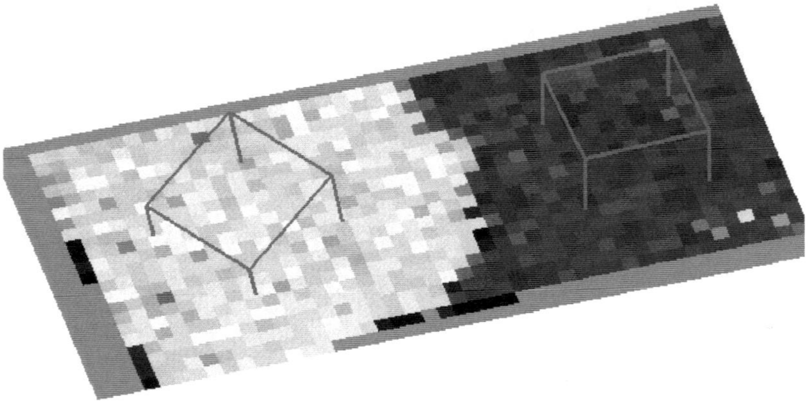

Figure 8. Velocity map with the grain orientation represented by a cube

5. Conclusions

We have presented the determination of grain orientation using spatially resolved acoustic spectroscopy. The SRAS instrumentation has been improved to allow faster data acquisition rates of 1200 points/second and a much smaller total footprint making the system easily transportable.
Obtaining the grain orientation is achieved by fitting the experimentally recorded velocity surface to a database of calculated velocity surfaces. The database is modified to take into account the response of the detector to different wave modes.

To date we can do this for any cubic material where the elastic constants are known and we are in the process of expanding this to other crystal types and optimising the fitting algorithm without comprising its robustness

6. References.

[1]Randle V 2008 *Microtexture determination and its applications* (Maney Publishing, London)

[2] Zaefferer S 2000 New developments of computer aided crystallographic analysis in transmission electron microscopy *Journal Applied Crystallography* **33** 10-25

[3] Sharples S D, Clark M and Somekh M G 2006 Spatially resolved acoustic spectroscopy for fast noncontact imaging of material microstructure *Optics Express* **14** 10435-40

[4] Sharples S D, Li W, Clark M and Somekh M G 2009 Microstructure imaging using frequency spectrum spatially resolved acoustic spectroscopy *Review of Progress in Quantitative Nondestructive Evaluation* **29** 279-86

[5] Farnell G W 1970 *Properties of Elastic Surface Waves (Physical Acoustics vol 6),* ed W P Mason and R N Thurston (Academic Press, New York and London) chapter 3 pp 109-66

Journal of Physics: Conference Series **353** (2012) 012004 doi:10.1088/1742-6596/353/1/012004

Evaluation of sound velocity inside underwater acoustic materials using test panel acoustic measurements

C Audoly
DCNS, Le Mourillon, BP 1306, 83076 Toulon Cedex, France

E-mail: christian.audoly@dcnsgroup.com

Abstract. Acoustic materials can be used as external coating to reduce radiated noise or target strength of immersed structures, or to improve performance of sonar systems. In order to optimise the design and the use of such coatings, it is necessary to have a good estimate of their intrinsic properties, in particular the complex sound velocity inside the materials. The type of material studied here consists in a viscoelastic slab containing a given proportion of micro-voids. Determination of the complex longitudinal velocity is not an easy task, and direct methods or experimental set-ups do not give sufficient results. The method presented here consists in determining the complex longitudinal velocity by solving an inverse problem, from acoustic measurements of the reflection and transmission coefficients of a test panel in a water tank. Best results, including attenuation, are obtained by over-determining the problem, using all data available. Results are consistent to the known physical behaviour of such materials, regarding influence of frequence and hydrostatic pressure.

1. Introduction

In the scope of underwater acoustics, some problems of interest are the reduction of radiated noise or target strength of immersed structures, or to improve sonar systems performances. Regarding design or choice of relevant materials, an approximate characterisation of the coatings can be done on planar structures, as shown on figure 1a and 1b.

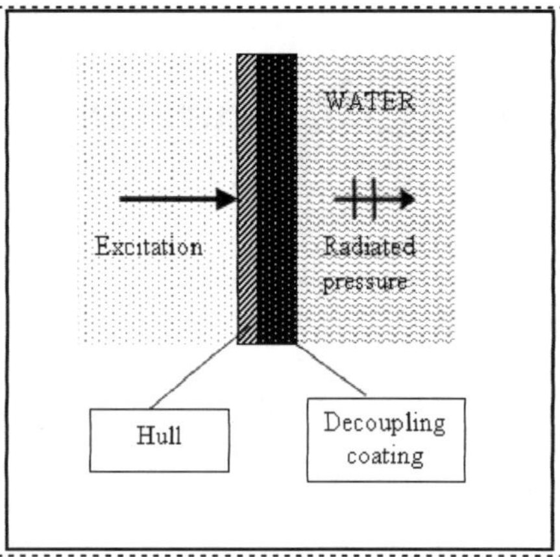

Figure 1a. Acoustic decoupling.

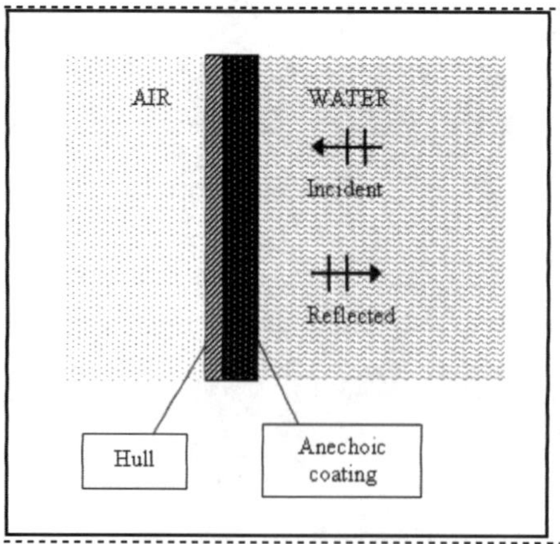

Figure 1b. Anechoism.

1.1. Type of material considered

On an engineering point of view, a possible solution is to realise composite materials consisting of a viscoelastic slab containing a given proportion of air in the form of micro-cavities or micro-inclusions. If ρ_l, K_l, G_l are respectively the density, dynamic bulk and shear moduli of the matrix, c the concentration of cavities, the composite material can be represented by an equivalent material with dynamic characteristics ρ_{eq}, K_{eq}, G_{eq}. As far as the cavities size is very small by comparison to the wavelength, these equivalent characteristics can be obtained using quasi-static approaches, such as Kuster-Toksöz method [1]. Under these conditions:

- density can be considered to be identical to static (physical density);
- resonance of cavities are not taken into account;
- frequency dependence of K_{eq} and G_{eq} appears through frequency dependence of K_l and G_l.

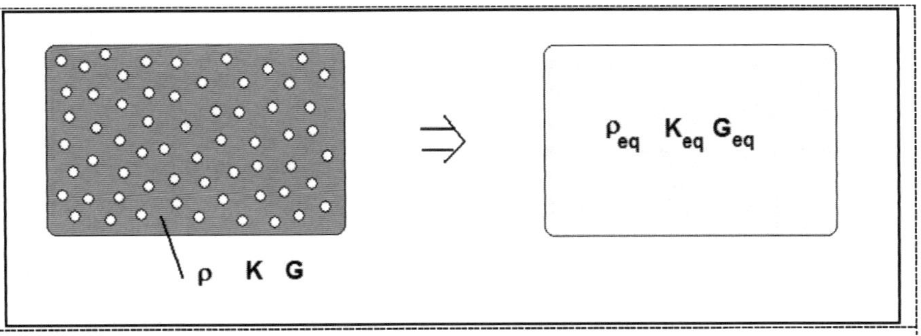

Figure 2. Representation of a composite material by an equivalent homogeneous material.

1.2. Characterisation

Two main criteria are introduced: the decoupling coefficient C_M and the anechoism coefficient C_A. It can be shown that these performance indicators are directly related to material density ρ_{eq}, material complex longitudinal sound velocity c_L, and coating thickness h [2]. Once K_{eq}, G_{eq} are known, the material is fully characterised by the density and the equivalent longitudinal and transversal sound velocities in the frequency range of interest, given by:

$$c_L = \left(\left(K_{eq} + \frac{4}{3}G_{eq}\right)\rho_{eq}^{-1}\right)^{1/2} \text{ and } c_T = \left(G_{eq}\rho_{eq}^{-1}\right)^{1/2} \tag{1}$$

Due to the attenuation, sound velocities are complex-valued, and can be represented, as follows, using a real part c'_L and a tangent loss η_L.

$$c_L = c'_L.(1 + i\eta_L) \tag{2}$$

That notation is used for simplicity, as propagation is best physically represented by using the complex-valued wavenumber given by $k_L=\omega/c_L$.

Direct determination of G_{eq} is possible using special devices such as a visco-analyser. On the other hand, direct determination of K_{eq} or c_L is difficult, all these methods having limitations [3], or cannot be used to evaluate the attenuation. The method proposed in this paper consists in solving an inverse problem, using acoustic measurement of the reflection and transmission coefficients of a test panel submitted to a plane wave in a water tank. Other authors have also developed similar approaches, some of them in the context of the study of acoustic metamaterials, for example [4]. However, [4] gives explicit formulae of equivalent densities and sound velocities, that could be very sensitive to measurement errors.

The purpose of the present paper is to improve the method developed in [5], and to show some practical results. Reduction of measurement errors is obtained thanks to an over-determination of the unknown quantities, using the more data possible.

2. Presentation of the method

2.1. Test panel measurements in a water tank

As a first step, the method requires measurements of the reflection R and transmission coefficient T of a test panel in a water tank, in the frequency range of interest as shown on figure 3. It is assumed that the method can provide the phases of R and T, as well as the amplitude.

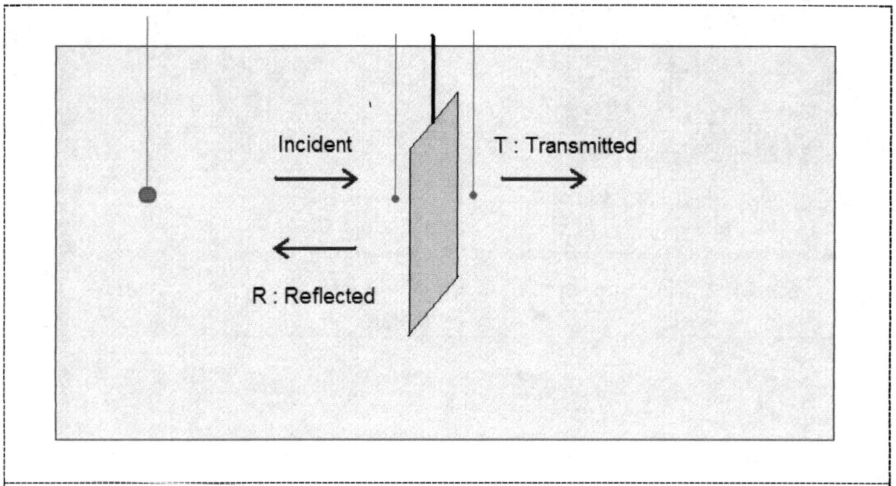

Figure 3. Set-up for test panel measurement in a water tank.

It is important to have at one's disposal good quality data. Some problems occur with this kind of measurement, in particular diffraction effects from the edges of the test panel. To reduce these phenomena, different methods are available, for example [6] and [7].

2.2. Evaluation of the complex longitudinal sound speed using test panel data

In the present case, i.e. simple layer materials in the quasi-static regime, the problem to be solved at a given frequency consists in the determination of two scalar values, which are c'_L and η_L, because it assumed that density ρ_{eq} has been determined using other means. Indeed, the complex reflection and transmission coefficients depend in that case only on c_L, through the following expressions:

$$T = \frac{2}{2\cos(k_L h) + i\left(\frac{Z_{eq}}{Z_0} + \frac{Z_0}{Z_{eq}}\right)\sin(k_L h)}, \quad R = \frac{i\left(\frac{Z_{eq}}{Z_0} - \frac{Z_0}{Z_{eq}}\right)\sin(k_L h)}{2\cos(k_L h) + i\left(\frac{Z_{eq}}{Z_0} + \frac{Z_0}{Z_{eq}}\right)\sin(k_L h)}, \quad (3)$$

where $Z_{eq}=\rho_{eq}\, c_L$ and $Z_0=\rho_0\, c_0$ are acoustic impedances of the material and the surrounding fluid medium (water), respectively.

The unknown values of c'_L and η_L are obtained by minimising numerically, in a two-dimensional space, the error between calculated and measured values of R and/or T. Different minimisation criteria have been considered:

- N°1 : $\left| |R_{calc}| - |R_{meas}| \right|$
- N°2 : $\left| |T_{calc}| - |T_{meas}| \right|$
- N°3 : $\left| |R_{calc}| - |R_{meas}| \right| + \left| |T_{calc}| - |T_{meas}| \right|$
- N°4 : $| R_{calc} - R_{meas} |$
- N°5 : $| T_{calc} - T_{meas} |$
- N°6 : $| R_{calc} - R_{meas} | + | T_{calc} - T_{meas} |$

In reference [5], minimising criterion N°5 has been used, which was giving acceptable results for some materials. With the present study, we recommend criterion N°6, because it is the one that uses the more information available, i.e. both amplitude and phase of reflection and transmission coefficients. As we have four scalar data with respect to two scalar unknowns, a more robust solution is obtained, as the problem is mathematically over-determined. That result is illustrated by the following example obtained on a test anechoic sample, 50 mm thick, at 10 kHz frequency. Figures 4 to 9 give two-dimensional plots of the errors between calculus and measurement, along c'_L and η_L, for criteria N°1 to N°6, respectively. Scale data are not shown for confidentiality reasons.

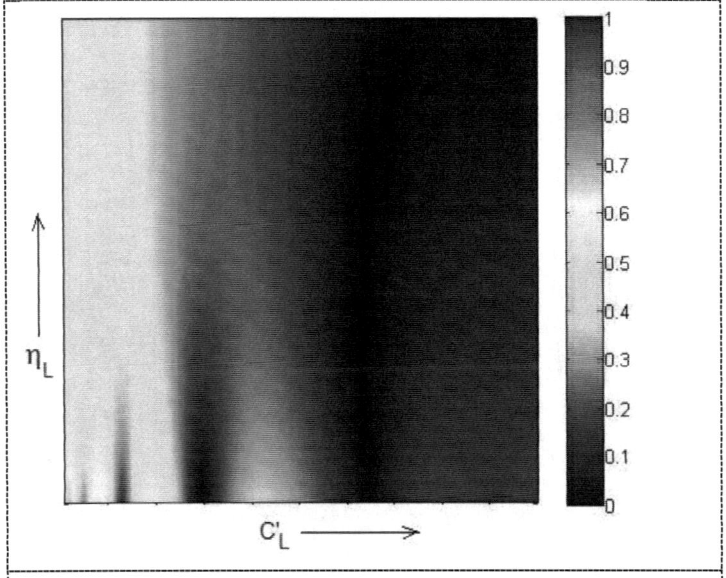

Figure 4. Test anechoic panel at 10 kHz – Two-dimensional plot of minimising criterion N°1.

Figure 4 shows different areas with a dark blue colour, which means that many different sets of (c'_L, η_L) values can correspond to a minimum. This is not surprising, because here we have used a criterion base on one scalar data only, as we look for two unknown scalar values. The problem is under-determined, leading to an unacceptable result. The same problem arises for criterion N°2, shown on figure 5.

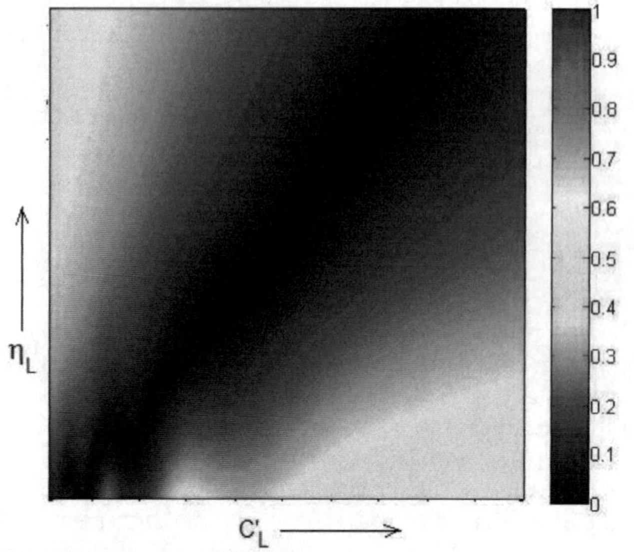

Figure 5. Test anechoic panel at 10 kHz – Two-dimensional plot of minimising criterion N°2.

Criterion N°3, shown on figure 6, gives much better results, because it combines the two previous ones. We have here two input data for two scalar values to determine. As the ill-determination using criteria N°1 and N°2 do not coincide, we obtain here a reasonable result, circled in white. However, some other values, circled in grey, could correspond to the solution.

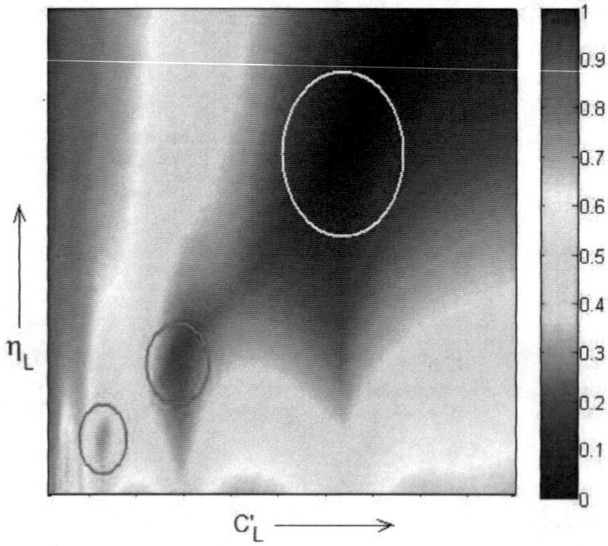

Figure 6. Test anechoic panel at 10 kHz – Two-dimensional plot of minimising criterion N°3.

When both amplitude and phase of panel coefficients are used, we obtain the results shown on figure 7 and 8. As previously, we can obtain a relevant estimate of (c'_L, η_L), but there is a doubt between different solutions

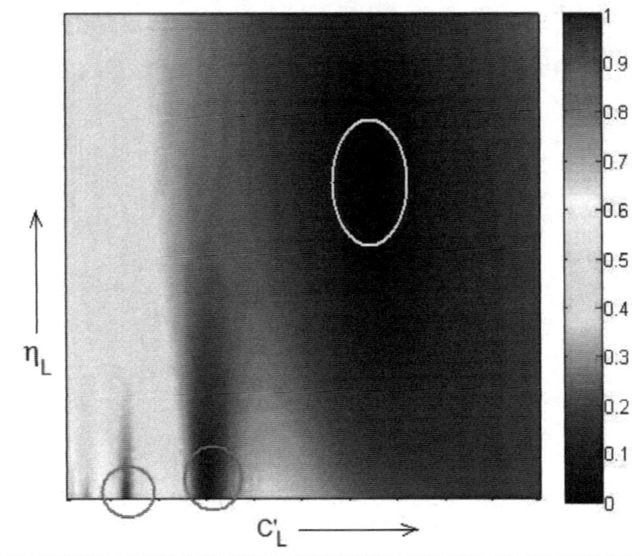

Figure 7. Test anechoic panel at 10 kHz – Two-dimensional plot of minimising criterion N°4.

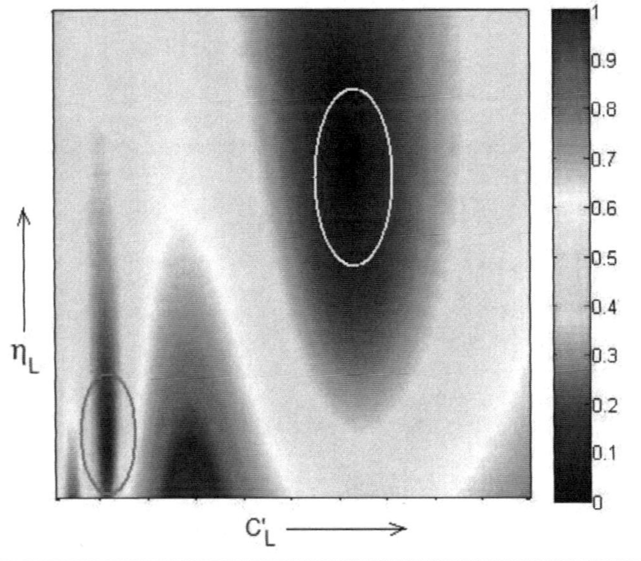

Figure 8. Test anechoic panel at 10 kHz – Two-dimensional plot of minimising criterion N°5.

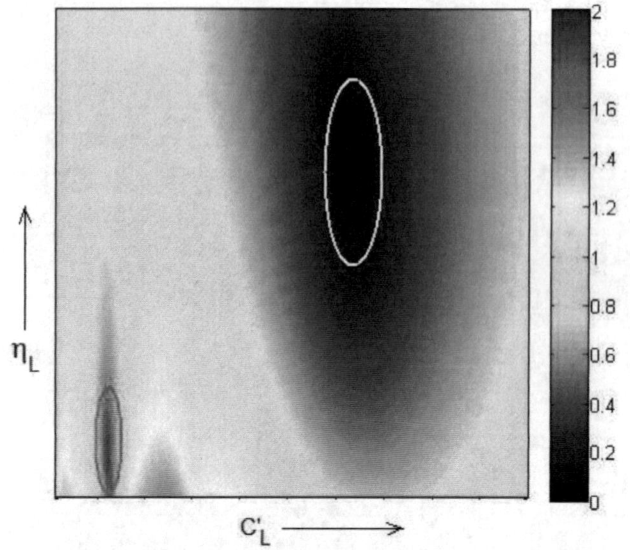

Figure 9. Test anechoic panel at 10 kHz – Two-dimensional plot of minimising criterion N°6.

Criterion N°6 on figure 9 gives the best result. There are still two possible solutions:

- one for a relatively high sound velocity and a high damping coefficient,
- an other one with lower velocity and damping.

The second solution can be discarded for different reasons:

- the minimum is weaker,
- the test material being anechoic, a strong damping coefficient is expected.

More analysis, based on numerical prediction of sound velocity of the composite, could be made to confirm the result.

3. Influence of frequency and hydrostatic pressure

This section presents more practical results on a 50 mm thick test anechoic material, in the frequency range 2 kHz – 20 kHz, and at different increasing hydrostatic pressures (P0 to P4).

3.1. Influence of frequency

Data at pressure P3 is considered here. Input information includes:

- density ρ_{eq} and thickness h at pressure P3,
- reflection and transmission coefficients versus frequency (amplitude and phase), shown on figures 10 and 11.

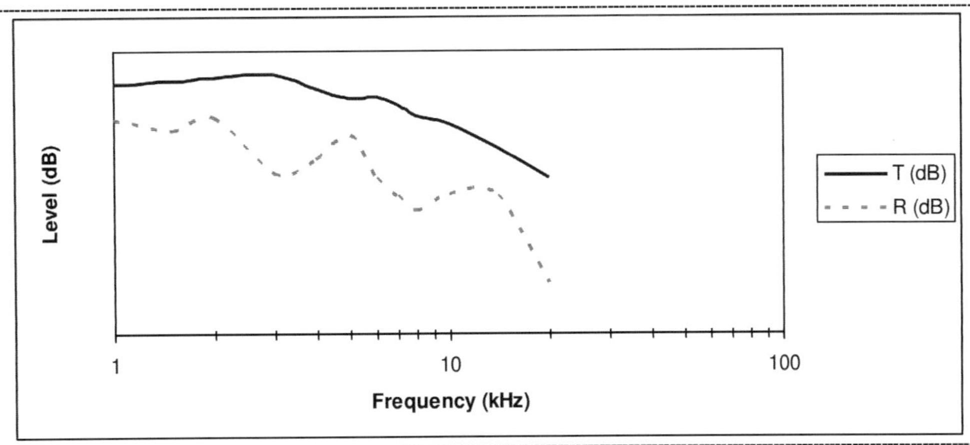

Figure 10. Test anechoic panel at pressure P3 – Level of transmission and reflection coefficients.

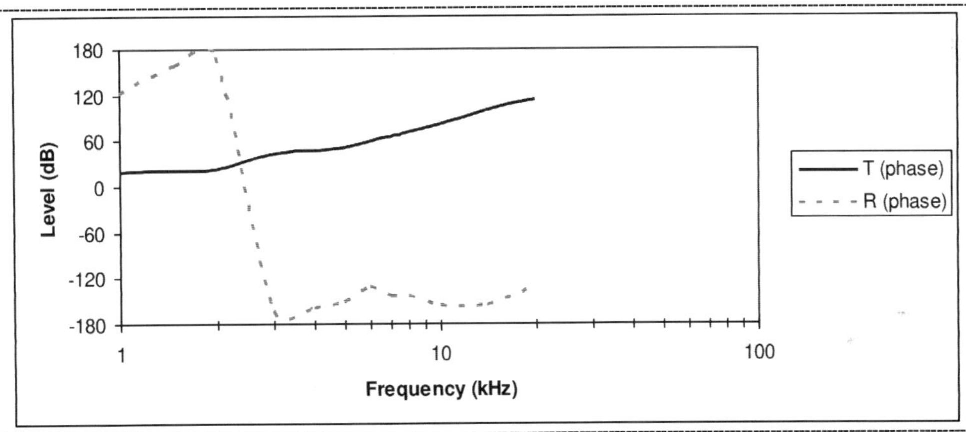

Figure 11. Test anechoic panel at pressure P3 – Phase of transmission and reflection coefficients.

The levels in dB on figure 10 are defined as $20 \log_{10} |T]$ and $20 \log_{10} |R|$. Scale is not shown for confidentiality reasons, the upper value in the scale being 0 dB, i.e. unit amplitude in natural value (corresponding to total transmission or total reflection). Not surprisingly, because of the relatively small thickness of the test panel compared to wavelength, and the panel being immersed with water on both sides, significant transmission is observed. Note also the material behaves like a partially soft reflector around 2.5 kHz, the phase of R being close to 180°.

Applying the process discussed in section 2 for each available frequency, we obtain the evolution of complex longitudinal velocity along frequency. Results are shown on figures 12 and 13. The following remarks can be made:

- results are consistent because sound velocity increases along frequency, as expected for this kind of material,
- some spurious points appear, due to error measurements on R and T,
- results diverge below a certain frequency, related to panel thickness,
- symmetrically, some attempts at higher frequencies did not give good results, the test panel being too opaque, acoustic measurement does not contain enough useful information.

For practical use, it is possible to fit raw data with smooth curves (dashed curves on the figures).

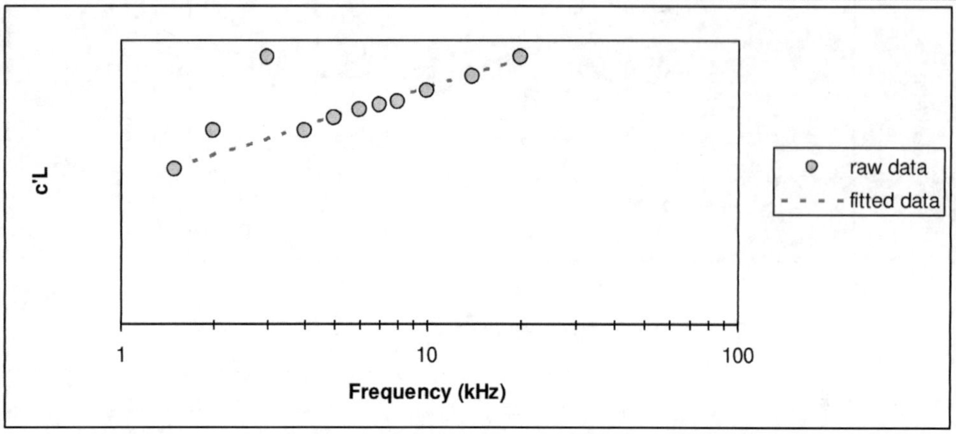

Figure 12. Test anechoic panel at pressure P3 – Estimate of sound velocity (real part).

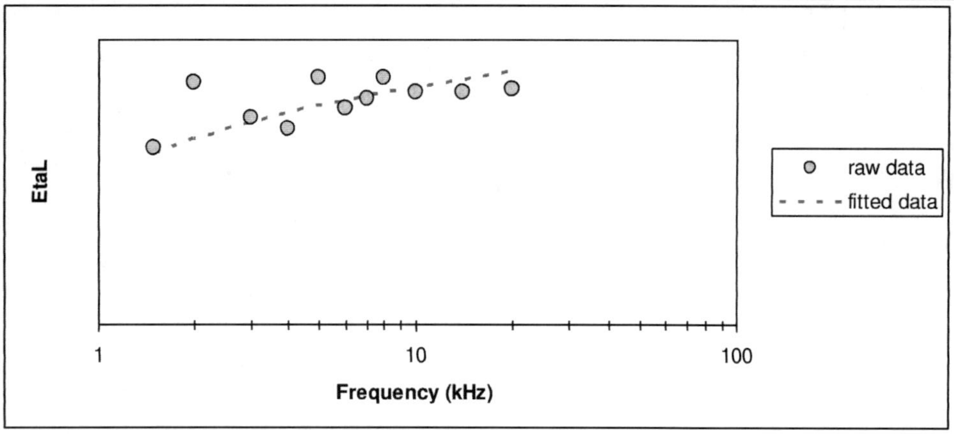

Figure 13. Test anechoic panel at pressure P3 – Estimate of sound velocity (damping).

3.2. Influence of hydrostatic pressure

Same analysis is done for hydrostatic pressures P0, P1, P2 and P4. Figures 14 and 15 show fitted curves of c'_L and η_L along frequency and their evolution with increasing hydrostatic pressure. When pressure increases, there is a variation in the actual concentration of voids in the material related to the collapse of the micro-balloons. However, results show an "abnormal" behaviour for the lowest pressures P0 and P1, as we could expect a lower sound velocity and a higher damping coefficient, the void concentration being larger. As a matter of facts, that behaviour is characteristic of materials made with a viscoelastic slab containing micro-balloons, as presented in reference [8].

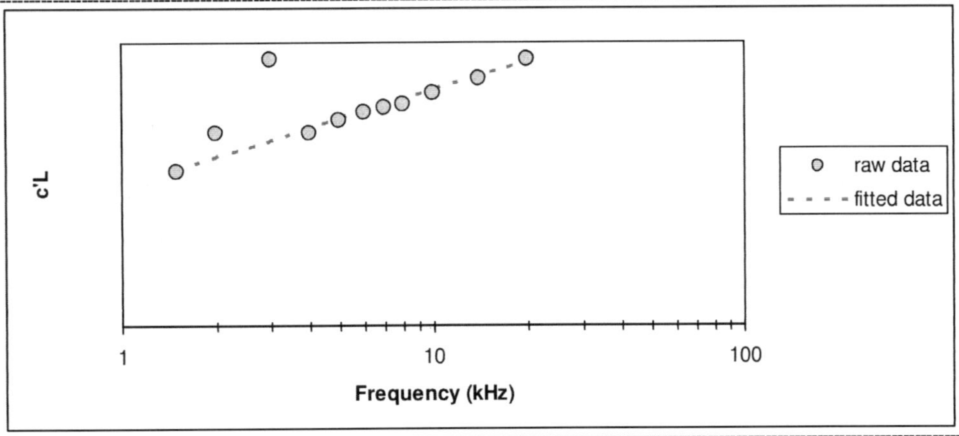

Figure 14. Test anechoic panel – Influence of hydrostatic pressure on longitudinal velocity (real part).

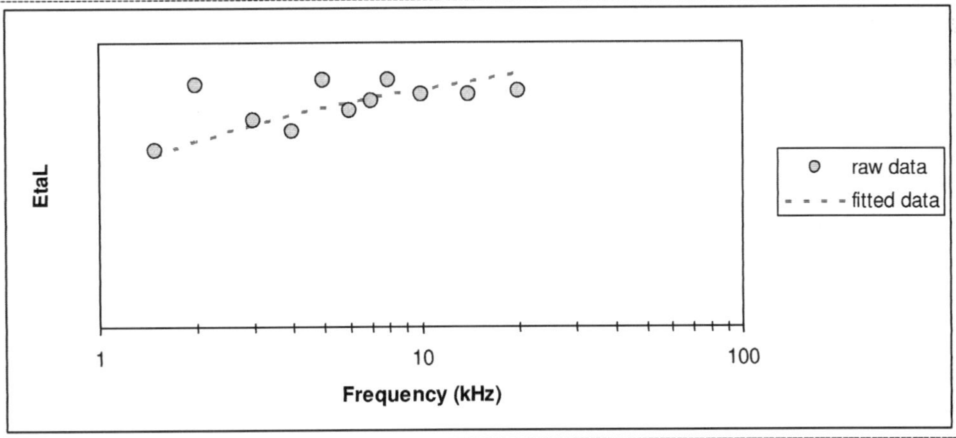

Figure 15. Test anechoic panel – Influence of hydrostatic pressure on longitudinal velocity (damping).

4. Conclusion

A method to evaluate both real part and damping of the longitudinal velocity of materials used in underwater acoustics has been presented, from test panel measurements in a water tank. Input data are material density, panel thickness and reflection and transmission coefficients of the test panel (amplitude and phase). As the problem is here mathematically over-determined, the method is less sensitive to error measurements than previous ones. It should be noted that the phase information in the data is important. Practical results have been obtained on a test anechoic panel. Dependency of longitudinal velocity along frequency and hydrostatic pressure is fully consistent with the known properties of such materials. The method suffers some frequency limitations both at low and high frequencies. However, the dimensions and thickness of the sample could be adapted to the user's frequency range of interest.

5. References

[1] Kuster G T and Toksöz M N 1974 Velocity and attenuation of seismic waves in a two-phase media *Geophysics* **39** 587

[2] Audoly C 1992 Global characterization of acoustic panels at normal incidence for underwater applications *Acustica* **76** 129

[3] Audoly C 1991 Characterization of underwater passive materials *Journal of Wave-Material Interaction* **5&6** 1

[4] Fokin V, Ambati M, Cheng Sun, Xiang Zhang 2007 Method for retrieving effective properties of locally resonant acoustic metamaterials *Physical Review* B **76**, 144302

[5] Audoly C and Giangreco C 1991 Determination of the complex sound speed of the longitudinal wave of materials using test-panel measurements *Journal of Wave-Material Interaction* **5&6** 225

[6] Humphrey V F, Robinson S P, Smith J D, Martin M J, Beamiss G A, Hayman G, Caroll N L 2008 Acoustic characterisation of panel materials under simulated ocean conditions using a parametric array source *J. Acoust. Soc. Am.* **124** 803

[7] Audoly C and Giangreco C 1990 Improvement of the measurement of the transmission coefficient of panels at normal incidence using surface receivers *J. Acoustique* **3** 369

[8] Beretti S 2010 Réponse acoustique d'élastomères micro-inclusionnaires soumis à la pression d'immersion *10ème Congrès Français d'Acoustique, Lyon, 12-16 Avril 2010*

Equation for acoustic propagation in inhomogeneous chemically active media with continuous distribution of relaxation losses

G.Vilensky[1], G. ter Haar[2] and N. Saffari[1]

[1] University College London, Deptartment of Mechanical Engineering, Torrington Place, London, WC1E 7JE, UK

[2] Institute of Cancer Research, 15 Cotswold Rd, Belmont, Sutton Surrey, SM2 5NG, UK

E-mail: g.vilensky@yahoo.co.uk

Abstract. The work focuses on the derivation of a lossy wave equation for ultrasound absorption in media with the power law of absorption, which plays an important role in bio-medical applications. Possible forms of nonlinear lossy wave equations are also discussed. The properties of the non-linear pressure fields resulting from the implementation of different non-linearity models are investigated. This part of the study is illustrated with the numerical computations of the related linear and nonlinear equations.

1. Introduction

The aim of this work is to extend the theory of acoustic absorption in chemically active media with a finite number of discrete relaxation mechanisms proposed in [1] to the case of a continuous distribution of relaxations mechanisms. Extension of this kind is needed for lossy media with "anomalous" laws of absorption such as, for example, the power law which finds wide applications in bio-medical and food industries.

In this paper we propose a rational method for derivation of lossy wave equations in such media.

According to the classical theory of sound absorption in reacting media (see [6, 5, 1],[7] and the references therein), the thermodynamic state of the propagation medium is described not only by local quasi-equilibrium variables such as the specific volume, the pressure and the specific internal energy, but also by a finite number of internal variables, also called relaxation parameters. Relaxation parameters describe unsteady physico-chemical processes taking place in the medium which has been perturbed from the state of its thermodynamic equilibrium. For example, these can be chemical reactions taking place in the media. As it is assumed that the perturbations from the state of equilibrium are small, each relaxation mechanism, is governed by the corresponding linear kinetic equation for the related relaxation parameter. Each such kinetic process is driven by the acoustic field applied to the medium and, in its turn, results in the additional compressibility of the reacting medium in excess of its equilibrium value of compressibility that would exist if the relaxation mechanisms were not excited. The total effect of the sum of these additional compressibilities on acoustic properties of the medium is manifested in the change of the phase speeds of sound waves and, more importantly, in the additional irreversible normal stresses and the related acoustic absorption, since part of the acoustic energy is used up to drive the involved physico-chemical reactions. These effects are reflected in the changes in the dispersion relation for the phase speed and absorption law, which describes the dependency of the absorption coefficient on the frequency of the related harmonic wave and can be registered experimentally.

When the number of such relaxation variables associated with the involved physico-chemical reactions is very large, the described classical scheme with a finite number of kinetic equations may become insufficient to give correct description of experimentally found laws of absorption. In this case its extension to the continuous distribution of internal parameters becomes welcome. In this paper we propose a rational method for derivation of lossy wave equations in such media.

A more detailed and an up-to-date discussion of the literature on the problem of ultrasound absorption in fluid media with complex internal structure can be found in the companion paper [8] on the treatment of various absorption laws available for biological media. It also tackles different aspects of the methodology proposed here, for instance, its relation to Kramers-Kronig equations, *etc.*

As in other mesoscopic non-equilibrium thermodynamic theories, [7], the main idea behind the proposed theoretical extension is to derive an evolution equation which describes the the kinetics of the distribution of the continuous relaxation variable (function p_r in the text of the paper) and, thus relates it behaviour to the experimentally measurable properties of the medium. In the present case such property is the absorption law. In this paper the required the evolution equation (19) appears as a natural extension of discrete kinetic equations of the classical theory with finite number of relaxations. This equation has one unknown coefficient before the forcing term, and the link between this coefficient and the experimentally measurable absorption law needs to be established. To achieve this we consider the limit of an infinite number of discrete relaxation processes. After that we express the sum of all inputs of each discrete relaxation mechanism into the expression for additional compressibility caused by the presence of each kinetic process in terms of the continuous relaxation parameter and introduce the required continuous approximation. Then, using a finite difference approach, we relate this additional compressibility to the relaxation variable governed by the generalized kinetic equation. This relation has the form of a convolution integral with the unknown kernel whose structure is obtained from consideration of the continuous analog of discrete dispersion relation. As a result, a single closed intergo-differential lossy wave equation is obtained for the acoustic pressure. In the end of the paper we compare it with related lossy wave equation recently reported in the literature and illustrate its usefulness for numerical modelling.

2. Problem formulation and governing equations

The starting point of the present work is the set of equations of linear acoustics for chemically active media derived in [1]:

$$\frac{\partial \rho'}{\partial t} + \nabla\left(\rho_o \underline{v}'\right) = 0, \tag{1}$$

$$\rho_o \frac{\partial \underline{v}'}{\partial t} = -\nabla p', \tag{2}$$

$$\frac{1}{c_e^2}\frac{\partial p'}{\partial t} - \left(\frac{\partial \rho'}{\partial t} + \underline{v}'\nabla\rho_o\right) + \sum_{k=0}^{N} \rho_o \kappa_k \frac{\partial\left(\triangle\xi_k\right)}{\partial t} = 0, \tag{3}$$

$$\left(\frac{\partial}{\partial t} + \frac{1}{\tau_k}\right)\left(\triangle\xi_k\right) = -\frac{\partial p'}{\partial t}. \tag{4}$$

Here ρ', p' , $\underline{v}'$ are the perturbations of the density ρ, the pressure p and the velocity vector $\underline{v}$ about their unperturbed values $\rho_o = \rho_o\left(\underline{r}\right)$, p_o and $\underline{v}_o$, where $p_o = \text{const}$ and $\underline{v}_o = 0$. It is assumed that a finite number N of molecular species are dissolved in the medium. The species k

is characterized by its affinity A_k and the number of molecules per unit mass n_k. Each quantity $\triangle\xi_k$ is proportional to the excess of n_k over its equilibrium value $n_k^e(p,T)$ in the absence of relaxation events, with T being the temperature. All $\triangle\xi_k$ satisfy the relaxation equations (4) with $k = 0,1,2,3,\ldots N$, where $\tau_k > 0$ represents the relaxation time for the k-th relaxation process. For convenience of further notation also define the relaxation frequencies $\varpi_k = 1/\tau_k$.

We hasten to note that the relaxation frequencies ϖ_k defined here differ from the modified relaxation frequencies $\tilde{\omega}_k$ defined in [1] in that the quantities ϖ_k depend only on just one relaxation mechanism, but not on all of them as in [1].

Each relaxation process is also characterized by its contribution κ_k into the total compressibility of the medium. As shown in [1] (see equations (42), (43) and (39) there), κ_k are related to the absorption coefficient $\alpha(\omega)$ and the phase velocity $c(\omega)$ of the harmonic wave $p' = p(\underline{r})\exp(-i\omega t)$ of frequency ω via the following equations:

$$\frac{2\alpha(\omega)}{c(\omega)} = \rho_o\omega^2\sum_{k=0}^{N}\frac{\kappa_k\varpi_k}{\varpi_k^2+\omega^2} \quad (5)$$

$$\left(\frac{\omega}{c(\omega)}\right)^2 - \alpha^2(\omega) = \rho_o\omega^2\left(K_\infty + \sum_{k=0}^{N}\frac{\kappa_k\varpi_k^2}{\varpi_k^2+\omega^2}\right), \quad (6)$$

where

$$K_\infty = \frac{1}{\rho_o c_e^2} - \sum_{k=0}^{N}\kappa_k. \quad (7)$$

$c_e = c_e(\underline{r})$ is the adiabatic speed of sound in the medium at rest and when there are no variations of relaxation parameters $\triangle\xi_k$. c_e, ρ_o and κ_k are functions of the position vector $\underline{r}$. Expressions for the attenuation (5) and for the wave speed (6) are the imaginary and real part of the complex dispersion equation obtained in [1].

The relationships (1), (2) and (3) are the linearized equations of continuity, momentum and energy in the absence of viscosity and heat conduction effects, which are neglected, for the sake of simplicity. System (1-4) coincides with the system (18-21) originally obtained in [1], where the details of its derivation and the related physical assumptions can be found.

The set of governing equations (1-7) will be the basis for the analysis of the present paper. It aims to derive a lossy wave equation for acoustic propagation in the media, where the number N of the relaxation frequencies ϖ_k is so large, and their distribution over the positive half axis $0 < \varpi_k < \infty$ is so dense, that N can be set equal to infinity, and the discrete sequence of ϖ_k can be approximated by a continuous spectrum.

3. Continuous distribution of relaxation parameters

3.1. Equations of motion

In what follows, instead of $\triangle\xi_k$, it will be more convenient to work with the non-dimensional relaxation parameters

$$\hat{p}_k = \kappa_k\triangle\xi_k.$$

Since κ_k is a time independent quantity, equations (3) and (4) can be rewritten in the form

$$\frac{1}{c_e^2}\frac{\partial p'}{\partial t} - \left(\frac{\partial\rho'}{\partial t} + \underline{v}'\nabla\rho_o\right) + \rho_o\frac{\partial}{\partial t}\sum_{k=0}^{\infty}\hat{p}_k = 0, \quad (8)$$

$$\frac{\partial\hat{p}_k}{\partial t} + \varpi_k\hat{p}_k = -\kappa_k\frac{\partial p'}{\partial t},\ k = 0,1,2,3\ldots+\infty. \quad (9)$$

Assuming that the relaxation frequencies ϖ_k fill in the positive half axis $\varpi_k > 0$ sufficiently densely, we substitute the discrete set of parameters $\{\varpi_k\}_{k=0}^{k=\infty}$ with a continuous variable ϖ which is allowed to vary in the semi-infinite interval $\varpi \geq 0$. Without the risk of an ambiguity, parameter ϖ will be also called the relaxation frequency in what follows.

Instead of the discrete set of values $\{\hat{p_k}\,(t, \underline{r})\}_{k=0}^{k=\infty}$, consider a continuous function $\hat{p}\,(t, \underline{r}, \varpi)$ which, apart from t and $\underline{r}$, depends on the additional argument ϖ and, in agreement with (9), satisfies the equation

$$\frac{\partial \hat{p}}{\partial t} + \varpi \hat{p} = -\kappa\,(\varpi)\,\frac{\partial p'}{\partial t}, \tag{10}$$

where $\kappa\,(t, \underline{r}, \varpi)$ is a smooth function, such that $\kappa\,(t, \underline{r}, \varpi = \varpi_k) = \kappa_k\,(t, \underline{r})$.

In order to derive the continuous analog of equation (8), the sum

$$J = \sum_{k=0}^{\infty} \hat{p}_k \tag{11}$$

in its third term needs to be appropriately rearranged. To do this, we introduce the Laplace originals $p_r\,(t, \underline{r}, \tau)$ and $\sigma\,(\underline{r}, \tau)$ of the functions $\hat{p}\,(t, \underline{r}, \varpi)$ and $\kappa\,(t, \underline{r}, \varpi)$, so that

$$\hat{p}\,(t, \underline{r}, \varpi) = \int_0^{\infty} p_r\,(t, \underline{r}, \tau) \exp\left(-\varpi\tau\right) d\tau, \tag{12}$$

$$\kappa\,(t, \underline{r}, \varpi) = \int_0^{\infty} \sigma\,(\underline{r}, \tau) \exp\left(-\varpi\tau\right) d\tau. \tag{13}$$

Here, in agreement with the definition of the Laplace transformation, functions $p_r\,(t, \underline{r}, \tau)$ and $\sigma\,(\underline{r}, \tau)$ satisfy the following restrictions

$$p_r\,(t, \underline{r}, \tau \leq 0) = \sigma\,(\underline{r}, \tau \leq 0) = 0, \tag{14}$$

The Laplace transformation variable τ has the dimension of time.

If we now assume that the discrete relaxation frequencies ϖ_k are distributed uniformly in the interval $\varpi \geq 0$, *i.e.*, $\varpi_k = k\varpi_*$, the sum (11) can be rewritten as

$$J = \sum_{k=0}^{\infty} \hat{p}_k = \sum_{k=0}^{\infty} \int_0^{\infty} p_r\,(t, \underline{r}, \tau)\, e^{-\varpi_k \tau} d\tau = \int_0^{\infty} p_r\,(t, \underline{r}, \tau) \left(\sum_{k=0}^{\infty} e^{-\varpi_k \tau} \right) d\tau$$

and, since

$$\sum_{k=0}^{\infty} e^{-\varpi_k \tau} = \sum_{k=0}^{\infty} \left(e^{-\varpi_* \tau}\right)^k = \frac{1}{1 - \exp\left(-\varpi_* \tau\right)} \approx \frac{1}{\varpi_* \tau}, \text{ when } \varpi_* \ll 1,$$

to leading order in ϖ_* , we get the following formula

$$J = \frac{1}{\varpi_*} \int_0^{\infty} p_r\,(t, \underline{r}, \tau)\, \frac{d\tau}{\tau}. \tag{15}$$

If the distribution of the relaxations frequencies ϖ_k is not uniform, but is arbitrary, the following general formula should be used to calculate the variable J:

$$J = \int_0^{\infty} p_r\,(t, \underline{r}, \tau)\, \varphi\,(\tau)\, d\tau, \tag{16}$$

Journal of Physics: Conference Series **353** (2012) 012005 doi:10.1088/1742-6596/353/1/012005

where $\varphi(\tau) = \sum_{k=0}^{\infty} e^{-\varpi_k \tau}$.

Note, that function $\varphi(\tau)$ is the result of summation over all individual relaxation frequencies ϖ_k. Consequently, $\varphi(\tau)$ does not depend on the particular values of the relaxation frequencies ϖ_k, but is determined by the parameters of their distribution as a whole, as in the case with formula (15), where there is only one such parameter equal to ϖ_*, that is,

$$\varphi(\tau) = 1/(\varpi_* \tau). \tag{17}$$

This results in the fact that J is a function of time t and position vector $\underline{r}$ only, but not of the relaxation frequency ϖ or the parameter τ. For this reason, the function p' also depends only on the time variable t and the position vector $\underline{r}$, but does not depend on the variable ϖ or τ, since none of the equations of mass conservation (1), momentum (2) and energy

$$\frac{1}{c_e^2}\frac{\partial p'}{\partial t} - \left(\frac{\partial \rho'}{\partial t} + \underline{v}'\nabla\rho_o\right) + \rho_o \frac{\partial}{\partial t} J(t, \underline{r}) = 0 \tag{18}$$

contain the variable ϖ or τ.

Consequently, we can conclude from (10) that the unknown function $p_r(t, \underline{r}, \tau)$ in the expressions for J must satisfy the following equation

$$\frac{\partial p_r}{\partial t} + \frac{\partial p_r}{\partial \tau} = -\sigma(\underline{r}, \tau) \frac{\partial}{\partial t} p'(t, \underline{r}). \tag{19}$$

Indeed, application of the Laplace transformation to the last equation results in the original relaxation equation (10).

Equation (19) is the continuous analog of the discrete set of kinetic equations (4). It plays the same role in the theory with continuous distribution of relaxation parameters which is proposed here as kinetic equations (4) in the conventional theory with a finite number of relaxation mechanisms.

The sought analog of the discrete governing equations (1,2,3,4) in the case of continuously distributed relaxation degrees of freedom is thus given by the set of equations (1),(2),(18),(19).

The variable J now is the functional of the function $p_r(t, \underline{r}, \tau)$ defined by either formula (15) or formula (16), depending on the type of distribution of the relaxation frequencies. However, in the absence of experimental data on the true distribution of relaxation frequencies ϖ_k, it makes sense to make use of the simplest type of their distribution, which results in (15).

If we assume that there was no motion in the medium before $t = 0$, *i.e.*,

$$p' = \underline{v}' = \rho' = p_r = 0 \text{ at } t \leq 0$$

and recall that the functions $p_r(t, \underline{r}, \tau)$ and $\sigma(\underline{r}, \tau)$ satisfy the initial condition (14) with respect to the variable τ, then equation (19) has the following solution

$$p_r(t, \underline{r}, \tau) = -\int_0^t \sigma(\underline{r}, t' - t + \tau) \frac{\partial}{\partial t'} p'(t', \underline{r})\, dt'.$$

This fact can be easily verified by direct substitution of the last formula into equation (19).

Substitution of this solution into formula (16) leads to the expression

$$\begin{aligned} J &= -\int_0^\infty \int_0^t \sigma(\underline{r}, t' - t + \tau) \frac{\partial}{\partial t'} p'(t', \underline{r})\, \varphi(\tau)\, dt' d\tau \\ &= -\int_0^t \frac{\partial}{\partial t'} p'(t', \underline{r}) \int_{t-t'}^\infty \sigma(\underline{r}, t' - t + \tau)\, \varphi(\tau)\, d\tau dt', \end{aligned}$$

since $\sigma(\underline{r}, t' - t + \tau) = 0$ when $\tau < t - t'$. After the change of the integration variable $\tau' = \tau + t' - t$ in the inner integral the expression for the function J takes the following form

$$J = -\int_0^t \frac{\partial}{\partial t'} p'(t', \underline{r}) \int_0^\infty \sigma(\underline{r}, \tau') \varphi(\tau' + t - t') \, d\tau' dt'.$$

If we now make use of formula (17) for the function $\varphi(\tau)$, then it follows from the previous equation for J that

$$J = -\int_0^t \frac{\partial}{\partial t'} p'(t', \underline{r}) \int_0^\infty \frac{\sigma(\underline{r}, \tau')}{\varpi_*} \frac{d\tau'}{\tau' + t - t'} dt'. \tag{20}$$

After some simple algebra equations (1),(2) and (18) can be reduced to a single equation for the perturbation pressure

$$\frac{1}{c_e^2} \frac{\partial^2 p'}{\partial t^2} - \rho_o \nabla \left(\frac{\nabla p'}{\rho_o} \right) = -\rho_o \frac{\partial^2 J}{\partial t^2}, \tag{21}$$

where the functional J is given by equation (20). It depends on the unknown function $\sigma(\underline{r}, \tau)$ which characterizes absorption properties of the medium and can be expressed via the absorption coefficient $\alpha(\omega)$ as will be shown in the next section.

3.2. Dispersion relation in media with continuous distribution of relaxations

In order to determine the function $\sigma(\underline{r}, \tau)$, consider discrete equation for attenuation (5) and dispersion relation (6). For $N = \infty$ and $\varpi_k = k\varpi_*$ the sum on the right-hand side of equation (5) can be rewritten as follows:

$$\sum_{k=0}^{\infty} \frac{\kappa_k \varpi_k}{\varpi_k^2 + \omega^2} = \sum_{k=0}^{\infty} \frac{\kappa(\varpi_k)}{\varpi_*} \frac{\varpi_k(\varpi_{k+1} - \varpi_k)}{\varpi_k^2 + \omega^2} \approx \int_0^\infty \frac{\kappa(\varpi)}{\varpi_*} \frac{\varpi d\varpi}{\varpi^2 + \omega^2},$$

where the rectangle quadrature rule was used to obtain the last formula. Substitution of this integral in place of the sum in equation (5) gives the following continuous version of this equation:

$$\frac{\rho_o c^2(\omega)\omega}{2} \int_0^\infty \frac{\kappa(\varpi)}{\varpi_*} \frac{\varpi d\varpi}{\varpi^2 + \omega^2} = \frac{\alpha(\omega)}{\omega / c(\omega)}. \tag{22}$$

Similar treatment of equations (6) and (7) results in the following formula:

$$1 - \left(\frac{\alpha(\omega)}{\omega / c(\omega)} \right)^2 = \frac{c^2(\omega)}{c_e^2} - \omega^2 \rho_o c^2(\omega) \int_0^\infty \frac{\kappa(\varpi)}{\varpi_*} \frac{d\varpi}{\varpi^2 + \omega^2}, \tag{23}$$

The right-hand side of equation (22) is the non-dimensional absorption coefficient which is equal to the ratio of the dimensional value of the absorption coefficient $\alpha(\omega)$ to the wave number $\omega / c(\omega)$. Normally, in situations of practical interest absorption of ultrasound waves over the distance equal to one wavelength $\lambda = c(\omega)/\omega$ is small, and the non-dimensional absorption coefficient is also small, that is,

$$\frac{\alpha(\omega)}{\omega / c(\omega)} \ll 1.$$

For this reason the non-dimensional quantity

$$\frac{\rho_o c^2(\omega) \kappa(\varpi)}{\varpi_*}$$

on the left-hand side of equation(22) must also be much less than unity. Consequently, it follows from equation (23) that, to leading order,

$$c(\omega) \approx c_e. \tag{24}$$

This formula could have been derived directly from formulae (5-7) and corresponds to the physical situation when absorption results in only slight dispersion of the sound field.

For a given medium, the absorption coefficient $\alpha(\omega)$ can be measured experimentally and can be considered known. Substitution of formula (24) into formula (22) then results in an integral equation for the function $\kappa(\varpi)$ with a known right-hand side.

In agreement with formula (13), function $\kappa(\varpi)$ is related to the unknown function $\sigma(\underline{r}, \tau)$ via the Laplace transformation. Thus, it will be possible to determine function $\sigma(\underline{r}, \tau)$, if the integral equation (22) has been solved for the function $\kappa(\varpi)$ (with $c(\omega) \approx c_e$).

Although it is possible to construct general solution of the mentioned integral equation with an arbitrary right-hand side, to avoid physically unessential algebraic complications the solution method will be illustrated for the power law of absorption

$$\alpha(\omega) = \alpha_o(\underline{r})\,\omega^{1+\gamma(\underline{r})}, \text{ where } \alpha_o(\underline{r}) \geq 0,\ 0 < \gamma(\underline{r}) < 1. \tag{25}$$

This law of absorption is widely used in problems of ultrasound propagation in biological media and is of considerable importance for bio-medical applications.

Equations (22), (24) and (25) lead to the following integral equation

$$\frac{\omega}{\varpi_*}\int_0^\infty \frac{\kappa(\varpi)\,\varpi d\varpi}{\varpi^2+\omega^2} = \frac{2\alpha_o\omega^\gamma}{\rho_o c_e}. \tag{26}$$

To construct its solution we take the inverse Laplace transformation of this equation with respect to ω. Its application to the left-hand side gives

$$\frac{1}{\varpi_*}\int_0^\infty d\varpi\kappa(\varpi)\,\varpi\frac{1}{2\pi i}\int_{a-i\infty}^{a+i\infty}\frac{\omega\exp(\omega x)\,d\omega}{\varpi^2+\omega^2} = \frac{1}{\varpi_*}\int_0^\infty \kappa(\varpi)\,\varpi\cos(\varpi x)\,d\varpi.$$

The inverse Laplace transformation of the right-hand side is

$$\frac{2\alpha_o}{\rho_o c_e}\frac{1}{2\pi i}\int_{a-i\infty}^{a+i\infty}\omega^\gamma\exp(\omega x)\,d\omega = \frac{2\alpha_o}{\rho_o c_e}\frac{x^{-(1+\gamma)}}{\Gamma(-\gamma)},$$

where $\Gamma(-\gamma)$ is the Euler gamma function. Equating the right-hand sides of the last two formulae we get the equation

$$\frac{1}{\varpi_*}\int_0^\infty \kappa(\varpi)\,\varpi\cos(\varpi x)\,d\varpi = \frac{2\alpha_o}{\rho_o c_e}\frac{x^{-(1+\gamma)}}{\Gamma(-\gamma)}.$$

Multiplying this equation by $\cos(yx)\,dx$, integrating it over all $x \geq 0$ and differentiating the result in y we obtain for its left-hand side

$$\frac{1}{\varpi_*}\frac{d}{dy}\int_0^\infty d\varpi\kappa(\varpi)\,\varpi\int_0^\infty \cos(\varpi x)\cos(yx)\,dx =$$

$$\frac{\pi}{2\varpi_*}\frac{d}{dy}\int_0^\infty \kappa(\varpi)\,\varpi\delta(\varpi-y)\,d\varpi = \frac{\pi}{2\varpi_*}\frac{d}{dy}(\kappa(y)\,y).$$

The right hand side of the resultant equation is

$$-\frac{2\alpha_o}{\rho_o c_e}\frac{1}{\Gamma(-\gamma)}\int_0^\infty x^{-1+\gamma}\sin(yx)\,dx = \frac{2\alpha_o\gamma}{\rho_o c_e}y^{-1+\gamma}.$$

Equating the right-hand sides of the last two formulae leads to the differential equation

$$\frac{1}{\varpi_*}\frac{d}{dy}\left(\kappa\left(y\right)y\right)=\frac{4\alpha_o\gamma}{\pi\rho_o c_e}y^{-1+\gamma}.$$

Integrating it once in y and making the substitute $y=\varpi$ we obtain the solution of the integral equation (26)

$$\frac{\kappa\left(\varpi\right)}{\varpi_*}=\frac{4\alpha_o\varpi_*}{\pi\rho_o c_e}\varpi^{-1+\gamma}. \tag{27}$$

Formula (27) makes it possible to calculate the increase in the phase speed of the harmonic wave of frequency ω from the dispersion relation (23). It can be shown after some algebra that in the lossy media with the power law of absorption (25) the phase speed is given by the following asymptotic formula:

$$\frac{c_e^2}{c^2\left(\omega\right)}=1-\frac{4\left(1-r\right)}{sin\left(r\pi\right)}\frac{\alpha\left(\omega\right)c_e}{\omega}+\ldots,$$

where dots stand for the quadratic and higher order terms in the non-dimensional absorption coefficient $\alpha\left(\omega\right)c_e/\omega\ll 1$.

4. Lossy wave equations

4.1. Linear equation

Formula (27) gives the following equation for the function $\sigma\left(\underline{r},\tau\right)$:

$$\int_0^\infty\sigma\left(\underline{r},\tau\right)\exp\left(-\varpi\tau\right)d\tau=\frac{4\alpha_o}{\pi\rho_o c_e}\varpi^{-1+\gamma}.$$

Hence, taking the inverse Laplace transformation, we get

$$\sigma\left(\underline{r},\tau\right)=\frac{4\alpha_o\varpi_*\tau^{-\gamma}}{\pi\rho_o c_e\Gamma\left(-\gamma\right)}. \tag{28}$$

It is now possible to obtain an explicit formula for the functional J, in order to formulate a closed problem for the acoustic pressure. To do this, we substitute expression (28) into formula (20) and carry out the following set of algebraic manipulations:

$$\begin{aligned}J&=-\int_0^t\frac{\partial}{\partial t'}p'\left(t',\underline{r}\right)\int_0^\infty\frac{4\alpha_o\tau'^{-\gamma}}{\pi\rho_o c_e\Gamma\left(-\gamma\right)}\frac{d\tau'}{\tau'+t-t'}dt'\\&=-\frac{4\alpha_o}{\pi\rho_o c_e\Gamma\left(-\gamma\right)}\int_0^t dt'\frac{\partial}{\partial t'}p'\left(t',\underline{r}\right)\int_0^\infty\frac{\tau'^{-\gamma}d\tau'}{\tau'+t-t'}\\&=-\frac{4\alpha_o}{\pi\rho_o c_e\Gamma\left(-\gamma\right)}\int_0^\infty\frac{z^{-\gamma}dz}{z+1}\int_0^t\frac{\partial}{\partial t'}p'\left(t',\underline{r}\right)\frac{dt'}{\left(t-t'\right)^\gamma}\\&=-\frac{4\alpha_o\Gamma\left(\gamma\right)}{\pi\rho_o c_e}\int_0^t\frac{\partial}{\partial t'}p'\left(t',\underline{r}\right)\frac{dt'}{\left(t-t'\right)^\gamma}.\end{aligned}$$

The wave equation (21) for the perturbation pressure then takes the following final form

$$\frac{1}{c_e^2}\frac{\partial^2p'}{\partial t^2}-\rho_o\nabla\left(\frac{\nabla p'}{\rho_o}\right)-\frac{4\alpha_o\Gamma\left(\gamma\right)}{\pi c_e}\frac{\partial^2}{\partial t^2}\int_0^t\frac{\partial}{\partial t'}p'\left(t',\underline{r}\right)\frac{dt'}{\left(t-t'\right)^\gamma}=0. \tag{29}$$

Equation (29) describes propagation of linear acoustic waves in a spatially inhomogeneous lossy media, whose absorption properties are governed by the power law.

Note, that the key feature of the presented derivation was the possibility to solve integral equation (26) analytically, which follows from the continuous version of the expression for attenuation (22). Since the proposed solution method can be repeated word for word for an arbitrary right-hand side of the equation for attenuation (22), wave equations governed by other absorption laws can be easily derived using the described approach.

A similar lossy wave equation, although not identical to equation (29), was proposed in [2] for the power law of absorption on the basis of a heuristic reasoning. It is easy to verify that our wave equation (29) can be rewritten in the form which is analogous to the *ad hoc* equation presented in [2]. If we integrate the third term of equation (29) by parts, differentiate it once with respect to time and use the fact that, to leading order,

$$\frac{\partial^2 p'}{\partial t^2} = \rho_o c_e^2 \nabla \left(\frac{\nabla p'}{\rho_o} \right)$$

in order to substitute the second time derivative of the pressure in the resultant integral, the following form of equation (29) is obtained:

$$\frac{1}{c_e^2}\frac{\partial^2 p'}{\partial t^2} - \rho_o \nabla \left(\frac{\nabla p'}{\rho_o} \right) - \frac{4\alpha_o \Gamma(\gamma)\, \rho_o c_e}{\pi} \frac{\partial}{\partial t} \int_0^t \nabla \left(\frac{\nabla p'(t', \underline{r})}{\rho_o} \right) \frac{dt'}{(t-t')^{\gamma}} = 0. \tag{30}$$

The main difference between this equation and its heuristic counterpart proposed in [2] is in that the coefficient $\alpha_o(\underline{r})$ appears outside the operator ∇.

In the present work, however, we have given a theoretical method of **systematic derivation** of acoustic wave equations in lossy media from first principles without the need to recourse to intuitive reasoning.

4.2. Nonlinear equation

In applications it is often of interest to take into account the nonlinear character of the sound field, when the nonlinearity is mainly caused by the equation of state, whereas all other second and higher order effects remain negligibly small in the medium. This situation is realized when the non-dimensional second derivative of pressure with respect to density, traditionally designated as B/A , is sufficiently large:

$$\frac{B}{A} = \frac{\partial^2 p}{\partial \rho^2}\frac{\rho_o}{c_e^2} \gg 1.$$

In this case the quadratic term in the Taylor series expansion of the density with respect to the pressure perturbations (or vice versa) needs to be retained:

$$\frac{\rho}{\rho_o} = 1 + \frac{p'}{\rho_o c_e^2} - \frac{1}{2}\frac{B}{A}\left(\frac{p'}{\rho_o c_e^2}\right)^2 + \ldots \tag{31}$$

In a more general context the adiabatic relation between the density and the pressure can be taken in the form

$$\rho = \rho_*(p'). \tag{32}$$

For example, along with equation (31) we can take Tait's equation of state, [4],

$$\frac{\rho}{\rho_o} = \left(1 + \gamma_L \frac{p'}{\rho_o c_e^2}\right)^{1/\gamma_L}, \tag{33}$$

where $\gamma_L = 1 - B/A$, or secant-modulus (original Tait's) equation of state, see [3] for more details:

$$\frac{\rho}{\rho_o} = \frac{1 + \left(1 + \frac{1}{2}\frac{B}{A}\right)\frac{p'}{\rho_o c_e^2}}{1 + \frac{1}{2}\frac{B}{A}\frac{p'}{\rho_o c_e^2}}. \tag{34}$$

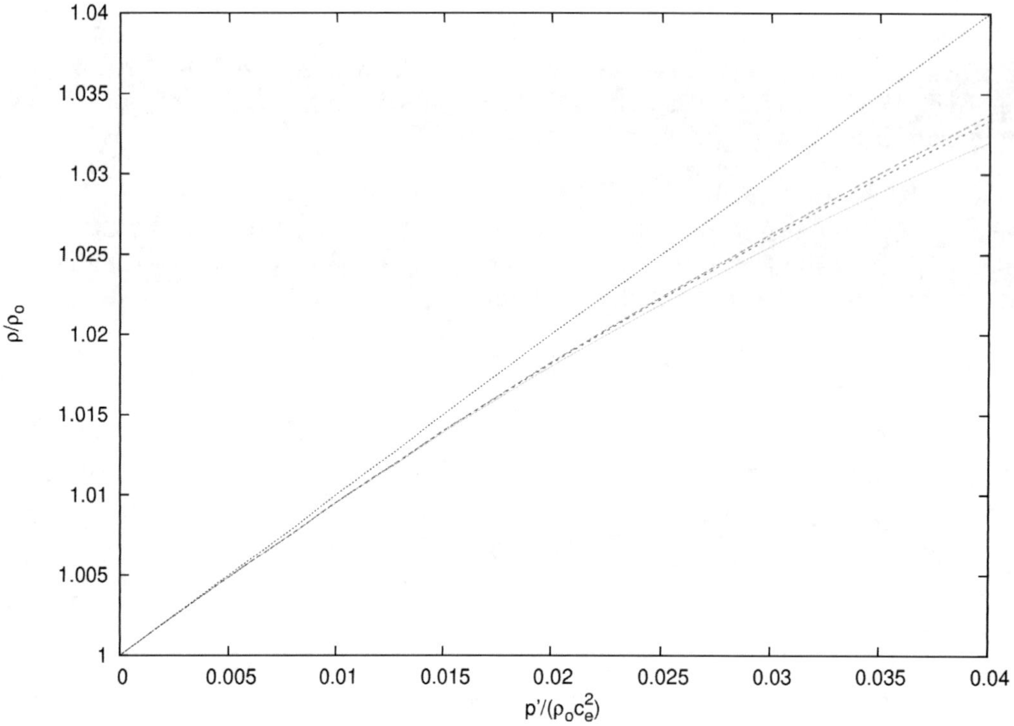

Figure 1. Adiabatic equations of state for $B/A=10$: green solid line - formula (31), red long dashed line - formula (33), blue short dashed line - formula (34), black straight dotted line - linear equation $\rho/\rho_o = 1 + p'/\rho_o c_e^2$.

Comparison of the equations of state (31), (33), (34) when $B/A=10$, which approximately corresponds to the case of human fatty tissue, [5], is presented in Figure 1. The dimensional range of pressure variations shown in this figure approximately corresponds to $0 \div 90$ MPa.

To derive the related nonlinear equations for the pressure field we need to consider the original nonlinear version of the linearized energy equation (3) obtained in [1]:

$$\frac{1}{\tilde{c}_e^2}\frac{dp}{dt} - \frac{d\rho}{dt} + \frac{\beta\rho T}{c_p}\sum_{k=0}^{N}\left(\frac{d}{dt}\frac{c_p\kappa_k}{\beta T}\triangle\xi_k\right) = \frac{\beta}{c_p}\sum_{k=0}^{N}\frac{\kappa_k}{\tau_k}\left(\triangle\xi_k\right)^2, \tag{35}$$

where $\tilde{c}_e$, c_p and β are the equilibrium values of the speed of sound, the specific heat at constant pressure and the coefficient of thermal expansion, respectively, $p = p_o + p'$, $\rho = \rho_o(\underline{r}) + \rho'$, $T = T_o + T'$, and the operator d/dt stands for the convective time derivative. As above, the subscript "o" refers to the unperturbed values of the pressure, the density and the temperature.

Since the second order effects are assumed negligible in comparison with the nonlinearity due to the equation of state (32), equations (1,2,4) remain unchanged, and equation (35) reduces to its original form (3). However, now the square of the equilibrium speed of sound c_e^2 for the medium at rest in its first term

$$\frac{1}{c_e^2}\frac{\partial p'}{\partial t}$$

must be substituted with the square for the adiabatic speed of sound calculated from the equation

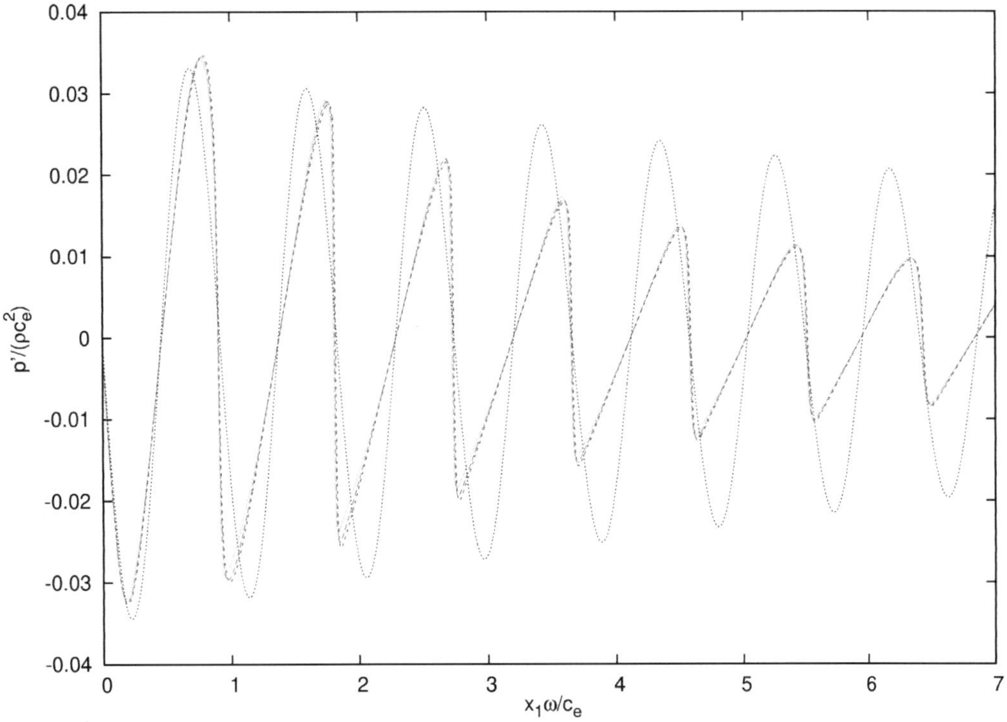

Figure 2. Pressure versus distance from the sound source: green solid line - formula (31), red long dashed line - formula (33), blue short dashed line - formula (34), black dotted line - linear model $\rho/\rho_o = 1 + p'/\rho_o c_e^2$.

of state (32):

$$\tilde{c}_e^2 = \frac{1}{d\rho_* (p') / dp'},$$

which is not a given function of the position vector only, but also depends on the pressure, that is, instead of equation (3) now we have the equation

$$\frac{\partial \rho_* (p')}{\partial t} - \left(\frac{\partial \rho'}{\partial t} + \underline{v}' \nabla \rho_o \right) + \sum_{k=0}^{N} \rho_o \kappa_k \frac{\partial (\triangle \xi_k)}{\partial t} = 0.$$

It is possible to show that this alteration in the original set of equations leads to the following nonlinear wave equation

$$\frac{\partial^2 \rho_* (p')}{\partial t^2} - \rho_o \nabla \left(\frac{\nabla p'}{\rho_o} \right) - \frac{4\alpha_o \Gamma (\gamma)}{\pi c_e} \frac{\partial^2}{\partial t^2} \int_0^t \frac{\partial}{\partial t'} p' (t', \underline{r}) \frac{dt'}{(t - t')^{\gamma}} = 0 \qquad (36)$$

where the function $\rho_* (p')$ defines the adiabatic equation of state, and can be taken, for example, in the form (31), (33) or (34).

To assess the influence of the nonlinear term given by different forms of the equation of state, and the influence of the absorption on the solution behaviour we carry out numerical integration of equation (36) for the case of uniform one-dimensional medium whose parameters approximately correspond to fatty tissue, [5]. They are as follows: $\rho_o = 1020$ kg/m^3, velocity of ultrasound

$c_e = 1440$ m/s, nonlinearity parameter $B/A = 10.0$, attenuation coefficient $\alpha = 7.5 \cdot \omega^{1.5}$dB/m. The sound wave is created by a piston which performs sinusoidal oscillations with the frequency ω equal to 1 MHz about its mean position $x_1 = 0$, and the pressure amplitude A_p which approximately corresponds to 73.4 MPa. The sound wave propagates along the axis x_1 in the positive direction.

Obtained numerical results are presented in Figure 2 which depicts the pressure fields calculated on the basis of the linear model (dotted line) and three nonlinear models corresponding to equations (31), (33) and (34). It can be inferred from Figure 2 that all three nonlinear models given by the equations of state (31), (33) and (34) are in reasonable agreement with each other, although the latter two models are in somewhat better agreement between each other than with the Taylor series expansion (31). This is particularly true for the region of negative pressures. The picture also illustrates the process of shock formation and the dramatic influence of the nonlinearity on the amplitude of the pressure wave.

5. Conclusion

The model for acoustic propagation in inhomogeneous chemically relaxing media originally presented in [1] has been extended to the case of lossy media whose absorption properties are governed by the power law of absorption. Since absorption laws of such type cannot be described within the framework of a finite number of kinetic equations, a rational theory based on the continuous distribution of relaxation mechanisms is required in this case. This work formulates such a theory.

The major results of the proposed analysis is linear lossy wave equation (29) and its nonlinear counterpart (36). Although both these equations are obtained for media with the power law of absorption, the method of their derivation can be used for other types of experimental absorption laws.

The obtained nonlinear equation allows for different types of nonlinearity in the adiabatic equation of state and has been shown to readily render itself for numerical solution.

Acknowledgement

The authors would like to acknowledge the support of the Engineering and Physical Sciences Research Council under grant number EP/F025750/1.

References

[1] Nachman, A.I., Smith, J.F., Waag, R.C. 1990 An equation for acoustic propagation in inhomogeneous media with relaxation losses. J. Acoust. Soc. Am. V. 88 (3). P. 1584-1595.

[2] Wismer,M.G., 2006 "Finite element analysis of broadband acoustic pulses through inhomogenous media with power law attenuation". J. Acoust. Soc. Am. 120 (6), 3493-3502.

[3] Hayward, A.T.J., 1967 Compressibility equations for liquids: a comparative study. Brit. J. Appl. Phys. V. !8. 965-977.

[4] Rossing, T.D. (Ed.) 2007. Springer Handbook on acoustics. 1182 p.

[5] Duck, F.A. 1990 Physical Properties of Tissue. A comprehensive reference book. Academic press. 346 p.

[6] Pierce, A.D. 1991 Acoustics. An introduction to its physical principles and applications. Published by the Acoustical Society of America. 500 Sunnyside Boulevard, Woodbury, New York 11797. 678 p.

[7] Lebon, G. , Jou, D., Casas-Vazquez, J. 2008 Understanding Non-equilibrium Thermodynamics. Foundations, Applications, Frontiers. Springer-Verlag Berlin Heidelberg. 325 p.

[8] Vilensky, G.,ter Haar, G., Saffari, N., 2011 A model of acoustic absorption in fluids based on a continuous distribution of relaxation times, Wave Motion, Article in press. Available online 11 August 2011, ISSN 01652125, 10.1016/j.wavemoti.2011.07.005.

Evaluation of adhesive bond Young's modulus during crosslinking using a mechanical method and an ultrasound method

B Mascaro, M K Budzik, M Castaings, J Jumel, M E R Shanahan

Université de Bordeaux, Institut de Mécanique et d'Ingénierie de Bordeaux (I2M), CNRS UMR 5295, Bât. A4, 351 cours de la Libération, 33405 TALENCE Cedex, FRANCE

Email : benoit.mascaro@u-bordeaux1.fr

Abstract. The strength and stability of adhesive bonded structures are related to polymer curing, when crosslinking occurs and leads to adhesive strength, stiffness and durability. Depending on the resin and curing agent used, cure time can vary from minutes to weeks. Methods based on dynamic mechanical analysis (DMA) or calorimetric techniques (DSC, DTA) are valuable for evaluating mechanical properties of adhesives, but are devoted specifically to the polymers alone, and not *in situ* in adhesive bonds. In this contribution, we have monitored - during crosslinking - the Young's modulus of a slow-curing DGEBA - PAMAM adhesive system, with two non-destructive, *in situ*, methods used for the characterisation of the adhesive in a bonded system. The first method is based on measurements obtained from strain gauges mounted on one bonded adherend. The second method uses an ultrasound technique based on the through-transmission. Both methods suggest the same curing kinetics.

1. Introduction

Adhesive bonding is a fast developing method of joining structural parts together, giving advantages of reducing weight by eliminating rivets or bolts, and reducing local stresses when using large bonding surfaces. Many structural adhesives (and also composite materials) are based upon epoxy resins. The mechanical properties and chemical stability of epoxies in their final, as-used, form are due to chemical crosslinking. This reaction is achieved using a variety of crosslinking agents of which amines (for room temperature curing), and imidazoles (for high temperature) are good examples. The essential problems of such cured adhesives are their fragility (relatively low fracture toughness) and, to some extent, their toxicity. Due to the high toxicity of amine and imidazole curing agents, substitutes are sometimes used. PolyAMidoAMines (PAMAM) curing agents can help with both the above-mentioned weaknesses of epoxy-based structural adhesives. However, one of the major disadvantages of PAMAM agents is their longer curing time compared to amines or imidazoles. PAMAMs are low-volatility, low-toxicity curing agents, which mix easily with standard Di-Glycidyl Ether of Bisphenol A (DGEBA, or BADGE) epoxy resins [1, 2]. DGEBA-PAMAM systems are cured at room temperature, resulting in low final shrinkage. The proportion of resin-to-agent can be widely changed in order to obtain the required reaction kinetics and final bulk properties.

We have monitored the crosslinking of a DGEBA-PAMAM system with two, essentially different, non-destructive methods. The first is based upon strain gauge measurements obtained from one of the bonded substrates. The second method is based on the measurement of ultrasonic transmission through a tested material and the use of transmission spectra to infer its mechanical properties. These methods present the impact of crosslinking time on important mechanical properties of bonded structures, particularly the adhesive Young's modulus. In addition, they are applied directly to adhesive joints, thus are potentially attractive for industry (e.g. for minimising the danger of leaving adhesive uncured in real, bonded structures).

2. Wedge test/strain gauge technique

The evaluation of mechanical properties of the adhesive, *in situ* in a bond, is made using the so-called wedge test (figure 1), in which a flexible, rectangular adherend is bonded to a second substrate, either also flexible or, in the present case, essentially rigid. A wedge, of thickness Δ_w, is inserted at the unbonded end (produced by prior insertion of a non-sticking spacer). Curvature of the elastic, bent member(s) leads to stored strain energy, and as a fracture front progresses, either within the adhesive or at an interface, this energy is liberated and becomes the fracture energy of the adhesive joint. In 1867, Winkler [3] studied the problem of a beam on an elastic foundation, in order to study effects of the loads acting on rails lying on different soils. Previously, some of us have derived and employed this model to analyse the adhesive wedge test [4,5]. An advantage of the Winkler treatment is that changes in the elastic properties of the adhesive layer may be monitored, since apparent changes in crack length are attributable to strain within the adhesive layer, if actual crack length is constant. The Winkler model is derived from two differential equations, corresponding respectively to the bonded zone and the free zone. The differential equation for the part of the beam in the bonded zone is:

$$\frac{d^4z}{dx^4}+\frac{\bar{k}}{EI}z=0, \quad a<x<+\infty, \tag{1}$$

where E is Young's modulus of the plate bent and I is the second moment of inertia of the beam section area ($I=bh^3/12$ in our case). To a good approximation, the foundation stiffness is given by

$$\bar{k}\approx\frac{E_A}{e}b \quad , \tag{2}$$

where E_A is Young's modulus of the adhesive, e its thickness and b is joint width. The general solution of differential equation (1) involves four arbitrary constants, but parts of the solution are physically unrealistic. Terms in $e^{\lambda(x-a)}$ are thus neglected, leading to:

$$z(x)\Big|_a^{+\infty}=e^{\lambda(a-x)}\left[A_1\cos\lambda(a-x)+B_1\sin\lambda(a-x)\right]. \tag{3}$$

In the free zone, the equation

$$\frac{d^4z}{dx^4}=0 \tag{4}$$

applies, since there is no reaction from the adhesive, and this has a straightforward polynomial solution:

$$z(x)\Big|_0^a=A_2x^3+B_2x^2+C_2x+D_2. \tag{5}$$

In both equations (3) and (5), capital letters refer to integration constants given explicitly in reference [4].

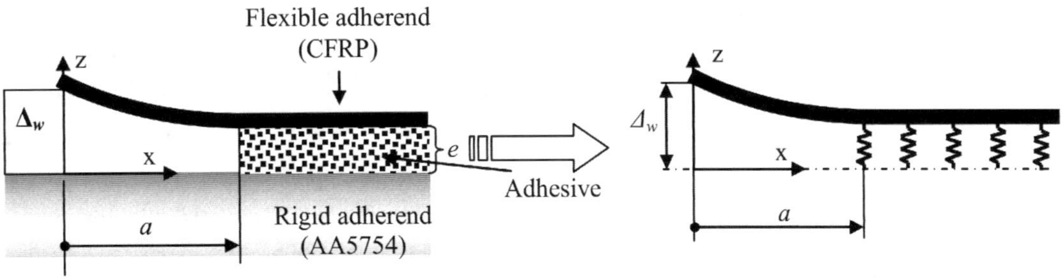

Figure 1. Geometry of asymmetric wedge test and schematic representation of adhesive joint using the Winkler model.

In the instrumented wedge test technique, an expression is required for the *upper* surface strain of the flexible adherend ε_S. For the bonded zone the (absolute value of) strain can be written [4] :

$$|\varepsilon_S(x)| = \frac{h}{2R} = \frac{h}{2}\left|\frac{d^2z}{dx^2}\right| = \frac{3\Delta h\lambda^2 e^{\lambda(a-x)}(-\sin\lambda(a-x) + \lambda a(\cos\lambda(a-x) - \sin\lambda(a-x)))}{(3+6\lambda a+6\lambda^2a^2+2\lambda^3a^3)}, \quad (6)$$

where R is the local radius of curvature of the adherend. This expression is derived from the 2nd derivatives of equations (3) and (5), using values of constants from reference [4]. In the free zone, the absolute value of strain is :

$$|\varepsilon_S(x)| = \frac{3\Delta h\lambda^3 x}{(3+6\lambda a+6\lambda^2a^2+2\lambda^3a^3)}. \quad (7)$$

In equations (6) and (7), the parameter λ, which we term the "wave number", is given by :

$$\lambda = \frac{\sqrt{2}}{2}\left(\frac{\bar{k}}{EI}\right)^{1/4}. \quad (8)$$

Thus, from equations (6) and (7), when a = constant, any change of the actual strain can only be caused by a change of λ, which in turn indicates a change of E_A (and possibly adhesive Poisson's ratio ν_A, but to a lesser extent), other parameters being constant (cf. equation (2)).

To take full advantage of the relations presented, in this Winkler Non-Destructive (ND) mode, six strain gauges were attached to the outer surface of the flexible plate: three in the free zone, and three in the bonded zone. The wedge test technique used here was asymmetric and employed in the newly introduced ND modes.

The thin, or flexible, adherend was made from 8-ply CFRP (carbon fibres in an epoxy matrix – Hexcel, Berkeley, CA, USA) composite with a [0°/90°]2s stacking sequence and 1.17 ± 0.02 mm thickness. In order to decrease flexural rigidity of the plate, thus increase strain recorded at the upper skin of the adherend, the composite plate was initially abraded using 320 grit paper. The final thickness achieved was h = 0.96 ± 0.02 mm. An aluminium plate of Hydronalium AA 5754, with thickness H = 5.4 ± 0.02 mm, constituted the rigid substrate. The flexible adherend of length, l = 120 mm, was bonded to the rigid one, of length L = 180 mm, along a bonded zone of length l_{adh} = 45 mm, making an unbonded zone length of 65 ± 0.2 mm. The width b of both adherends was 25 mm.

Prior to bonding, the aluminium surface was abraded with 320 grit emery paper, sand blasted with carborundum (SiC) particles of average diameter 60 μm, and washed in an ultrasound bath (in ethanol). The CFRP plate was cleaned with ethanol. Subsequently, the plates were bonded with an epoxy-based adhesive resin : Araldite GY784BD (modified DGEBA, epoxy equivalent weight 196 - 208 g/eq.) crosslinked with Aradur 125 high viscosity polyamidoamine (PAMAM) hardener, supplied by Huntsman (Huntsman, Fullertone, CA, USA). The adhesive was prepared by mixing resin and hardener in a ratio of 2:1 by weight, respectively, following the supplier's recommendation. Ambient conditions during the test, as for the crosslinking, were ca. 23 ± 2°C, ca. 55 ± 5 % relative humidity and 100kPa pressure. Two PTFE strips, of thickness 0.2 mm, placed on the bonded zone extremities, were used to assure sharply defined bonded zone edges (sharp crack front with a constant value of crack length). To provide a homogenous bondline thickness, the epoxy resin was filled (by the supplier) with glass beads of 0.2 mm diameter. Their presence, as well as that of the PTFE spacers placed at the extremity, resulted in a bondline of thickness $e = 0.2 \pm 0.01$ mm as measured with a digital microscope system (Dino - Lite Pro-IS Production S.A., 01633 St. Genis Pouilly, France).

The procedure used for the ND wedge test can be summarised as follows. A tapered asymmetric wedge of thickness, $\Delta_w = 0.9$ mm, is inserted to the desired position at the open bond end. To assure that the value of initial crack length is constant in all tests, two wedge positioning (stopping) pins are adjusted in the set up (placed at 61.5 mm from the crack front towards the unbonded zone). Since the aluminium wedge used has a tapered attacking edge (to allow smooth insertion), the crack length, defined as the distance between the line of contact of the wedge with the flexible adherend and the crack front was measured as $a = 62.0 \pm 0.05$ mm. Measurements being made, then the wedge is removed after each test. Overall, the time taken for each of the measurements was a maximum of ca. 10s. No crack propagation was observed in this short time, as verified by microscopic examination of the side of the sample.

3. Ultrasonic air-coupled through-transmission technique

Ultrasonic measurements represent a useful and easy way to characterize material properties of a single component [6], as well as layered structures [7, 9]. Here, a through-transmission method was employed to infer elastic properties of the adhesive during crosslinking. Air-coupled transducers are used to follow advancing crosslinking of the adhesive joint at atmospheric pressure and ambient temperature. Transmitted waveforms through the sample are collected for several wave incidence angles θ and frequencies f (figure 2). In the classic, direct problem of plane wave transmission, the complete transmitted acoustic field, including internal echoes, is a function of the incident angle θ and frequency f. The transmitted spectra modulus at each incidence angle $|A_T(f,\theta)|$ are calculated by multiplying the plane wave transmission coefficients $T(f,\theta)$ by the frequency spectrum $A_I(f)$ of the real wave measured with no sample between the transducers:

$$|A_T(f,\theta)| = |T(f,\theta) A_I(f)|. \tag{9}$$

In this manner, the effect of noisy experimental data, being in the wings of the frequency distribution due to the limited frequency bandwidth of the transducers, will be considerably reduced during the inversion process. The transmission coefficient is predicted using the surface impedance matrix method [8], which permits both fast and stable computations. This model takes into account the anisotropy and viscoelasticity of materials by using the concept of heterogeneous plane waves and complex viscoelasticity moduli, $C_{ij} = C'_{ij} + iC''_{ij}$ [9]. The plane wave condition considerably simplifies computations. Finite fields, on the other hand, would require the summation of plane waves over an angular aperture corresponding to that of ultrasonic transducers. As a result, the plane wave consideration has an important advantage for solving the inverse problem. The inversion scheme determines values of the viscoelastic coefficients, C_{ij}, of the material giving the best fit between theoretically predicted and measured transmission frequency spectra (figure 3), by using the Simplex method. The problem and procedure are covered in references [9, 10] and the method has been applied to infer both elastic and viscoelastic properties of materials. However, this application to adhesive joints, specifically when change of the adhesive properties is expected with time, is, to our knowledge, novel. The main advantage over the method already used to study changes in the state of bonding agents [11], is gaining values of engineering constants (such as Young's modulus). Therefore, the ultrasonic method presented is potentially both a qualitative and a quantitative tool.

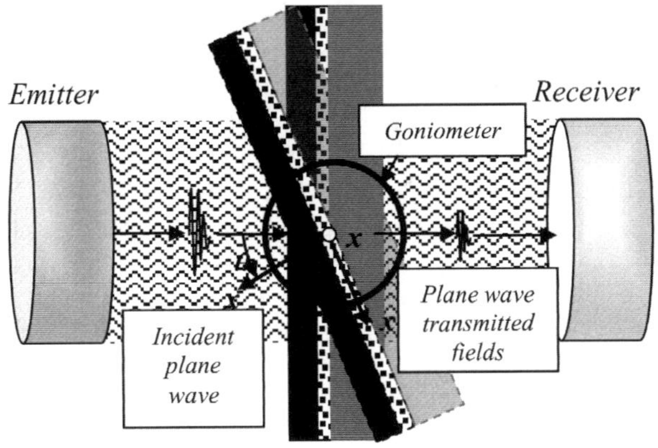

Figure 2. Schematic representation of air-coupled ultrasound through-transmission method for characterisation of adhesive elastic coefficients.

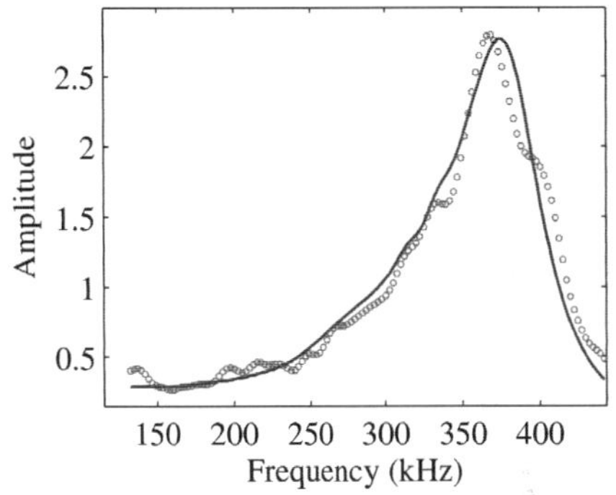

Figure 3. Optimisation of C_{ij} coefficients by minimising cost function between theoretical (—) and experimental (○) spectra of transmitted waveform.

In the adhesive joint, shown schematically in figure 2, two of the constituents, the adherends (in the present case, composite CFRP and aluminium AA5754), are preliminary characterised separately by using through-transmission measurements and the inversion technique. Also, the required thicknesses and densities of the assembled plates were measured before adhesive bonding was realised. These characteristics are used for calculations of the theoretical spectra of transmitted waves through the adhesive assembly. Therefore, only the viscoelastic properties of the adhesive are unknown in the theoretical model, and only their change in time can lead to changes in the transmitted wave field. In this paper we consider only the real part of the C_{ij} moduli in order to compare results with the wedge test method. The adhesive material is assumed to be isotropic and can be characterised by elastic coefficients C_{11} and C_{66}.

3.1. Ultrasound setup

In the case of ultrasonic inspection, the sample was made from the same materials as in section 2, whilst the geometry used was necessarily different. The CFRP composite plate, the same as for the wedge test, but now of dimensions 150 x 150 x 1.17 mm^3, was bonded to a rectangular aluminium alloy plate 150 x 150 x 3 mm^3. Surface preparation for the aluminium was kept as before. However, preparation of the CFRP in this case was limited to cleaning with ethanol and drying before bonding. Since a vertical orientation of the plate is required in this test, special spring jigs were used to prevent any relative movement between the composite and aluminium plates. PTFE spacers at the joint extremities were not used here. This had, as a consequence, a reduced bondline thickness, compared to the wedge test. Measured bondline thickness was $e = 0.16 \pm 0.01$ mm (calliper gauge and dial gauge).

An air-coupled transmitter was excited with a chirp frequency range of 250 kHz to 450 kHz. The reference signal is transmitted directly to an air-coupled receiver without a sample in between. This was measured with low voltage amplification on the transmitter and a low gain on the receiver, to

Journal of Physics: Conference Series **353** (2012) 012006 doi:10.1088/1742-6596/353/1/012006

prevent saturation of the signal. The bonded sample was placed between both transducers on a motorised goniometer, and parallelism between sample plate and transducers was carefully adjusted. Wave coupling between air and the solid medium is expected to be very poor and consequently acoustic energy transmitted through the sample to be very low. Therefore, both amplification voltage excitation and reception gain must be increased significantly to get signals transmitted through the sample (this can be done with the assumption that the frequency response of the instrumentation is constant and independent of the gain). In addition, signals were averaged over 2000 acquisitions in order to remove any additional noise from instrumentation and obtain a high signal-to-noise ratio.

Subsequently, through-transmission signals were collected at different times during the crosslinking period of the epoxy adhesive, and at several angles of incidence, θ, between 0° and 12° with a 2° step, in order to optimise values of the elastic moduli C_{11} and C_{66}. In the optimisation process, transmission spectra are normalised to overcome the gain discrepancy between the reference signal and through-sample signals (low gain for the reference signal and high gain for through-sample signals). Values of elastic moduli are optimized so that predicted transmitted fields fit at best those measured, over every incidence angles.

4. Results and discussion

4.1. Crosslinking evolution

Crosslinking occurs in three basic stages, during which the polymer transforms from a viscous liquid to a rigid solid viz. resol-resitol-resit [12]. Resol is a polymer in which no crosslinking has yet occurred and is within its gel time. Resitol corresponds to a partially crosslinked, partially unreacted polymer. Finally, resit consists of (mostly) crosslinked adhesive, since completing crosslinking at

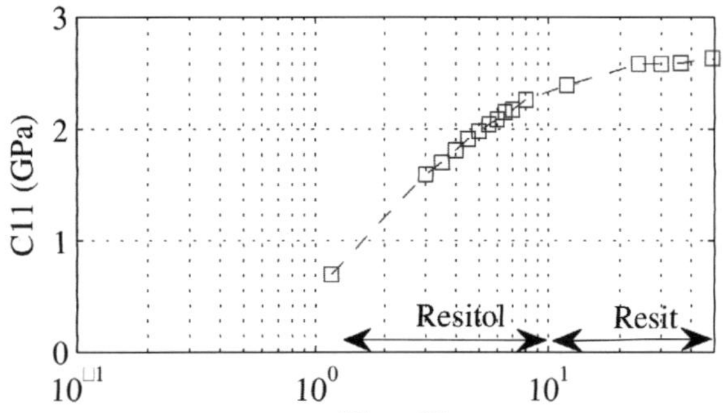

Figure 4. Data obtained from ultrasound technique representing change in C_{11}, on a logarithmic time scale.

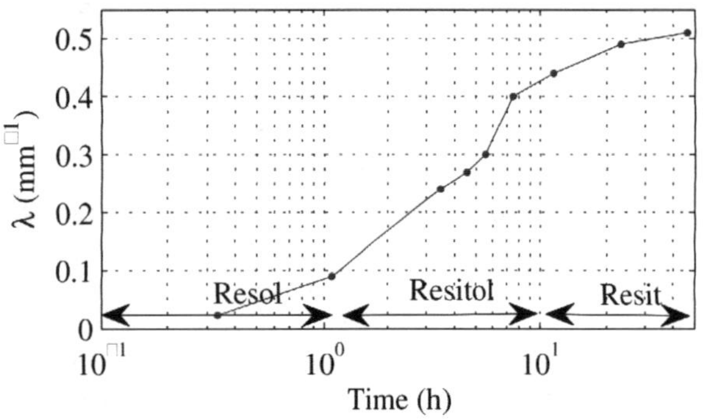

Figure 5. Variation of characteristic parameter λ, during crosslinking on a logarithmic time scale. Points indicate optimised λ values from experimental results and using the Winkler model. Three stages can be identified.

room temperature is very hard to achieve because the reaction of secondary (or tertiary) amine groups may be blocked (correct mixing ratio is crucial), and leads to residual, unreacted epoxy groups, therefore the final resit state of the adhesive is rarely achieved.

We address first the influence of the evolution of crosslinking on parameters λ and C_{11} from both the wedge and ultrasound measurements. Evolution of the measured parameters appears smooth with no particular discontinuity occurring. As can be seen, the results obtained with the two different approaches present the same trend of crosslinking kinetics. The three expected stages can be identified from the wedge test data (figure 5) : an initial, slow reaction (resol), acceleration (resitol), and deceleration (resit) leading to (the same) final curing time. From the ultrasound data (figure 4) the resol stage cannot be identified, because early crosslinking data are not available. Regarding the resitol and resit stages, data from both techniques show the same trend of deceleration and plateau reaching, which is associated with the decreasing rate of the crosslinking reaction. Therefore, the use of the wedge test and Winkler model, which shows the change in elastic properties of the adhesive with time at constant crack length, seems to be a suitable and meaningful way of estimating curing time as well as monitoring adhesive properties.

4.2. Comparison of Young's modulus obtained from both methods

Although parameters C_{11} and λ are both sensitive to crosslinking kinetics, and follow a similar trend, they cannot be compared *directly* to a mechanical property. From both techniques presented in this paper, it is possible to obtain the Young's modulus, which is a standard parameter related to mechanical properties of the bonded joint.

To obtain Young's modulus of the adhesive E_A, from the ultrasound method, the modulus C_{66} (which is equal to the shear modulus G_A) should also be estimated, because it is required together with C_{11} [13], since Young's modulus can be expressed as

$$E_A = C_{66}\frac{3C_{11}-4C_{66}}{C_{11}-C_{66}}. \tag{10}$$

The normal incidence signal is used for the characterisation of the C_{11} coefficient, since the longitudinal wave is insensitive to C_{66} for that angle. Angles of incidence above 0° are expected to generate both longitudinal and shear wave modes inside the sample, shear wave modes being used for characterisation of C_{66}.

From the wedge test method, having established the value of the wave number λ, it is then possible to estimate Young's modulus of the adhesive. Wave number is essentially a function of one variable during crosslinking, the rest remaining constant: $\lambda = f(E_A) \propto E_A^{1/4}$ (see equations (8) and (2)).

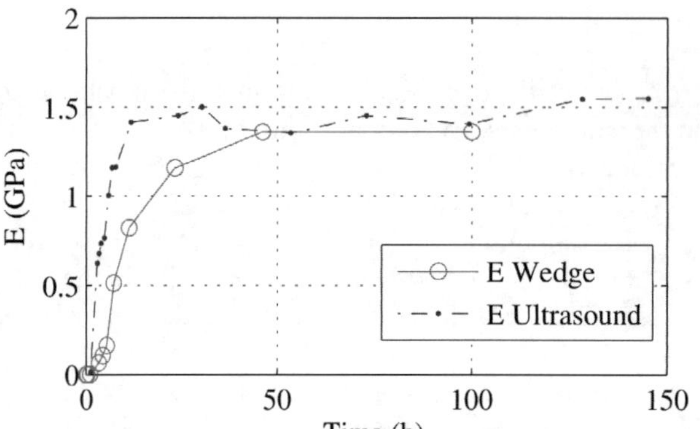

Figure 6. Young's modulus obtained during monitoring of crosslinking, from wedge method and ultrasound method

Values of Young's modulus calculated from both techniques are compared in figure 6. Clearly, there is good agreement between the two methods for values of modulus obtained towards completion of chemical reaction, but this is less good near the start of the crosslinking. The discrepancy between ultrasound method and wedge method during the first phase of crosslinking could be explained by a low sensitivity of ultrasound due to high viscosity at the first stages of crosslinking. As a result, the uncured adhesive has a low shear stiffness, and we observe a poor sensitivity of the transmission spectra when optimising C_{66} values, this unsensitivity probably being reinforced by the very low thickness of the adhesive bond. High uncertainty on C_{66} values results directly from high uncertainty on Young's modulus. In addition, high frequencies used with the ultrasound method (dynamic regime) are expected to give a higher value of Young's modulus compared to the wedge Young's modulus (quasi-static regime), this is observed during crosslinking. When the rate of crosslinking is maximal and reaches a plateau, we observe that both methods give a very good agreement of $E_{wedge} = 1.36$ GPa and $E_{US} = 1.40$ GPa at $t = 100$ hours. The ultrasound method also gives the shear modulus $G = C_{66} = 0.51$ GPa. For the characterisation of the cured adhesive, which represents the most frequent cases, both methods give consistent results and present a great interest for non-destructive evaluation of mechanical properties of adhesive bonds.

5. Conclusions

The curing of a DGEBA-PAMAM adhesive has been studied with two non-destructive approaches. One was based on an instrumented, asymmetric, wedge test technique, where changes of strain measured on the bent substrate during wedge insertion (without causing fracture) reveal the state of the adhesive. The second was based on an ultrasound through-transmission method, where changes of optimized transmitted spectra yield the adhesive properties. The main conclusions are as follows:

- Application of the Winkler model to the adhesive joint allows quantitative estimation of the adhesive properties, changing with time, as well as curing time itself.
- The ultrasonic through transmission method using inverse problem solution allows estimation of curing time and adhesive properties during crosslinking, when an actual adhesive joint is used.
- The final properties of the DGEBA-PAMAM system in question, as obtained with wedge test and ultrasonic experiments, are: $E_A = 1.38$ GPa as an average from both methods, and $G_A = 0.51$ GPa given by the ultrasound method.

The problem of non-destructive evaluation of curing time is particularly important in structural applications. Thus, these methods could allow in situ, non-destructive measurements to be made on a structure in service. On the contrary, standard techniques, such as thermogravimetry (TG or TGA), calorimetric spectroscopy (CSC) and dynamic mechanical and thermo-mechanical analysers (DMA or DMTA) tend to be destructive and/or do not give the opportunity for use on working structures. In addition, they are devoted to polymers alone. This study shows the potential of the two techniques presented to be developed together, or separately, for applications in non-destructive testing (NDT), and for monitoring the development of strength in *situ* of structural adhesive bonding.

Acknowledgments
We would like to thank the company Rescoll for free supplying of the adhesive components. The work on the ultrasound method has been done in the framework of INMAT2 project.

References

[1] Mika T F and Bauer R S 1988 In *Epoxy resins: chemistry and technology* (ed. C. A. May) 2nd edn New York: Marcel Dekker Inc. 456-550
[2] Czub P, Bończa-Tomaszewski Z, Penczek P and Pielichowski J 2002 In *Chemistry and Technology of epoxy resins*, Warsaw: WNT
[3] Winkler E 1867 Die Lehre von der Elasticitaet und Festigkeit. Teil 1, Verlag H. Dominicus, Prague (As noted in L Fryba 1993 History of Winkler Foundation *Vehicle System Dynamics Supplement* **24** 7-12)
[4] Budzik M K, Jumel J, Imielinska K and Shanahan M E R 2011 *J. Adhes. Sci. Technol.* **25** 131-149
[5] Budzik M K, Jumel J and Shanahan M E R 2011 *Inter. J. Adhes. Adhes.* (accepted for publication)
[6] Chimenti D E 1997 *Appl. Mech. Rev.* **50** 247-84
[7] Cawley P, Pialucha T P and Zeller B D 1996 *Proc. R. Soc. Lond. A* 452 1903-26
[8] Hosten B and Castaings M 2003 *Ultrasonics* **41** 501-507
[9] Castaings M 2009, in *Materials and Acoustics Handbook*, Under supervision of M Bruneau and C Potel, Ed. ISTE-WILEY, London ISBN 9781848210745 chap. 20 561-576
[10] Castaings M and Hosten B 2000 *Ultrasonics* **38** 781-786
[11] Saksena T K and Babbar N K 1979 *Ultrasonics* **17** 3 122-124
[12] Brydson J A 1999 *Plastics Materials*, 7th edn, London: Butterworth Scientific 641-642
[13] Royer D and Dieulesaint E 2000 *Elastic waves in solids 1 : Free and guided propagation*, Springer

Application of ultrasonic tomography to characterize the mechanical state of standing trees (*Picea abies*)

L Brancheriau[1], A Ghodrati[2], P Gallet[1], P Thaunay[1], P Lasaygues[3]

[1] Production and Processing of Tropical Woods, CIRAD, Montpellier, France
[2] Islamic Azad University, Science and Research Branch, Tehran, Iran
[3] Laboratory of Mechanics and Acoustics, UPR CNRS 7051, Marseille, France

E-mail: loic.brancheriau@cirad.fr

Abstract. Wood is a biological growth medium. It is orthotropic with longitudinal, radial and tangential axes. Furthermore, standing trees adapt themselves to environmental growth conditions, and their material properties vary with age. These changes result in variations that are much more complex than anisotropy. Studying wood quality and intraspecific variability is useful for clonal selection and for the genetic improvement of plantations. In this study, two logs of *Picea abies* underwent transmission tomography. The mean diameter was 16 cm (26-year-old tree) and the moisture content was 22%. The effect of the presence of bark and artificial defects was investigated. The tomographic device was specifically built for tree imaging. The imaging process was automatic with 900 ultrasonic acquisitions in 40 minutes (emission at 55 kHz with 5 periods of square wave form). The main conclusions were: speed near the bark is higher than in the centre because of the existence of juvenile wood combined with the moisture content gradient (moisture content lower near the bark). Likewise, damping near the bark is lower than in the centre. A significant relationship was observed between slowness and attenuation (R^2=0.50); when the speed increased, damping decreased. No clear effect of the presence of bark was shown on the tomographic images. The bark was thin (3 to 5 mm thick) compared to the wavelength (26 mm). The 10, 20 and 50 mm artificial holes were clearly visible on the tomographic images. However, quantitative tomography does not enable the precise location of defects.

1. Introduction

Wood is a biological end-product generated during cambial growth over successive years [1]. During the growth process, cambium cells convert most photosynthesized products and nutrients into various biopolymers for use in the formation of woody tissues (annual ring and cork formation). Wood can thus be considered as a record of the responses of trees exposed to different growing conditions, but also to changing environments. The properties of an annual ring describe the morphological structure of the species and quantify the reaction of trees to variations in their environment.

In the particular case of plantations, species are usually selected for their fast growth, but the quantity of wood produced does not necessarily mean quality. With a view to improving breeding programmes and genetic selection adapted to new forestry practices, non-destructive techniques need to be directly employed in plantations. The challenge is thus to use accurate techniques capable of quantifying a large number of properties in a few minutes; the apparatus should also be easy to move around the plantation.

The use of elastic waves (ultrasound or sound waves) to detect decay in trees has been explored and reported by many researchers [2, 5, 8, 16]. The concept of detecting decay by this method was initially based on the observation that wave propagation is sensitive to the presence of degradation in wood. In fact, elastic waves appear to travel more slowly in deteriorated wood than they do in normal wood [15]. The first generation of equipment used involved two-probe systems that measured wave TOF in a single path. The suitability of a single-path approach for tree-decay detection has proved to be limited, as stress-wave velocity across tree stems varies substantially even for intact trees, and a standard reference velocity for data interpretation is not readily available [17].

More recently, tomography techniques developed for medical and industrial applications, or geophysical prospecting, have been assessed for their applicability in standing trees. Different imaging techniques were reported by Bucur in 2003 [4]. In that review article, the author detailed the possibilities of using ionizing radiation (X-rays and gamma rays), thermography, microwaves, ultrasonics and nuclear magnetic resonance. A tomograph using ionizing radiation was designed for standing trees [12]. Commercial devices were also developed and were based on determining the time of flight (TOF) of elastic waves induced by an instrumented hammer (acousto-ultrasound tools). These devices were presented in detail by Schubert in 2007 [13]. The three tools were PiCUS®, ARBOTOM® [11] and FAKOPP3D® [7]. In contrast to tomography based on X-rays and gamma-rays, tomographic methods based on electrical resistance, microwaves, or elastic waves are not dangerous. Nicololli [10] compared these three approaches by calculating the tomograms of decayed trunks. These tomographic methods consist in inverting the electrical resistance, or TOF, of electromagnetic and elastic waves by algorithms from the field of geophysics. Tomography based on elastic waves has given the best results in terms of depicting the geometry and location of inner wood-decay. However, times of flight are inverted by filtered back-projection, which assumes isotropic material behaviour and straight rays (PICUS® and FAKOPP3D®). Straight rays are appropriate in isotropic media, otherwise inversion should account for curved rays [6, 14]. Socco (2004) used an iterative TOF inversion method (root mean square optimization method) developed by geophysicists. This inversion method can be used with straight or curved rays [14]. In 2006, Maurer also proposed a correction procedure to take into account the anisotropy of wood. This correction assumed slight anisotropy and took into account a radial to tangential velocity ratio in a standard isotropic reconstruction procedure [9].

With the challenge of using an accurate and fast, non-destructive technique in plantations, the objectives of this study were to design an automatic tomograph and compare how the presence of bark and drilled holes affect the ultrasonic parameters: slowness and attenuation. In the early stages of developing the tomograph, two 300-kHz ultrasonic probes were used [3]. The probes consisted of a wheel in which the emitter was placed, and the coupling medium was formed by an elastomer surrounding the wheel. After several tests on green and dry woods in the transverse and longitudinal directions, those probes were replaced because transmission depth at 300 kHz was between 3 cm and 10 cm, depending on the test configuration (without the presence of bark). The anisotropy of wood was not taken into account and a classic filtered back-projection algorithm was used to compute the tomography maps.

2. Experimental methodology

2.1. Materials

Two spruce (*Picea abies*) logs were harvested from different trees near the city of Nowshahr (Iran). The sampling height was between 0.80 m and 1.30 m. The corresponding mean diameter was between 15.0 cm and 16.5 cm. The cross-section of the logs was almost circular and the logs also had a straight vertical appearance. The trees were 26 years old. Samples were taken in September 2010 and shipped to France to be tested in October; the samples were wrapped in plastic film to minimize the effect of wood drying.

2.2. Experimental protocol

The first log was cut into two 5-cm thick disks and one 15-cm thick disk. The disks were taken from the middle section of the log to avoid cracks due to wood drying, and to have a high moisture content. The 5-cm thick disks were used to determine the mean moisture content, along with the moisture content profile along the radius (figure 1). The moisture content was determined before and after ultrasonic testing. The samples in figure 1 were weighed then dried in an oven (103°C) for 48 hours, then weighed again. Equation 1 was used to compute the moisture content.

$$MC\% = 100\frac{M_H - M_0}{M_0} \qquad (1)$$

Where MC is the moisture content, M_H the green weight and M_0 the oven-dry weight.

Figure 1. Sampling pattern for moisture content determination (dimensions of the small samples: 2 x 3 x 5 cm^3).

The 15-cm thick disk underwent tomography testing with 3 sample states: with bark, without bark and drilled (2 holes measuring 1 cm and 2 cm in diameter, see figure 2). The tomography tests were carried out with two replications and an angular step of 30° to estimate measurement uncertainties, and with an angular step of 10° for tomographic imaging.

Figure 2. Tomography tests on the first log (from left to right: with bark, without bark and drilled with two holes measuring 1 cm and 2 cm in diameter).

The second log was cut into two 15-cm thick disks. The disks were taken from the middle section of the log. In one disk, a hole measuring 5 cm in diameter was drilled in the middle. In the other disk, a hole measuring 5 cm in diameter was drilled off-centre (figure 3). The two disks underwent tomography testing with an angular step of 10°.

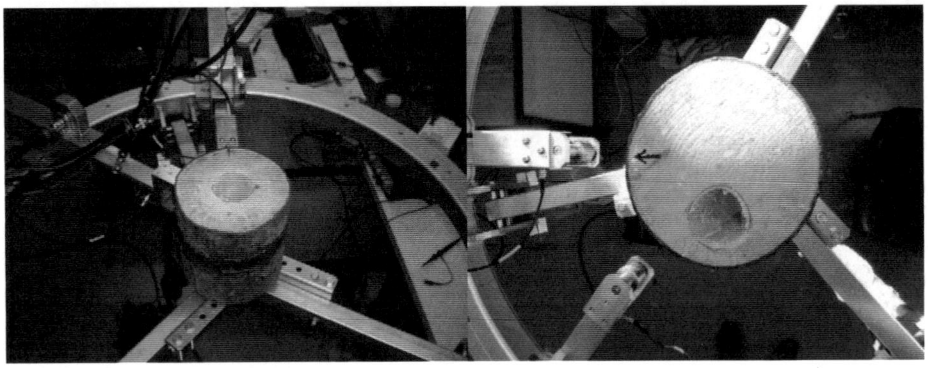

Figure 3. Tomography tests on the two disks of the second log (5-cm hole diameter).

2.3. Ultrasonic tomography

The ultrasonic tomograph used was an aluminium ring supported by a tripod (figure 4) [3]. In order to be able to test small disks, a specific support was added and mounted inside the tripod (figure 3). This device was designed to test standing trees. Thus, the height of the ring can be adjusted from 1 m to 1.6 m but the typical height is 1.3 m. If the apparatus is mounted on a trunk (figure 4) the two parts of the ring are first assembled, and then the tripod is fixed. The apparatus is then placed against the trunk using two narrowing elements. Each probe is mounted on a trolley and the trolleys are slid onto the ring. Finally, the last section of the ring is mounted and the third narrowing element is applied.

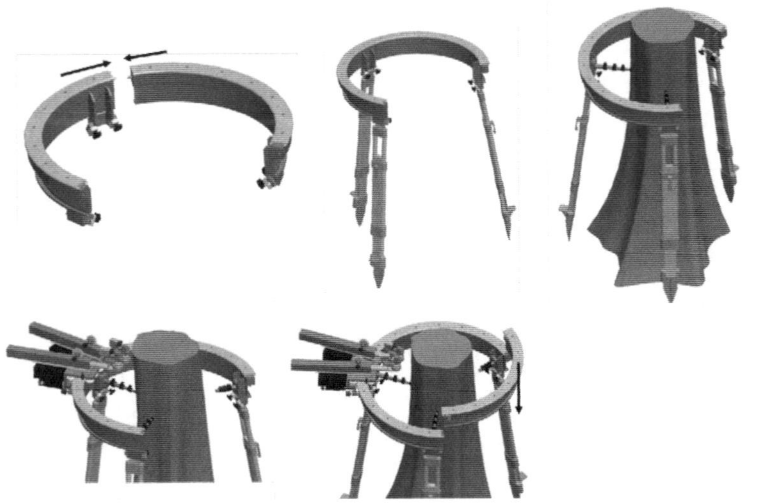

Figure 4. Assembly steps for the ultrasonic tomograph.

The probes had a frequency of 55 kHz and were designed for acoustic emission purposes. They were mounted on a specific support with a wheel, and grease was used as the coupling medium. Emission consisted in 5 square periods of 480 V. The output analog signal was amplified by 80 dB. Acquisition was via a converter with 16-bit resolution and a sampling frequency of 5 MHz. The acquisition time was set at 410 μs. The output signal was filtered beforehand by a Morlet wavelet at 55 kHz with a bandwidth of 20 kHz for denoising (figure 5).

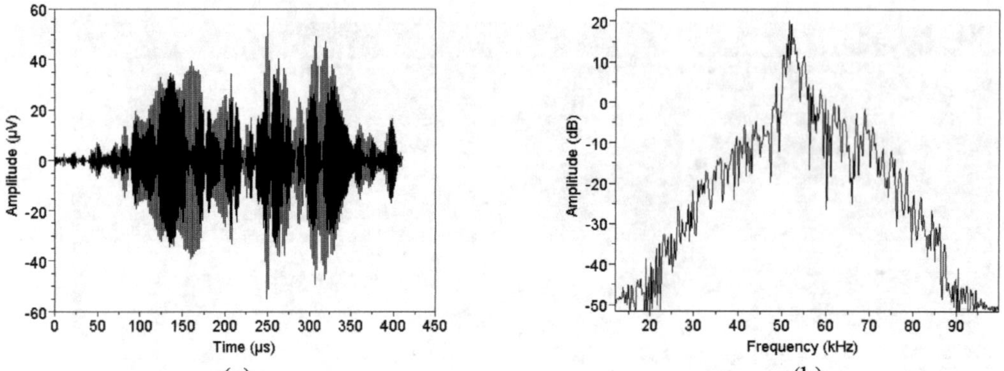

(a) (b)

Figure 5. Ultrasonic signal filtered by a Morlet wavelet at 55 kHz with a bandwidth of 20 kHz (a) and associated Fourier transform (b).

The ultrasonic parameters used for image reconstructions were slowness (s/m) and attenuation (dB/m at 55 kHz). Slowness, *Sl,* was computed using cross-correlation between the input and the output signals (equation 2). The ratio of received energy to emitted energy was used to determine attenuation *At* (equation 3).

$$Sl = \frac{1}{d} Max\left(\int e(\tau)\, s(t-\tau)\, d\tau\right) \quad (2)$$

$$At = \frac{1}{d} 10\log_{10}\left(\frac{\int s^2(t)\,dt}{\int e^2(t)\,dt}\right) \quad (3)$$

where *Sl* is slowness, *At* attenuation, *d* the distance between the emitter and the receiver, *e(t)* the input signal and *s(t)* the output signal.

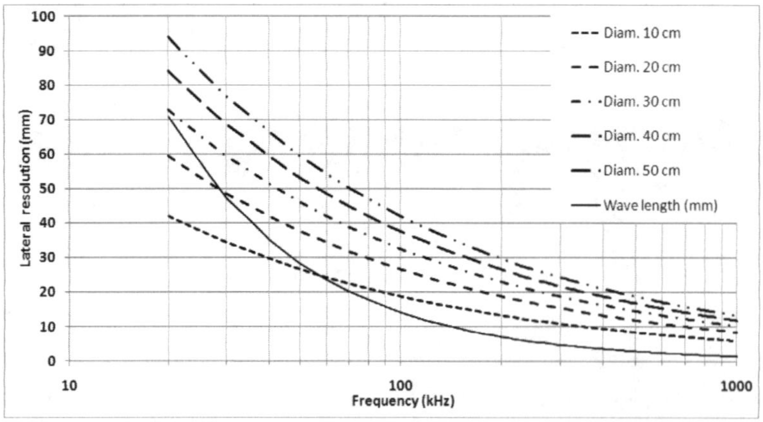

Figure 6. Lateral resolution and wavelength as a function of the frequency and depending on the sample diameter (the acoustic wave speed was set at 1400 m/s).

The tomographic images were computed using an angular step of 10°, which led to 900 acquisitions (36 equidistant acquisition points). A complete scan was completed in 40 minutes with the apparatus. In comparison, three well-trained operators were asked to carry out their own scan with 18 measurement points (306 acquisitions) and the duration of the scanning process was never less than one hour.

For a projection, the emitter remained motionless while the receiver moved around the ring. The next projection was initiated by an angular step motion of the emitter followed by another motion of the receiver all the way around the ring (acquisition took place on each angular step). The reconstruction algorithm used was based on the inverse Radon transform in fan beam geometry (linear interpolation) with a modified Shepp-Logan filter. Considering the medium to be isotropic, the elastic waves were assumed to propagate in a straight line. The velocity (and therefore the attenuation) of waves was then obtained by measuring travel-time (of energy) differences by inverting the Radon transform.

The theoretical upper limit of the number of computed pixels was 450 (mesh of 39 mm², equation 4). However, the physical resolution was about 26 mm (axial resolution: wavelength) and 31 mm (lateral resolution: Fresnel zone, equation 5, figure 6). This resolution led to a mesh of 806 mm²; the images were thus 2D smoothed by a Blackman filter of the wavelength size (the errors induced by the use of the modified Shepp-Logan filter were also greatly reduced). It is important to point out that the redundancy of the information (39 mm² compared to 806 mm²) made it possible to decrease the effect of erroneous experimental values in the imaging process.

$$N_{Pixels} = \frac{N_{\mathrm{Projections}} \times N_{Sample}}{2} \tag{4}$$

where N_{Pixels} is the number of pixels, $N_{\mathrm{Projections}}$ the number of projections (36) and N_{Sample} the number of acquisitions per projection (25).

$$R_F = \sqrt{\frac{S_d C}{2f}} \tag{5}$$

where R_F is the Fresnel zone radius, S_d the scanned depth (the rotation radius of the emitter or receiver transducer), C the wave celerity and f the frequency of the transducer.

3. Results and discussion

In the following paragraphs, slowness is presented in 10^3 (s/m) or its inverse, speed in (m/s). Attenuation (in dB/m) was normalized by the diameter and thus presented in (dB).

3.1. Moisture content determination

The mean moisture content was 23% before the tests and 21% at the end of the experiment. The moisture content profile along the radius is presented in figure 7. The difference in moisture content from the pith to the bark was 5% before the test and 10% after the test. Despite the logs and disks being wrapped in plastic film, the wood samples tested were not in a green state and the natural drying effect increased while testing (a major phenomenon after bark removal). Small cracks appeared around the edge of the disks due to drying (figure 2, right-hand image).

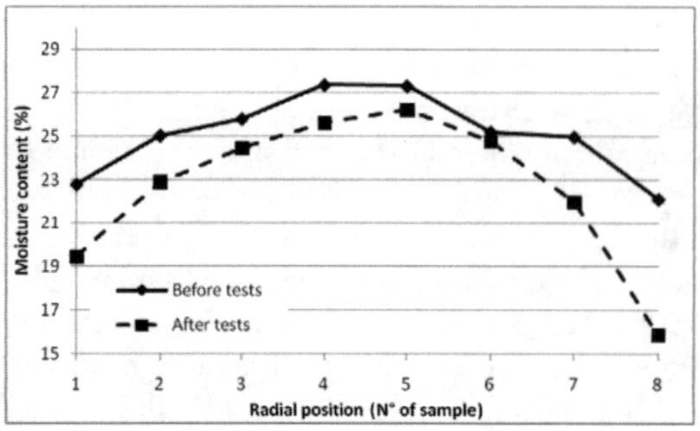

Figure 7. Moisture content (%) depending on the radial position before and after the ultrasonic tests (samples from the first log).

3.2. Measurement uncertainties

Two tomography tests were conducted with an angular step of 30° (108 acquisitions) on the same disk (with bark, without bark and drilled). The results are shown in figure 8 for the disk with bark only, as the trends were very similar after bark removal and drilling. Significant linear relationships were found for speed ($R^2 = 0.54$, $F = 105$, $p < 0.001$, $N = 108$) and for attenuation ($R^2 = 0.93$, $F = 1373$, $p < 0.001$, $N = 108$). The presence of outliers in the scatter plots for speed (figure 8, a) was caused by errors in the travel time estimation. The standard error of speed determination was thus ±420 m/s. Attenuation had a low measurement error compared to the speed estimation, with a standard error equal to ±1.7 dB.

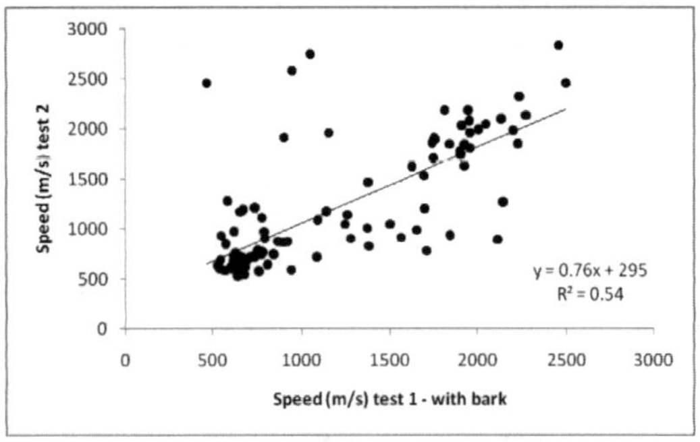

(a)

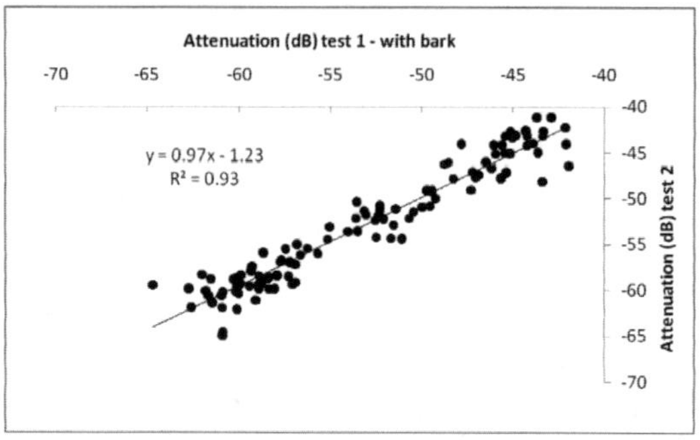

(b)

Figure 8. Relationship between the two replications for (a) speed and (b) attenuation (disk with bark, N=108).

In actual practice, defects detection in trees like fungal decay depends on the ability of measuring acoustic impedance variations. It was reported that the weight loss due to fungal activity was linked to the decrease of transverse acoustic velocity [13]. Taking into account the speed measurement error of the tomograph (±420 m/s), the threshold value of weight variation was computed and equal to 13%. A weight loss of 13% is thus linked to the minimum variation of velocity detectable by the tomograph. This threshold can be associated to a moderate degradation state of the material.

3.3. Effect of bark and holes on the ultrasonic parameters

Mean comparisons were made to examine how bark and holes affected the ultrasonic parameters (figure 9). There were 108 acquisitions with an angular step of 30°. The statistical tests were firstly a

Student's t-test on paired samples (table 1). The *t*-test provides an exact test for the equality of the means of two normal populations with unknown, but equal, variances. The test statistic is based on a Student's distribution *t* with the degrees of freedom *d.f.* (table 1). The associated *p*-value is the probability that the means are the same.

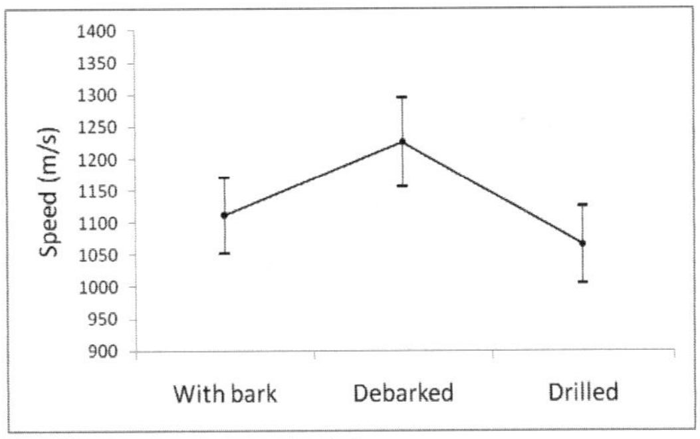

(a)

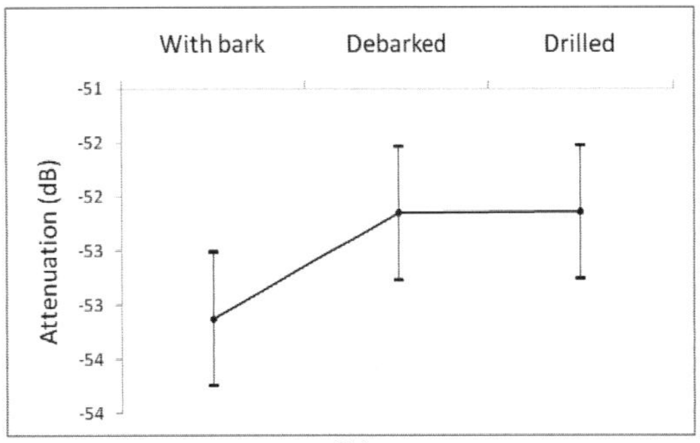

(b)

Figure 9. Means and standard deviations of speed (a) and attenuation (b) computed on 108 acquisitions for the disk with bark, without bark and drilled.

Table 1. *t*-Test and Wilcoxon test on paired samples for speed and attenuation between the disk with bark, without bark and drilled.

Data sets		t	d.f.	*p*-value (bilateral)	Wilcoxon *p* (bilateral)
With bark - without bark					
	Speed	-1.934	107	0.056	0.496
	Attenuation	-5.888	107	< 0.001	< 0.001
Drilled – Without bark					
	Speed	2.196	107	0.030	0.126
	Attenuation	-0.059	107	0.953	0.998

The *t*-test results, presented in table 1, were biased by the non-normal distribution of the parameters. To highlight this phenomenon, two histograms were plotted (figure 10) after the tomographic tests with an angular step of 10° (N=900). The bimodal distribution is shown in this figure for speed, with two modes between 1200 (m/s). The same type of distribution was also found for attenuation between -51 (dB). Equivalent findings were made for the disks without bark and the drilled disks.

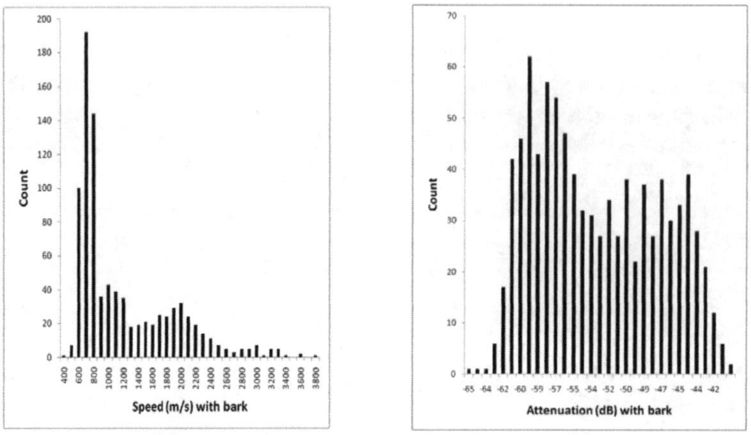

(a) (b)

Figure 10. Histograms of speed (a) and attenuation (b) of acoustic waves at 55 kHz computed for the disk with bark (N=900).

The results in table 1 were then completed by non-parametric Wilcoxon tests (valid when the normality assumption is not satisfied or the sample size is too small). The Wilcoxon tests agreed with those of the Student's t test for attenuation but did not confirm the Student's t results for speed (table 1). The mean attenuation measured on the disk with bark was thus significantly different from that for the disk without bark. The mean attenuation was higher with bark than without bark. No significant difference was found between the mean attenuation of the disk without bark and that of the drilled disk. For the speed measurements, differences existed between disks with bark, without bark and drilled disks, but the probability values were too high to be absolutely affirmative (figure 9 and table 1).

3.4. Relationship between slowness and attenuation

Figure 11 shows a scatter plot between slowness and attenuation for the disk with bark. A significant linear relationship was found with an R^2 equal to 0.51 (F=933, $p < 0.001$, N=900). Attenuation increased when slowness increased or speed decreased. However, the degree of fit should be read with care, as figure 11 clearly shows two populations (associated with the distributions of figure 10). Outliers were also visible due to the uncertainty of slowness determination, but they were not associated with particular positions of the probes.

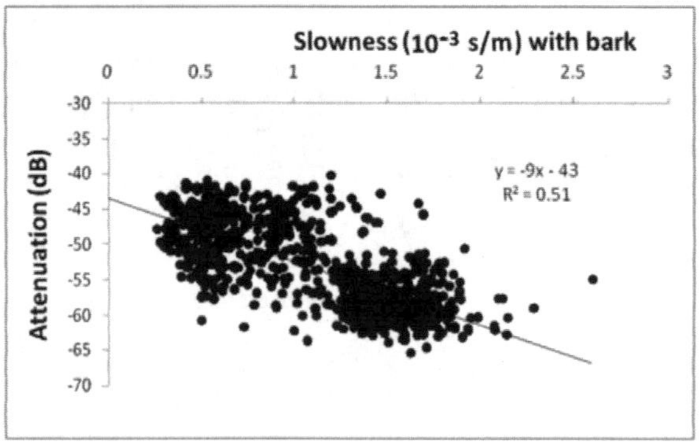

Figure 11. Relationship between slowness and attenuation for the disk with bark (N=900).

3.5. Tomographic images

The computed maps of slowness and attenuation for the first log with and without bark are presented in figure 12. The images were displayed in 87 x 87 pixels and based on a 10° angular step (900 acquisitions). There was no visible effect of the presence of bark on the outlines of the maps (figure 12). This phenomenon was due to the limited thickness of the bark (3 to 5 mm) compared to the wavelength (26 mm), combined with the smoothing effect of the 2D filter.

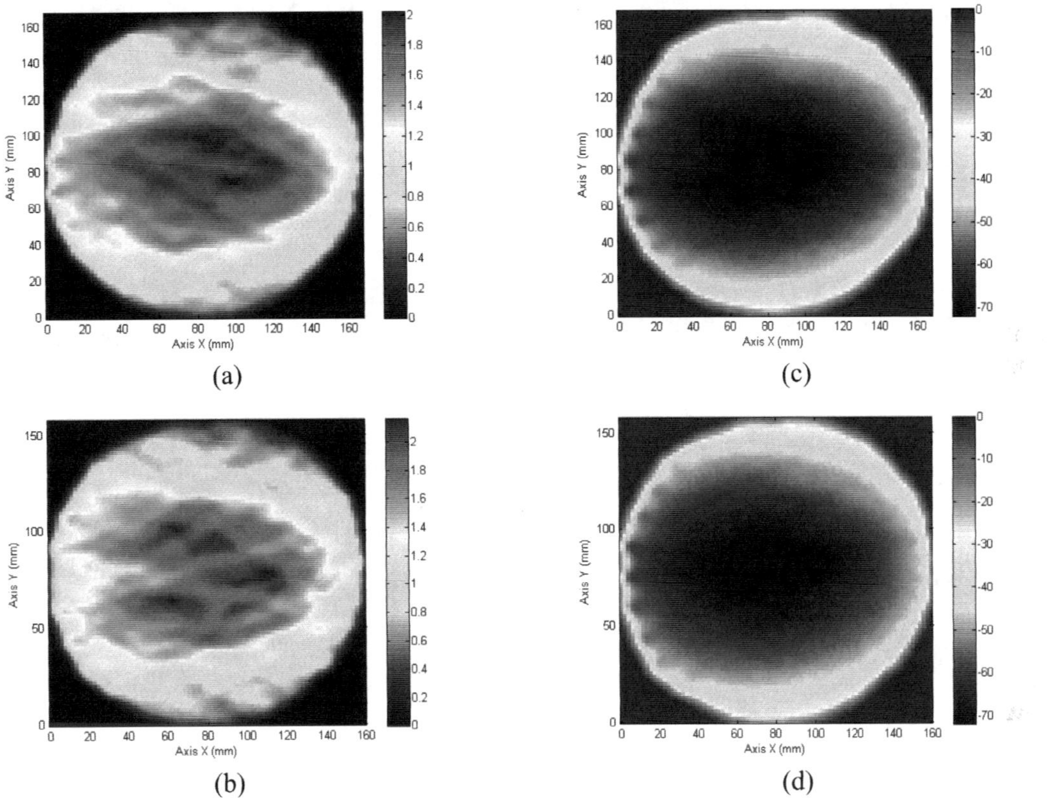

Figure 12. Tomographic maps of slowness (10^{-3} s/m) (a, b) and attenuation (dB) (c, d) for the disk of the first log, with bark (a, c) and without bark (b, d).

However, figure 12 shows two different areas from the pith to the periphery. These areas are highlighted in figure 13 in which the paths were drawn if slowness was lower than 1.2×10^{-3} s/m and if attenuation was over -51 dB. The periphery was thus characterized by higher speed and lower damping than those of the pith. The reverse phenomenon was found by Socco and interpreted as an effect of wood anisotropy [14]. The shorter paths (periphery) were more influenced by the tangential behaviour of wood, while the radial behaviour had a greater effect for longer paths (centre). However, the radial velocity of wood is known to be greater than the tangential velocity. In this specific case (figure 12), this observation was explained by the transition from juvenile to mature wood, combined with the effect of moisture distribution (figure 7).

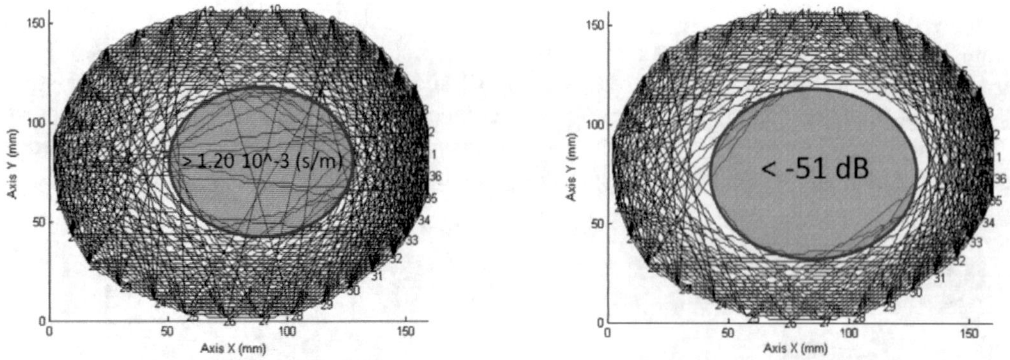

(a) (b)

Figure 13. Paths from emitter to receiver with slowness under 1.2x10^{-3} s/m (a) and attenuation over -51 dB (b) for the disk of the first log with bark.

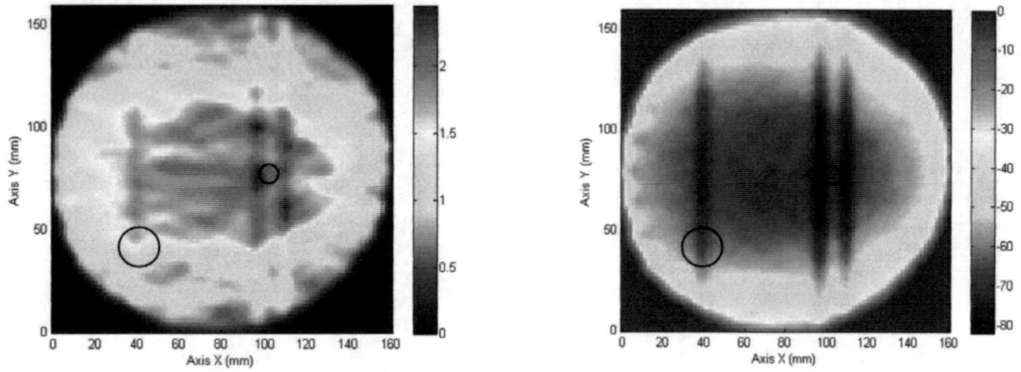

(a) (b)

Figure 14. Tomographic maps of slowness (10^{-3} s/m) (a) and attenuation (dB) (b) for the drilled disk of the first log. The black circles indicate the positions of the drillings.

The tomographic images obtained for the drilled disks of the two logs are presented in figure 14 and figure 15. The effect of the holes can be seen on the maps by areas with high slowness and also high damping. This observation was valid even for the hole with a diameter (10 mm) half that of the wavelength (26 mm). It seems that the position of the holes in the disk had an effect on the reconstructed pattern; the holes close to the pith were characterized by two elliptic patterns unlike those close to the periphery. The shape of the patterns associated with the holes was not a circle but an ellipse and several hypotheses were advanced: anisotropy was not taken into account, some cracks due to natural drying occurred, the reconstruction procedure was not well adapted for missing data (the dimension of the probe supports did not enable scanning of the total angular area but only 250° for each projection), the mounting of the probes (figure 5, right-hand image) induced back-waves which altered the received signals. These hypotheses will be investigated in a future study to improve the tomographic process.

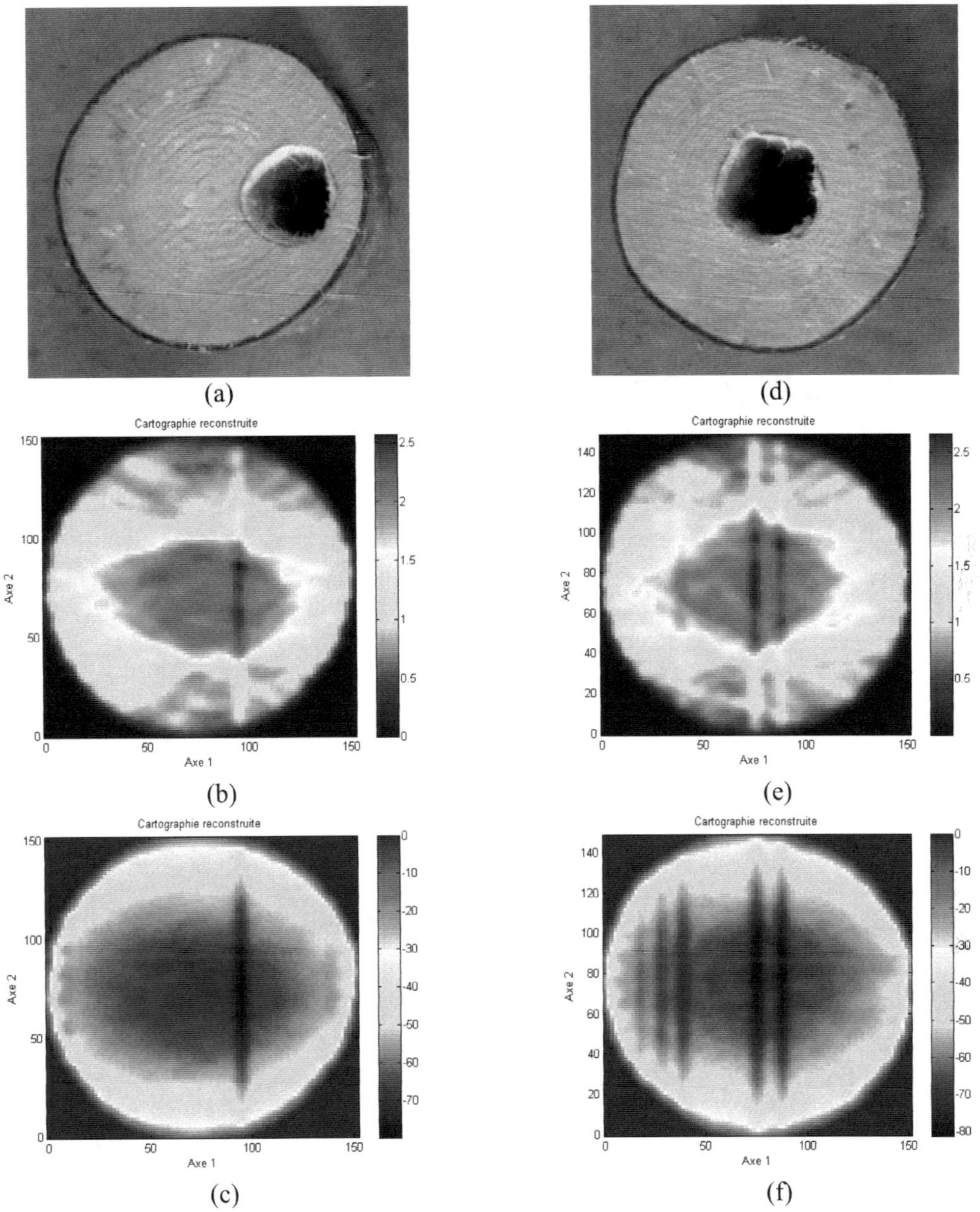

Figure 15. Tomographic maps of slowness (10^{-3} s/m) (b, e) and attenuation (dB) (c, f) for the two disks of the second log (a, b).

4. Conclusions

An automatic tomograph device specifically designed for standing trees is presented. The scanning process employed constitutes an improvement compared to previous work. This improvement has resulted in a fast process with only two probes giving scans with an angular precision of 10 degrees without any intervention by an operator during the process. This tomograph was used on two spruce (*Pica abies*) logs to study the effect of bark, and also the effect of defects, such as holes, on the reconstructed images. The tests were not carried out in the green state, but with a mean moisture content of 22% and a gradient of 7% between the pith and the periphery. The ultrasonic parameters used for wood imaging were slowness and attenuation. The use of attenuation has not been reported in earlier work and its determination was found to be more accurate compared to that of slowness. Time of flight determination was found to be less accurate, even when cross-correlation was employed. This phenomenon was due to the dispersion and dissipation of wood material, leading to high attenuation combined with modifications in the incident waves. The means of the ultrasonic parameters were not sufficient for determining the existence of defects in trees, but the presence of bark significantly contributed to signal attenuation. For the test samples, it was shown that speed near the bark was greater than in the pith, due to the existence of juvenile wood, combined with the moisture gradient (natural drying occurred during the tests). Likewise, damping near the bark was lower than in the centre. A significant relationship was shown between slowness and attenuation, but the population was bimodal due to the presence of juvenile wood. However, it was possible to see that a high speed was coherent with low damping. The presence of bark was not visible on the tomographic images. This was probably due to its limited thickness compared to the wavelength. However, the presence of holes was clearly visible, but artifacts prevented precise shape reconstruction and location. Removing those artifacts will be the objective of future work.

5. Acknowledgments

The BIOGMID project was supported by a grant from the French National Research Agency (BLAN07-1_183692, "BioGMID").

References

[1] Barnett J and Jeronimidis G 2003 *Wood Quality and its Biological Basis* (Oxford: Blackwell)
[2] Beall F C 2002 *Wood Sci. Technol.* **36** 197–212
[3] Brancheriau L, Gallet P, Thaunay P and Lasaygues P 2009 *6th Plant Biomechanics Conf.* 285-88
[4] Bucur V 2003 *Meas. Sci. Technol.* **14** 91-8
[5] Deflorio G, Fink S and Schwarze F 2008 *Wood Sci. Technol.* **42** 117–32
[6] Dikrallah A, Kabouchi B, Hakam A, Brancheriau L, Baillères H, Famiri A and Ziani M 2010 *C. R. Mecanique* **338** 107-12
[7] Divos F and Divos P 2005*14th Int. Symp. on Nondestructive Testing of Wood* 309-14
[8] Lin C, Kao Y and Lin T , Tsai M, Wang S, Lin L, Wang Y and Chan M 2008 *Int. Biodeterioration & Biodegradation* **62** 434-41
[9] Maurer H R, Schubert S, Baechle F, Clauss S, Gsell D, Dual J, and Niemz P 2006 *Holzforschung* **60** 567–73
[10] Nicolotti G, Socco L V, Martinis R, Godio A and Sambuelli L 2003 *Journal of Arboriculture* **29** 66–78
[11] Rinn F 2004 *Neue Landschaß* **7** 44-7
[12] Schmoldt D L 1996 *24th Annual Hardwood Symp.* 69-80
[13] Schubert S 2007 *Acousto-ultrasound assessment of inner wood decay in standing trees: possibilities and limitations* (Zürich: Diss ETH #17126)
[14] Socco L V, Sambuelli L, Martinis R, Comino E and Nicolotti G 2004 *Research in Nondestructive Evaluation* **15** 31-54
[15] Wang X, Allison R B, Wang L and Ross R J 2007 *U.S. Forest Products Laboratory* FPL-RP-642
[16] Wang X, Divos F, Pilon C, Brashaw B K, Ross R J and Pellerin R F 2004 *U.S. Forest Products Laboratory* FPL-GTR-147
[17] Wang X, Wiedenbeck J, Ross R J, Forsman J W, Erickson J R, Pilon C and Brashaw B K 2005 *U.S. Forest Products Laboratory* FPL–GTR–162

Approximations for acoustically excited bubble cluster dynamics

D Sinden, E Stride and N Saffari

Department of Mechanical Engineering, Roberts Building, Torrington Place, London, WC1E 7JE, United Kingdom

E-mail: d.sinden@ucl.ac.uk

Abstract. In this paper the effect of interaction on the expansion of a bubble in a regular monodisperse cluster is investigated. By a geometric construction a two-dimensional ordinary differential equation with an exact expression for first-order bubble interactions is derived for an n-bubble model. An approximate equation is derived for the rapid expansion of the bubble which can be solved yielding an analytic expression for the collapse of a bubble which undergoes inertial cavitation. It is then demonstrated that the maximum volume of a bubble in a cluster is considerably less than that of a single bubble. This result is of significance as typically the dispersion relationship, the wave speed and the co-efficient of attenuation are calculated using a single bubble model and summed for the total number of bubbles to yield the void fraction. Furthermore it is shown that the maximum radius of a bubble in the cluster is considerably smaller than that of a single bubble, yet the duration of the collapse phase is only weakly dependent on the number of bubbles. Hence, it is conjectured that the likelihood of fragmentation due to Rayleigh–Taylor instability is reduced. The results from the analysis are in good agreement with full numerical simulations of multi-bubble dynamics, as well as experimental observations

1. Introduction

It has long been known that when bubbles interact with each other, their dynamics are often significantly different from those of isolated bubbles, for example bubbles may attract or repel one another [1] or their collapse phases may synchronize [2]. There is a wide range of studies, both experimental [3, 4] and theoretical devoted to multi-bubble dynamics, by direct numerical simulations of few-bubble systems [5–9] or averaged cloud models [10–12].

There have been several studies, in which bubbly flows at nondilute, finite void fraction were investigated through effective equations intended to capture wave phenomena. There are essentially two approaches to the problem. The first approach was pioneered by Foldy [13], who derived a dispersion relation by treating wave propagation in a bubbly mixture as a problem of multiple scattering by randomly distributed isotropic scatterers representing the spherical bubbles.

Alternatively, van Wijngaarden [14] introduced volume-averaged quantities which matched the microscopic conservation laws for each phase in order to remove the local fluctuations due to scattering and then derived the averaged equations based on heuristic, physical reasoning. Caflisch *et al.* [15, 16] rigorously extended the work of van Wijngaarden and proved that the relative motion between the bubbles and the liquid does not play a significant role, justifying the assumptions. Linear equations in the limits of weak applied forcing and small void fraction recovered the results of Foldy, by essentially replacing the complex distribution of gas bubbles and liquid with a continuous effective medium. From these linear equations Commander and Prosperetti [17] derived a dispersion relation.

Typically the dispersion relation and the co-efficient of attenuation are calculated using a single bubble model and summed for the total number of bubbles. As Leighton *et al.* [18] highlight, attenuation and wave speed are often measured in order to infer bubble size distributions. However, the derivation of the attenuation and wave speed are based on linear theory [17]. Thus accuracy in the measurement of the bubble size distribution is implicitly limited by the underlying assumptions that the bubbles must not collapse inertially, which is dependent on the bubble sizes. Ida [7] computed radius-time curves in coupled polydisperse bubble systems and observed that the maximum radius of smaller bubbles was reduced. This was attributed to the expansion of a bubble being suppressed by the pressure waves which larger neighbouring bubbles emit during their expansion.

The first estimations of the maximum radius of a single bubble were derived by Apfel and Holland [19, 20]. In a heuristic manner, the authors assumed that initially the bubble growth was linear, then dealt with the subsequent growth by equating the potential and kinetic energy of the bubble. Löfstedt *et al.* [21] and Hilgenfeldt *et al.* [22] demarcate the radius-time profile of a single bubble into distinct regions of expansion, collapse and afterbounce. Within each regime the dominant effects on the bubble were identified and reduced governing equations in each regime derived. In this paper, using geometric arguments [23] to reduce a $2n$-dimensional system of coupled interacting Rayleigh–Plesset type equations to an exact two-dimensional modified Rayleigh–Plesset equation, it will be demonstrated that for inertially collapsing bubbles the time-averaged volume of an n-bubble cluster will be significantly less than that of n isolated bubbles. Hence a more accurate approximation of a time-averaged void fraction may be derived. This is significant as experiments have shown that for water-air mixtures void fractions as low as 0.5% can lead to a drop in sound speed from 1500m/s to 160m/s, which is actually lower than the sound speed in air [24]. The explanation is that the sound speed depends on both the compressibility and the density of the medium. As the former is determined by the density of gas bubbles and the latter by the properties of the liquid, the resulting sound speed can vary drastically from that of either the liquid or the gas [25].

With the assumption of local homogeneity, Rubinstein [26] extended the approach derived in [15, 16] to a nonlinear analysis of periodic lattice multi-bubble system by homogenisation and derived an modified Rayleigh–Plesset equation, which includes the effects of interactions. Beylich and Gülhan [27] used a correction factor for the bubble-bubble interaction effect by fitting the result of a model equation to experimental data for a bubbly shock flow. Kubota *et al.* [28] modelled the bubble-bubble interaction around a hydrofoil by adding the velocity potential of other bubbles. As Seo *et al.* [29] state most of the proposed models for the bubble-bubble interaction have not been critically validated and the proposed correction factors for bubble-bubble interaction vary dramatically, not only in magnitude but also in sign, from study to study. In [29] the authors also used a locally averaged volume averaged Rayleigh–Plesset equation and compared the results against direct numerical simulations using a front tracking method. The effects of interaction became noticeable at void fractions 1% and became significant beyond 10%. Reasonable agreement between averaged equations and direct numerical simulation was found for void fractions up to 13%. Yasui *et al.* [30] investigated inertial collapse within a bubble cloud and found that shock waves do not lead to broadband noise if the bubbles are shape-stable, but broadband noise can be attributed to temporal fluctuations in the number of bubbles within a cloud.

The outline of this paper is as follows, in section 2 the Rayleigh–Plesset equation is introduced and in subsection 2.1 the reduced set of governing equations based on geometrically exact constructions is presented. In section 3 an analytical expression which approximates the expansion phase of a bubble in cluster is derived. Finally, in section 4 the results are summarised and the implications discussed.

2. The Rayleigh–Plesset Equation

The Rayleigh–Plesset equation is an nonlinear ordinary differential equation which models the oscillations of a single gas filled spherical bubble of radius $R = R(t)$ whose centre is stationary. The Rayleigh–Plesset equation can be derived by balancing the energy supplied to the bubble by the incident pressure and the surrounding fluid to the kinetic energy of the bubble oscillations [21].

The interaction between two bubbles, R_i and R_j, a distance D_{ij} apart from their centres, due the re-radiated pressure field is given by Mettin *et al.* [31]. It is derived by matching the velocity field for an incompressible liquid with the pressure field using the conservation of momentum. On computing the interaction to first-order, neglecting terms of $\mathcal{O}\left(1/D^5\right)$ and $\mathcal{O}\left(1/cD\right)$, for n-bubbles the governing equation is given by the $2n$-dimensional equation

$$\rho\left(R_i\ddot{R}_i + \frac{3}{2}\dot{R}_i^2 + \sum_{\substack{j=1,\\ j\neq i}}^{n}\frac{R_j}{D_{ij}}\left(R_j\ddot{R}_j + 2\dot{R}_j^2\right)\right) = p_g\left(R_i\right) + p_v - p_\infty - \frac{2\sigma}{R_i} - \frac{4\mu\dot{R}_i}{R_i} + p\left(t\right). \tag{1}$$

Here ρ is the density of the surrounding (irrotational, infinite, incompressible) liquid, p_g is the internal pressure of the gas in the bubble given by

$$p_g\left(R\right) = \left(p_\infty - p_v + \frac{2\sigma}{R_0}\right)\left(\frac{R_0^3 - a^3}{R^3\left(t\right) - a^3}\right)^{\kappa}, \tag{2}$$

with p_∞ is the hydrostatic pressure, p_v is the vapour pressure, $R_0 = R(0)$ is the initial bubble radius, σ is the surface tension, κ is the polytropic exponent of the gas within the bubble, a is the hard-core van der Waals radius and μ is the shear viscosity of the liquid. The gas is assumed to be ideal, thus the internal pressure is a function of the bubble radius only, hence p_v is constant. The applied acoustic pressure takes the form

$$p(t) = p_a \sin(\omega t + \varphi_0) \tag{3}$$

where p_a is the magnitude of the pressure, ω is the driving frequency and φ_0 is the phase of the applied field. The internal conditions, such as vapour content and polytropic index and external conditions, such as the fluid density, viscosity, surface tension and ambient pressure are all assumed to be identical for each bubble.

In the derivation of the Rayleigh–Plesset equation it is assumed that the wavelength of the sound field is large in comparison with the radius of the bubble and that the motion of the bubble wall is slow in comparison with the speed of sound in the liquid and the gas. Diffusive, thermal and chemical effects are all neglected. Discussions of the underlying assumptions and limitations of the classical Rayleigh–Plesset equation are given in [21, 22].

In the case of interacting bubbles some further assumptions are necessary. Firstly, the fluid is assumed to be incompressible so that the interactions between bubbles are instantaneous. Secondly, the distances between bubbles are assumed to be sufficiently large so that the bubbles remain spherical [32]. For violently collapsing bubbles a shock-like pressure wave may be released [33] containing small-amplitude high-frequency components whose wavelengths may be of a comparable size to the bubbles. As the focus of this paper is on the behaviour of interacting bubbles during expansion this situation will not arise and so the assumptions underpinning the Rayleigh–Plesset equation are still valid.

2.1. Geometric Averaging

Although it is possible to generalise the Rayleigh–Plesset equation (1) to an n-bubble model, it is not possible to express the natural frequencies of the system analytically for an arbitrary n-bubble model if $n \geq 5$. However, if the system of n-interacting equally sized bubbles forms a complete graph, i.e. are arranged in a regular polygon planar to the incident wave, the Rayleigh–Plesset equation can be simplified by exploiting the invariance of the dihedral symmetry group. The physical interpretation is that although the distances between pairs of bubbles are different, and thus the magnitude of the pressure waves due to interactions different, the sum of the distances between a bubble and all of its neighbours will be the same for every bubble, as is illustrated in Fig. 1. Thus the total incident pressure wave due to interaction will be the same for every bubble. Hence the Rayleigh–Plesset equation can be generalised to incorporate the multiple bubble interactions via

$$R_j \mapsto R \quad \text{and} \quad \sum_{\substack{j=1,\\ j\neq i}}^{n} \frac{R_j}{D_{ij}} \left(R_j \ddot{R}_j + 2\dot{R}_j \right) \mapsto \frac{\alpha_n}{D} R \left(R\ddot{R} + 2\dot{R}^2 \right) \tag{4}$$

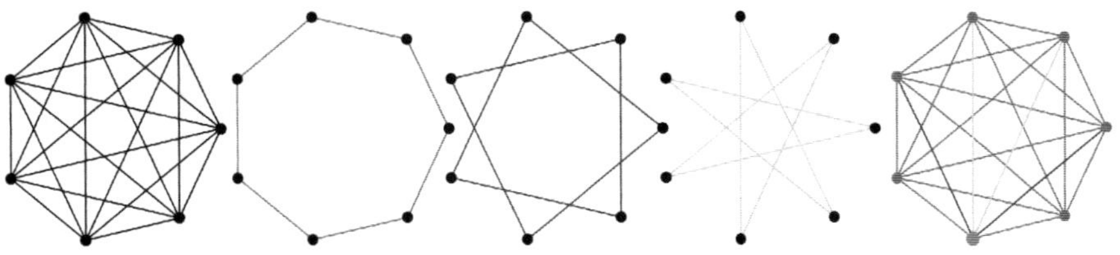

Figure 1: The interactions of a set of equally sized bubbles arranged in a heptagon ($n = 7$) form a complete graph which can be decomposed into three star polygons which connect with the first, second and third nearest neighbour. Each star polygon has a unique length scale λ_i, $i = 1, 2, 3$. The total interaction force on a bubble is given a sum of the length scales weighted by the total number of interactions per length scale.

where α_n is a dimensionless weighting parameter for the interaction amongst the n-bubbles and D is the distance between a bubble and its nearest neighbour. The dimension of the governing equation is thus independent of the number of bubbles and can be reduced from a $2n$-dimensional system to a two-dimensional system. Hence, for a single bubble $\alpha_1 = 0$ as there is no interaction, for a pair of bubbles $\alpha_2 = 1$, for three bubbles arranged in an equilateral triangle $\alpha_3 = 2$, for a square array of bubbles

$$\frac{1}{D} + \frac{1}{D} + \frac{1}{\sqrt{2}D} = \frac{1}{D}\left(2 + \frac{1}{\sqrt{2}}\right) \Rightarrow \alpha_4 = \frac{1 + 2\sqrt{2}}{\sqrt{2}} \tag{5}$$

and for a regular pentagon the sum of the internal lengths gives

$$\frac{1}{D} + \frac{1}{D} + \frac{1}{D\left(1+\sqrt{5}\right)/2} + \frac{1}{D\left(1+\sqrt{5}\right)/2} \Rightarrow \alpha_5 = \frac{2\left(3+\sqrt{5}\right)}{1+\sqrt{5}} \tag{6}$$

and so on. Fig. 1 illustrates the length scales of the heptagon, which is the first regular polygon whose length scales can not all be expressed by real numbers.

Rather than by repeatedly applying trigonometric formulae for each regular array, the procedure can be generalised by noting that the sum of all the interactions forms a complete graph which can be decomposed into star polygons. The ith-star polygon has an associated length scale, λ_i, based only on the total number of bubbles and the minimum distance between any two bubbles. Each of the length-scales may be found via the roots of Chebyshev polynomials [23]. The total interaction force on a bubble is given by the sum of the length scales, λ_i, weighted by the total number of interactions per length scale, which is denoted by ϕ_i. The parameter α_n can expressed as

$$\alpha_n = \sum_{i=1}^{n} \frac{D}{D_i} - 1 = \sum_{i=1}^{n} \phi_i \lambda_i - 1 \tag{7}$$

where the subtraction is due to the fact that when $n = 1$ there will be no interaction. When $n = 2k + 1$ the length scales λ_i are the n positive roots of the polynomial $u_n = u_n\left(1 - \lambda^2/2\right)$ where $u_n\left(\cdot\right)$ is an nth-order Chebyshev polynomial of the second kind. When $n = 2k$ then λ_i are found as the positive roots of $v_n = v_n\left(1-\lambda^2/2\right)$ where $v_n = u_n + u_{n-1}$. For the case of $n = 2k+1$ equally sized bubbles there will be $2k$ interactions so each length scale will be equally weighted as there will be two interactions every length scale. However for $n = 2k$ equally sized bubbles there will be $2k-1$ interactions, where, on the largest length scale, the bubble interacts with a single bubble. The parameter ϕ_i is given by the following relation

$$\text{if} \quad n = \begin{cases} 2k & \text{then} \quad \phi_i = 2 \\ 2k+1 & \text{then} \quad \phi_i = \begin{cases} 2 & \text{for} \quad i = 1, \ldots, 2k \\ 1 & \text{for} \quad i = 2k+1. \end{cases} \end{cases}$$

Hence, for regular configurations, the dimension of the system is reduced to a two-dimensional modified Rayleigh–Plesset equation of the form

$$\rho\left(R\ddot{R} + \frac{3}{2}\dot{R}^2 + \frac{\alpha_n R}{D}\left(R\ddot{R} + 2\dot{R}^2\right)\right) = p_g\left(R\right) + p_v - p_\infty - \frac{2\sigma}{R} - \frac{4\mu\dot{R}}{R} + p\left(t\right) \tag{8}$$

which, on linearisation, yields a single natural frequency, given by

$$\omega_0 = \sqrt{\frac{1}{\rho\left(1 + R_0\alpha_n/D\right)}\left(\frac{3\kappa\left(p_\infty - p_v\right)R_0}{R_0^3 - a^3} + 2\sigma\left(\frac{3\kappa}{R_0^3 - a^3} - \frac{1}{R_0^3}\right)\right)}. \tag{9}$$

An increase in the number of bubbles increases α_n and, therefore, reduces the natural frequency of the system, in agreement with experimental observations [34]. However, in the limit $n \to \infty$ the formulation is of limited applicability as the configuration approaches a circle of infinite radius.

The averaging of the bubble interactions has previously been attempted by Yasui *et al.* [4, 30] and Kubota *et al.* [28], who assumed the bubble cloud to be monodisperse and made a "crude" approximation based on the averaging of the geometric configuration of the bubble cloud for the coupling strength S

$$R_j \mapsto R \quad \text{and} \quad \sum_{\substack{j=1,\\ j\neq i}}^{n} \frac{R_j}{D_{ij}}\left(R_j\ddot{R}_j + 2\dot{R}_j\right) \mapsto SR\left(R\ddot{R} + 2\dot{R}^2\right),$$

where the coupling strength is approximated as

$$S = S_i = \sum_{j=1}^{n-1} \frac{1}{D_{ij}} \approx 4n\pi \int_{l_{\min}}^{l_{\max}} r\, dr \approx 2n\pi l_{\max}^2 \tag{10}$$

with $l_{\min} \ll l_{\max}$ the distances to the nearest and furthest bubbles for a bubble in the cluster. Yasui *et al.* estimated that the coupling strength could be of the magnitude 10^6/m in the homogeneous bubble cloud found in ultrasonic cavitation at large pressures, 2.9atm.

While the assumption of geometric regularity is somewhat unphysical, the primary benefit of the assumptions is the reduction in the dimension of $2n$-dimensional system to a two-dimensional system, allowing many analytical techniques from single bubble theory to be carried over. However, numerical simulations on the two-dimensional Rayleigh–Plesset equation (8) suggest that the geometric formulation is robust under perturbation in that if the bubbles are arranged in a regular array and the location of each bubble is perturbed slightly in the plane of the wave, each radius-time profile will be perturbed form the radius-time profile of the bubbles regular array by a small amount. However, whether the spatial perturbation is robust for consequent translational motion under the primary Bjerknes force is not yet known.

If compressibility was incorporated in the Rayleigh–Plesset equation then delays would be included in the bubble-bubble interaction but the system would still be two-dimensional, where the delays would be given by $\tau_i = D_i/c = \alpha_i D/c$.

Although the likelihood of nucleation occuring in a regular array in an uncontrolled environment is highly unlikely, nucleation in a regular array can be performed in a controlled and repetable manner. Kroon *et al.* [35] induced cavitation using an array of laser foci by shaping a pulsed laser beam using a digital hologram and focusing it into a small volume of liquid. Bremond *et al.* [36] used a hydrophobic surface patterned with micro-cavities as nucleation sites when the liquid pressure is lowered. In both cases the authors showed that the dynamics of the clusters were in good agreement with analytical models if the bubbles remained spherical. However, if the separation distance was beyond the threshold derived by Zavtrak [32], the interaction would lead to non-spherical collapse and jetting towards the centre of the arrays.

Figure 2 illustrates a radius-time profile for a one, two and three bubble clusters, where density is given by $\rho = 1000\text{kg/m}^3$, hydrostatic pressure $p_\infty - p_v = 1\text{atm}$, surface tension $\sigma = 0.073\text{kg/m}^2$, polytropic index $\kappa = 5/3$, van der Waals radius $a = 8.85$, frequency $\omega = 2\pi \cdot 27\text{kHz}$ and applied pressure $p_a = 1.36\text{atm}$. The figure clearly shows that the maximum radius decreases as the number of bubbles increases for a bubble in the cluster.

3. Approximating the Expansion Phase

By geometric arguments, the dimension of the multi-bubble system is reduced from a $2n$-dimensional system to a two-dimensional system and is thus far more amenable to analysis. In this section an analytical estimate for the expansion phase of the bubble is derived. As Fig. 2 illustrates, the expansion phase of a bubble will account for a significant part of the time-averaged volume of a bubble. Thus an estimation of the bubble radius during this phase has utility in predicting the attenuation for example.

During the expansion phase, the governing equation is dominated by the inertial terms and the driving pressure. The expansion phase of a bubble is determined by the time in which the applied acoustic pressure is negative. From Fig. 3 let

$$t_- = \frac{1}{\omega}\sin^{-1}\left(\frac{p_\infty - p_v}{p_a}\right) \quad \text{and} \quad t_+ + t_- = \frac{\pi}{\omega} \tag{11}$$

so that (t_-, t_+) is the interval in which the total acoustic pressure on the bubble is negative. Note that an inertially collapsing bubble may continue to expand when the applied pressure field is positive due to its own momentum, before collapsing at time $t_{\min}$ [37]. Thus, the inertial terms in the Rayleigh–Plesset equation, and the applied

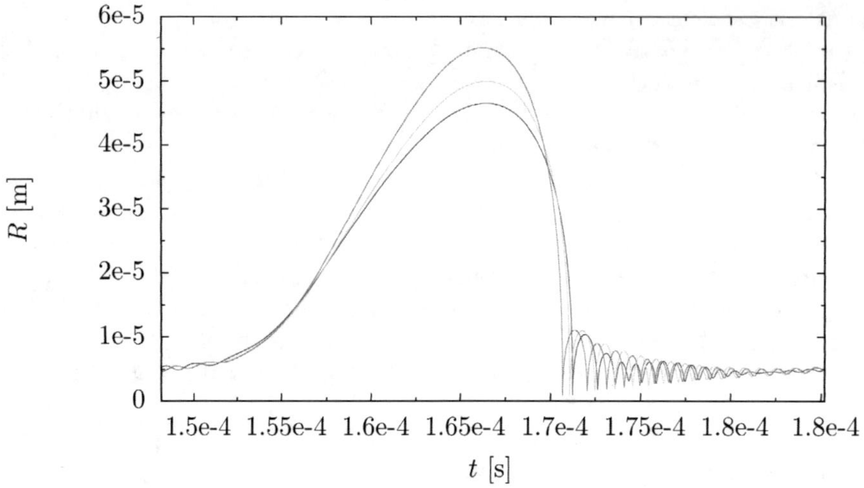

Figure 2: Radius-time profiles for a single bubble (red), two (green) and three (blue) bubbles of initial radius $R_0 = 5\mu$m with $\rho = 1000$kg/m^3, $p_\infty - p_v = 1$atm, $\sigma = 0.073$kg/m^2, $\kappa = 5/3$, $a = 8.85$ under acoustic field with $\omega = 2\pi \cdot 27$kHz and $p_a = 1.36$atm. Note that the maximum amplitude decreases as the number of bubbles increases. The radius-time profiles displayed are from the fifth acoustic cycle, so that transient oscillations have passed.

acoustic pressure are the significant terms during the expansion and beginning of the collapse phase.

The inertial terms may be written as

$$R\ddot{R} + \frac{3}{2}\dot{R}^2 = \frac{3}{4}\frac{\mathrm{d}^2}{\mathrm{d}t^2}R^2 - \frac{1}{2}R\ddot{R} \tag{12a}$$

or as

$$R\ddot{R} + \frac{3}{2}\dot{R}^2 = \frac{1}{2}\frac{\mathrm{d}^2}{\mathrm{d}t^2}R^2 + \frac{1}{2}\dot{R}^2. \tag{12b}$$

Numerical simulations [22] show that, as observed in [19], in the interval $t \in (t_-, t_+)$ the velocity of the bubble is the dominant effect as the growth is almost linear. Whereas in the interval $t \in (t_+, t_{\mathrm{min}}]$, that is when the bubble reaches it maximum radius and collapses, it is the acceleration of the bubble which is the dominant effect. Thus in the two domains, the weaker contributions to the expansion can be neglected.

The complete effect of interaction can be included in both regimes by exploiting the fact that the interaction is an exact second derivative, i.e.

$$\frac{\alpha_n}{D}R\left(R\ddot{R} + 2R\dot{R}^2\right) = \frac{\alpha_n}{3D}\frac{\mathrm{d}^2}{\mathrm{d}t^2}R^3. \tag{13}$$

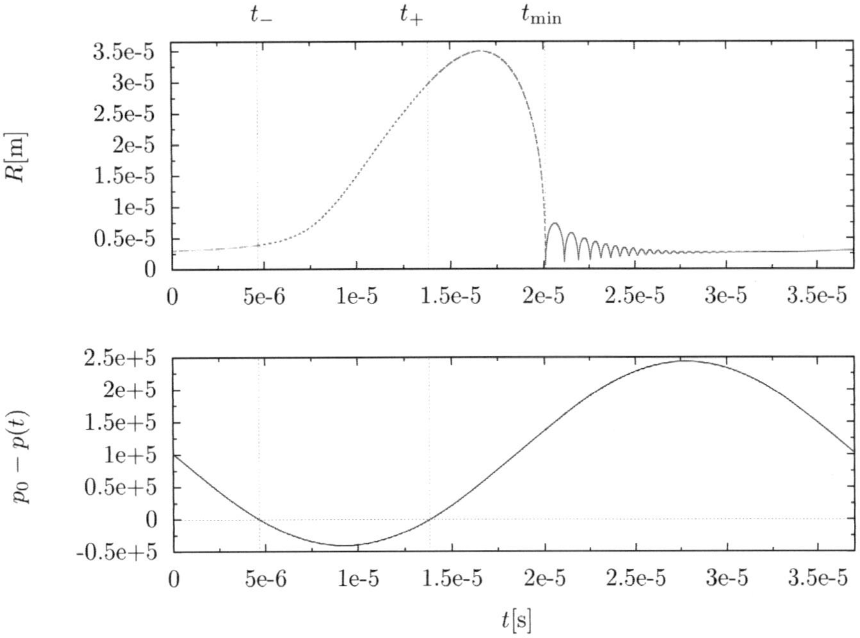

Figure 3: Regimes in which expansion phase of bubble is modelled, showing the effective applied acoustic wave and the times which determine the bubble expansion (11). In this figure $R_0 = 3\mu$m and $p_a = 1.36$atm. In the radius-time profile, the numerical solution on the domains (t_-, t_+) and $(t_-, t_{\min}]$ are shown by the red dotted and red dashed lines respectively.

Hence, following [22] and [38], the expansion phase can be formulated as

$$\frac{3\rho}{4}\frac{\mathrm{d}^2}{\mathrm{d}t^2}\left(R^2\left(1+\frac{4\alpha_n}{9D}R\right)\right) = p_a \sin(\omega t) - A + \mathcal{O}\left(R\ddot{R}\right) \quad \text{on} \quad (t_-, t_+) \quad \text{as} \quad \dot{R}^2 \gg R\ddot{R} \tag{14a}$$

and

$$\frac{\rho}{2}\frac{\mathrm{d}^2}{\mathrm{d}t^2}\left(R^2\left(1+\frac{2\alpha_n}{3D}R\right)\right) = p_a \sin(\omega t) - A + \mathcal{O}\left(\dot{R}^2\right) \quad \text{on} \quad (t_+, t_{\min}] \quad \text{as} \quad \dot{R}^2 \ll R\ddot{R} \tag{14b}$$

where A is given by

$$A = p_0 - \frac{2\sigma}{R_0}\frac{1}{K(p_0, p_a)} \quad \text{with} \quad p_0 = p_\infty - p_v \tag{15}$$

and K is a constant effective surface tension, dependent only on the ambient and applied pressures. The effective surface tension shall be determined later.

Denoting the solution over the range (t_-, t_+) as $R_-(t) = R(t)$, the first equation can be solved on the domain by integrating twice, with the initial conditions $R_-(t_-) = \zeta R_0$

and $\dot{R}_-(t_-) = \omega\zeta R_0$, where $\zeta = 1.6$ is a matching parameter inferred from full numerical simulations. Thus on (t_-, t_+) the governing equation is given by

$$R_-^2\left(1+\frac{4\alpha_n}{9D}R_-\right) = \frac{4}{3\rho\omega^2}\left(p_0 - p_a\sin\omega t + p_a\omega\left(t-t_-\right)\cos\omega t_- - \frac{A}{2}\omega^2\left(t-t_-\right)^2\right) + \zeta^2 R_0^2\left(1+\frac{4\alpha_n}{9D}\zeta R_0 + 2\omega\left(1+\frac{2\alpha_n}{3D}\zeta R_0\right)\left(t-t_-\right)\right). \tag{16a}$$

Solving the equation on the next domain $(t_+, t_{\min}]$ for $R_+(t) = R(t)$, by enforcing that R is derivative is continuous at t_+, i.e. $R_-(t_+) = R_+(t_+)$ and $\dot{R}_-(t_+) = \dot{R}_+(t_+)$, yields

$$R_+^2\left(1+\frac{2\alpha_n}{3D}R_+\right) = \frac{2}{\rho\omega^2}\left(p_0 - p_a\sin\omega t + p_a\omega\left(t-t_+\right)\cos\omega t_+ - \frac{A}{2}\omega^2\left(t-t_+\right)^2\right) + c_1\left(t-t_+\right) + c_2 \tag{16b}$$

where the constants

$$c_1 = 2\dot{R}_-\left(t_+\right)R_-\left(t_+\right)\left(1+\frac{\alpha_n}{D}R_-\left(t_+\right)\right) \quad \text{and} \quad c_2 = R_-^2\left(t_+\right)\left(1+\frac{2\alpha_n}{3D}R_-\left(t_+\right)\right) \tag{17}$$

ensure continuity. Closed form expressions for the radius during expansion can be found by solving a cubic equation, but are overly complicated and are omitted for brevity. However, for a single bubble the solution can be written as

$$R_+^2(t) = \frac{2}{\rho\omega^2}\left(p_0 - p_a\sin\omega t + p_a\omega\left(\frac{1}{3}t + t_+\right)\cos\omega t_+ - \frac{A}{2}\omega^2\left(t^2 + t_+^2 + \frac{2}{3}tt_+\right)\right) + \zeta^2 R_0^2\left(1 + 2\omega\left(t+t_+\right)\right) \tag{18}$$

It is important to note that at the instance of collapse the model breaks down, with all solutions unphysical. That is, the roots of the cubic are a pair of complex conjugates and negative real root. The instance of collapse can be crudely approximated by setting the right-hand side of Eq. (16b) equal to zero and solving for $t_{\min}$. Although this provides a reasonably accurate approximation for the collapse time, during the latter part of the collapse phase the model neglects effects such as internal pressure which become comparable to the inertial terms when arresting the collapse of the bubble.

The maximum radius $R_{\max} = R_+(t_{\max})$ can be found by setting $\dot{R}_+(t) = 0$ to yield an implicit expression for the time $t = t_{\max}$, which can be substituted into Eq. (16b). The time at which the radius of the bubble is maximal is found via numerical solutions to the transcendental equation

$$\frac{1}{\rho\omega^2}\left(p_a\cos\omega t_{\max} + p_a\cos\omega t_+ - A\omega\left(t_{\max}-t_+\right)\right) + c_1 = 0. \tag{19}$$

There is only one root to equation (19) in the domain $(t_+, t_{\min}]$, as in the limit $t \to t_{\min}$ the radius tends to zero, $R_+(t) = 0$.

The transition radius R_{trans} is the initial radius which demarcates weakly oscillating bubbles from bubbles which undergo rapid expansion. Hilgenfeldt [22, Fig. 7] showed numerically that for an inertially collapsing bubble, that is when $p_a \gg p_0$, the ratio of the maximum radius to the initial radius was constant when the initial radius was close to the transition radius. Thus

$$\frac{\partial}{\partial R_0}\left(\frac{R_{\max}(p, R_0)}{R_0}\right)\bigg|_{R_0=R_{\text{trans}}} = 0 \quad \text{with} \quad R_{\text{trans}} = \frac{4\sigma}{3(p_0 - p_a)}. \tag{20}$$

Following Hilgenfeldt's analysis it is simple to show that the transition radius is independent of the number of bubbles. Hence, the effective surface tension can now be approximated by matching the transition radius with the maximum of the nondimensional function $R_{\max}/R_0$.

The maximum radius was found by solving Eq. (19), without dependence on ζ or the effective surface tension K, that is $A = p_0$. Trigonometric terms can be expanded to first-order to give

$$t_{\max}^{(0)} = \frac{1}{\omega p_0}\left(p_a + \frac{1}{3}\left(p_a \cos\omega t_+ - p_0 \omega t_+\right)\right) \tag{21}$$

which depends only on the applied and ambient pressures. In [38] the procedure was extended to higher-orders. Again, neglecting the effects of surface tension, equation (19) is solved and trigonometric terms expanded to second-order as

$$t_{\max}^{(1)} = \frac{1}{\omega}\left(\frac{p_a}{p_0} + \sqrt{\frac{2}{p_a}\left(\omega p_0 t_{\max}^{(0)} + 3\omega^2 \rho c_1 + \frac{p_0^2}{2p_a}\right)}\right) \tag{22}$$

which now depends on R_0 and through c_1 on α_n, as well as the applied and ambient pressures. However, it is assumed that the dependence on R_0 is weak, so that $\partial t_{\max}^{(1)}/\partial R_0$ can be neglected. Thus expression (22) can be substituted in Eq. (16b) to give the maximum radius and the condition (20) applied to yield an expression for the effective surface tension. When $\alpha_n = 0$ the effective surface tension can be written in the simple form

$$K(p) = \frac{3t_{\max}^2 + 3t_+ + 2t_{\max}t_+}{g(p, t_{\max})}\frac{3}{4\sqrt{3}}(p-1) \quad \text{with} \quad t_{\max} = t_{\max}^{(1)} \tag{23}$$

where

$$g(p,t) = \frac{2}{\rho\omega^2}\left(p_0 - p_a \sin\omega t + p_a\omega\left(\frac{1}{3}t + t_+\right)\cos\omega t_+\right). \tag{24}$$

However, when interaction is considered the expression for the effective surface tension becomes far more complicated.

From the analysis, the time when the radius of the bubble is largest is very weakly dependent on the number of bubbles, which is also illustrated by the full numerical simulations of the Rayleigh-Plesset equation (8) in Fig. 2. The maximum radius is strongly influenced by the number of bubbles. The maximum radius $R_{\max}$ is found by substituting the value of the collapse time into the expression (16b) and solving the cubic equation for $R_{\max}$. Typically, there is a positive real root and a pair roots those real part is always negative. The positive real root of the cubic is a monotonically decreasing function with respect to α_n/D. Thus effect of interaction decreases the time-averaged size of the bubble.

Numerical calculations, illustrated in Fig. 4 clearly show that the approximations are in good agreement with the full numerical simulations up to the point of collapse.

The maximum radius for small multi-bubble clusters is shown in Fig. 5. The reduction in the maximum radius due to the influence of interaction is clear, in good agreement with those observed in [7]. Note that the calculations hold over a portion of a single acoustic cycle, so that if a bubble displays characteristics of chaotic behaviour, such as a period-doubling and sensitivity to initial conditions, then these will not be directly observed in the analysis. However, if a bubble is oscillating chaotically this can be incorporated into the analysis by letting ζ be a random variable for each acoustic cycle.

Towards the end of the inertial collapse phase while the acceleration is positive, that is while the gas is still accelerating into the fluid, any non-spherical perturbations will grow exponentially and shape instability may develop during this brief time. This mechanism of instability is known as the Rayleigh–Taylor instability [39]. As the times at which the bubble reaches its maximum and minimum amplitudes of oscillation are weakly influenced by the number of bubbles, yet the maximum amplitude decreases noticeably as the number of bubbles increases, the analysis suggests that the influence of the interaction may soften the collapse. Thus, due to the influence of the interactions the likelihood of Rayleigh–Taylor instability should be reduced.

4. Conclusion

In this paper the interaction of a cluster of equally sized bubbles is investigated. If the bubbles are arranged in a regular array, planar to the applied acoustic field, then for each bubble the sum of the re-radiated pressure fields are equal, thus the governing equation for a bubble cluster can be reduced from an arbitrarily large $2n$-dimensional system to a single second-order modified Rayleigh–Plesset differential equation. Furthermore, the first-order interactions of bubbles arranged in a regular array can be calculated exactly using Chebyshev polynomials of the second kind. As the number of bubbles increases, the natural frequency of the cloud decreases, in agreement with existing experimental observations. The expansion phase of the bubble in a multi-bubble model is approximated by exploiting the observation that the first-order interaction is an exact differential. From the approximate equation, the expansion phase for a bubble in the cluster is found explicitly. Thus a time-averaged void fraction can easily be derived. The maximum radius can be found by solving for the time at which the bubble radius is stationary. An expression for the time at which the bubble radius is maximal is derived and solved numerically. It is observed that the time of maximum expansion is weakly dependent on the number of bubbles, which is also seen from full numerical simulations of the Rayleigh–Plesset equation. On substituting the time of maximum expansion into the expression for the expansion, the maximum radius can be calculated. This expression is influenced by the number of bubbles, again in good agreement with full numerical simulations of the Rayleigh–Plesset equation. As the maximum amplitude of oscillation

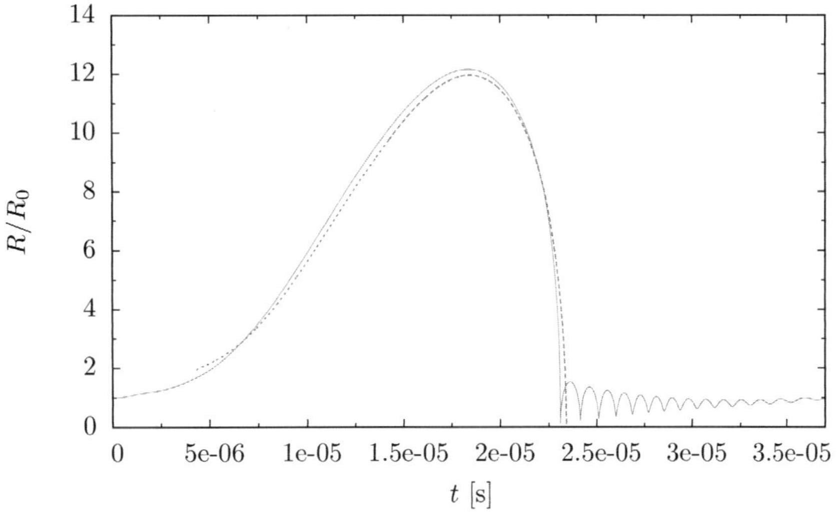

Figure 4: Comparison of analytical solution given by Eq. (16) and full numerical simulation (8) for a single bubble of size $R_0 = 5\mu$m with parameters given as in Fig. 2.

is reduced but the time between maximum to minimum radii essentially remains the same, it is conjectured that the likelihood of Rayleigh–Taylor instability developing may be reduced.

The analysis is complicated by the fact that the interaction is an exact differential yet approximations are made for the inertial terms of the bubble over two time domains. For a single bubble the expansion phase was modelled by a pair of differential equation whose inhomogeneous forcing terms had different co-efficients. Now, in the two domains, the third-order differentials also differ, cf. (16). An alternative would be to make the same assumptions for the interaction as are made for the inertial terms of the bubble, and thus approximate the interaction. However, this does not simplify the analysis as closed form expressions for the expansion are still given by the roots of a cubic equation, with a decrease in the accuracy of approximation.

Acknowledgments

This work was supported by the EPSRC under grant EP/F025750/1 and is gratefully acknowledged.

References

[1] Harkin A, Kaper T J and Nadim A 2001 *J. Fluid Mech.* **445** 377–411

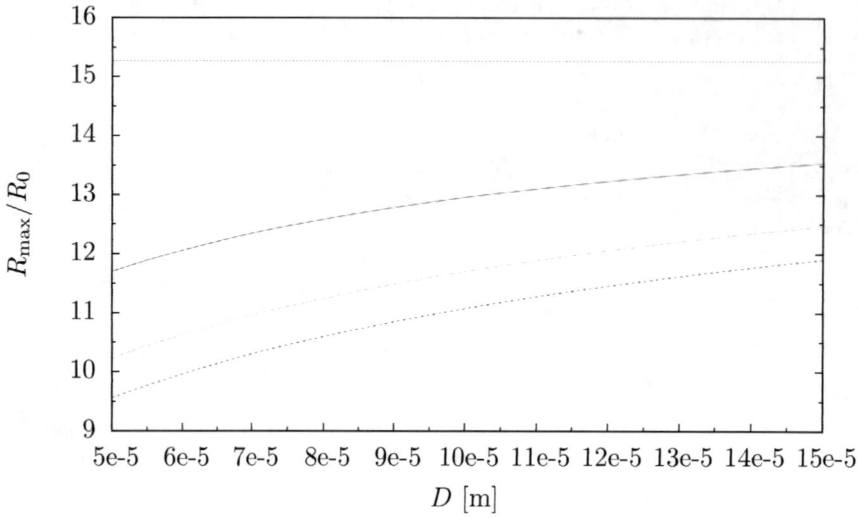

Figure 5: Ratio of maximum amplitude of the bubble radius and the initial radius for single (dotted), two (red solid), three (green/long-dash) and four (blue/short-dash) bubble clusters as the separation distance between the bubbles varies.

[2] Tervo J T, Mettin R and Lauterborn W 2006 *Acust. – Acta Acust.* **92** 178–180 ISSN 1610-1928

[3] Hatanaka S, Yasui K, Kozuka T, Tuziuti T and Mitome H 2002 *Ultrasonics* **40** 655–660

[4] Yasui K, Iida Y, Tuziuti T, Kozuka T and Towata A 2008 *Phys. Rev. E* **77** 16609

[5] Takahira H, Yamane S and Akamatsu T 1995 *JSME Int. J., Ser. B* **38** 432–439

[6] Ida M 2009 *Phys. Rev. E* **79** 16307

[7] Ida M 2009 *Phy. Fluids* **21** 113302

[8] Macdonald C A and Gomatam J 2006 *Proc. IMechE. C: J. Mech. Eng. Sci.* **220** 333–343

[9] Chong K J Y, Quek C Y, Dzaharudin F, Ooi A and Manasseh R 2010 *J. Sound Vib.* **329** 687–699

[10] Akhatov I, Parlitz U and Lauterborn W 1994 *J. Acoust. Soc. Am.* **96** 3627–3635

[11] Akhatov I, Parlitz U and Lauterborn W 1996 *Phys. Rev. E* **54** 4990–5003

[12] Parlitz U, Mettin R, Luther S, Akhatov I, Voss M and Lauterborn W 1999 *Phil. Trans. R. Soc. A* 313–334

[13] Foldy L L 1945 *Phys. Rev.* **67** 107–119

[14] van Wijngaarden L 1972 *Annu. Rev. Fluid Mech.* **4** 369–396

[15] Caflisch R E, Miksis M J, Papanicolaou G C and Ting L 1985 *J. Fluid Mech.* **153** 259–274

[16] Caflisch R E, Miksis M J, Papanicolaou G C and Ting L 1985 *J. Fluid Mech.* **160** 1–14

[17] Commander K W and Prosperetti A 1989 *J. Acoust. Soc. Am.* **85** 732–746

[18] Leighton T G, Meers S D and White P R 2004 *Proc. R. Soc. A* **460** 2521–2550

[19] Apfel R E 1986 *IEEE Trans. Ultrason., Ferroelect., Freq. Contr.* **33** 139–142

[20] Holland C K and Apfel R E 1989 *IEEE Trans. Ultrason., Ferroelect., Freq. Contr.* **36** 204–208

[21] Löfstedt R, Barber B P and Putterman S J 1993 *Phy. Fluids* **5** 2911–2928

[22] Hilgenfeldt S, Brenner M P, Grossmann S and Lohse D 1998 *J. Fluid Mech.* **365** 171–204

[23] Savio D Y and Suryanarayan E R 1993 *Am. Math. Monthly* **100** 657–661

[24] Smereka P 2002 *J. Fluid Mech.* **454** 287–325

[25] Sangani A S 1991 *J. Fluid Mech.* **232** 221–284

[26] Rubinstein J 1985 *J. Acoust. Soc. Am.* **77** 2061–2066

[27] Beylich A E and Gülhan A 1990 *Phy. Fluids* **2** 1412–1428

[28] Kubota A, Kato H and Yamaguchi H 1992 *J. Fluid Mech.* **240** 59–96

[29] Seo J H, Lele S K and Tryggvason G 2010 *Phy. Fluids* **22** 063302

[30] Yasui K, Tuziuti T, Lee J, Kozuka T, Towata A and Iida Y 2010 *Ultrason. Sonochem.* **17** 460–472

[31] Mettin R, Akhatov I, Parlitz U, Ohl C D and Lauterborn W 1997 *Phys. Rev. E* **56** 2924–2931

[32] Zavtrak S T 1987 *Akust. Zh.* **33** 240–242

[33] Matula T J, Hallaj I M, Cleveland R O, Crum L A, Moss W C and Roy R A 1998 *J. Acoust. Soc. Am.* **103** 1377–1382

[34] Kozhevnikova I N and Bjørnø L 1992 *Ultrasonics* **30** 21–25

[35] van der Kroon I, Quinto-Su P A, Li F and Ohl C D 2010 *Phys. Rev. E* **82** 066311

[36] Bremond N, Arora M, Ohl C D and Lohse D 2006 *Phys. Rev. Lett.* **96** 224501

[37] Flynn H G 1975 *J. Acoust. Soc. Am.* **57** 1379–1396

[38] Louisnard O 2008 *Phys. Rev. E* **78** 36322 ISSN 1550-2376

[39] Lin H, Storey B D and Szeri A J 2002 *Phy. Fluids* **14** 2925

Models Comparison for the scattering of an acoustic wave on immersed targets

Bo Lu[1], Michel Darmon[1], Catherine Potel[2,3], Victor Zernov[4]

[1] CEA-LIST, CEA-Saclay, 91191 Gif-sur-Yvette, France;

[2] Laboratoire d'Acoustique de l'Université du Maine (LAUM), UMR CNRS 6613, 72085 Le Mans, France ;

[3] Fédération Acoustique du Nord-Ouest (FANO), FR CNRS 3110, France;

[4] Sound Mathematics Ltd., Cambridge, CB4 2AS, U.K.

E-mail: michel.darmon@cea.fr

Abstract. Ultrasonic telemetry techniques consist in locating various immersed structures (for instance, components in the main vessel of fast breeder reactors). The interactions beam-targets give rise to different scattering phenomena: tip diffraction of boundaries and edges of the different parts, specular reflection, and corner effect. In order to conceive and design such imaging techniques, simulation tool needs to account for these effects. Classical methods have been studied for such problems. The diffraction coefficients based on the Geometrical Theory of Diffraction (GTD) fail in the transition regions adjacent to shadow and reflection boundaries. The uniform diffraction theory provides continuous solutions in these regions, but with more sophisticated formulation. Another simple approximation based on the integral equation, widely used for scattering problems, is the so-called Kirchhoff approximation. The Kirchhoff approximation has good performance in the specular reflection zone but fails at predicting amplitude of diffracted waves by edges. A refinement of the Kirchhoff approximation which is based on the Physical Theory of Diffraction (PTD) and combines GTD and Kirchhoff edge diffraction coefficients has been studied. This refined Kirchhoff approximation provides a simple formulation and correct results for all scattered directions, which will be illustrated in the case of a rigid halfplane or wedge.

1. Introduction

Monitoring and inspection of nuclear reactor are stringent requirements from operator and safety authorities. The sodium-cooled fast reactor (SFR) is one of the perspectives chosen for the 4th generation reactor. The characteristics exhibited by sodium, such as its opacity, have led the designers to devise specific monitoring and inspection techniques. Consequently ultrasonic techniques are seen as suitable candidates. Two approaches are being followed: the core monitoring where transducers are directly immersed in sodium near the reactor's core and the outside inspection with transducers located along the wall of the main vessel (outside sodium medium).

Ultrasonic telemetry is one of the core monitoring techniques that allows checking the position of the various objects contained inside the main vessel and the possible detection of defects inside these objects. The distance between the transducer and the immersed targets can be determined by measuring the time of flight of backscattering acoustic waves generated by the transducer installed inside. While in-service the flow of sodium creates turbulence that leads to temperature inhomogeneities, which convert into ultrasonic velocity inhomogeneities. These variations of velocity could impact directly upon the location tolerance by introducing times of flight variations. Different scattering phenomena can also be produced while the interaction between the acoustic beam radiated by the probe and the immersed targets: specular reflection, tip diffraction from boundaries and edges of the different parts and corner effect. Thus various variables will influence this technique behavior. In order to optimize the probes parameters that are being developed and to predict the probes performances, a simulation tool is necessary to assist the design of each element of the ultrasonic telemetry.

A wave propagation model has been developed in a previous work to calculate the ultrasonic field radiated in an inhomogeneous medium [1]. Calculations have shown that the inhomogeneous characteristic of the fluid doesn't impact much the time of flight and amplitude of the acoustic wave propagating but leads to beam deviations. In this paper, we will consider the field scattered from the immersed targets; different models describing the scattering phenomenon have been studied and compared. Firstly the scattered acoustic field can be modeled using the high-frequency asymptotics known as the Geometrical Acoustics (GA) and Geometrical Theory of Diffraction (GTD) [2], all based on the ray theory. The former describe incident and reflected waves and the latter, wave diffracted by obstacle edges (the so-called edge waves). The regions that support different kinds of waves are classified as either geometrical regions (illuminated region and shadow region) or transition zones that are the boundaries between an illuminated region and a shadow region. The sum of GA and GTD gives a perfectly adequate description of geometrical regions but fails inside transition ones. A more sophisticated uniform GTD is required to complete the description.

In some NDE applications, another approach, the so-called Kirchhoff approximation (KA) [3], is widely used in high-frequency scattering problems, particularly when dealing with obstacles of a complicated shape. The fundamental principle of this method is the use of Green's function representing in a given region Σ the solution to the Helmholtz equation. The KA provides a correct description of the reflected wave and the fields inside the transition region. The integral formulation of the KA solution enables description of the field in more intricate regions, such as focusing areas, shadow boundaries of edge waves, where the known GTD procedures are no longer applicable. This approximation leads to qualitatively correct description of edge waves, but with incorrect amplitudes. To eliminate the deficiencies of KA and GTD and combine their advantages, a model called here the refinement of KA is proposed to modify KA by employing GTD diffraction coefficients: this approach is based on the Physical Theory of Diffraction (PTD) [4].

The targets to be inspected by telemetry are steel structures immersed in liquid sodium, thus, the assumption of perfect rigid boundary condition should be not suitable here. In order to take into account the real boundary condition of our case, a modified GTD model applied to nearly-rigid wedge has been implemented. The reflection coefficient of this nearly-rigid interface can also be taken into account in the KA integral creating the rigorous KA formalism valid for arbitrary boundary condition.

2. Geometrical Theory of Diffraction

2.1. Non uniform asymptotic solution for scattered waves

Consider a simple example: a plane wave scattered by a two-dimensional halfplane as shown in Figure 1. Presence of this halfplane in a plane incident wave field (with θ_0: incident angle) gives rise to a shadow region of incident wave, a reflected wave and the related shadow region. Thus two light-shadow boundaries can be identified which are function of the incident angle: $\theta_1 = \pi - \theta_0$, $\theta_2 = \pi + \theta_0$. The incident and reflected wave are given by Geometrical Acoustics and these solutions are discontinued on light-shadow boundaries jumping to zero in the shadow region. These discontinuities of the field will be eliminated by adding the diffracted fields. Therefore the scattered field can be written as

$$U^{\mathrm{Scat}} = U^{\mathrm{GA}} + U^{\mathrm{GTD}}. \tag{1}$$

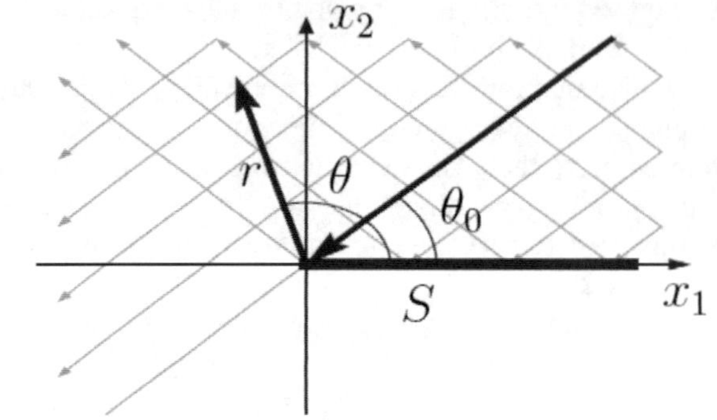

Figure 1: scattering of a plane acoustic wave by a halfplane.

The geometrical acoustic fields can be written as

$$U^{\mathrm{GA}} = U^{\mathrm{Inc}} + U^{\mathrm{Ref}} = e^{-ikr\cos(\theta-\theta_0)} \pm e^{-ikr\cos(\theta+\theta_0)}, \tag{2}$$

the plus sign is taken for the Neumann boundary condition (hard case) and the minus sign for the Dirichlet boundary condition (soft case). We'll treat in the following only the Neumann's case (perfect rigid target).

In the problem under consideration the diffraction field is a cylindrical wave generated by the obstacle edge. According to the Geometrical Theory of Diffraction [2], the "main-order" term with respect to large kr in the diffraction field has the form

$$U^{\mathrm{GTD}} \simeq \frac{e^{ikr}}{\sqrt{kr}} D^{\mathrm{GTD}}(\theta,\theta_0), \tag{3}$$

where θ_0 is the incidence angle of the plane wave, r and θ are the polar coordinates of the observation point, k is the wavenumber and $D^{\mathrm{GTD}}(\theta,\theta_0)$ is the diffraction coefficient [2] given by

$$D^{\mathrm{GTD}}(\theta,\theta_0) = -\frac{e^{i\pi/4}}{2\sqrt{2\pi}}\left(\sec\frac{\theta-\theta_0}{2} + \sec\frac{\theta+\theta_0}{2}\right), \tag{4}$$

where the angles θ and θ_0 are measured with respected to the illuminated surface of the halfplane (S^+) as shown in Figure 1. From eqn.(4) it follows that the diffraction coefficient grows without bound if the observation point is at a light-shadow boundary ($\theta = \pi - \theta_0$ or $\theta = \pi + \theta_0$) as shown in Figure 2. Clearly this diffraction coefficient is inapplicable in the vicinity of light-shadow boundaries where their poles are located. Hence GTD fails on the light-shadow boundaries.

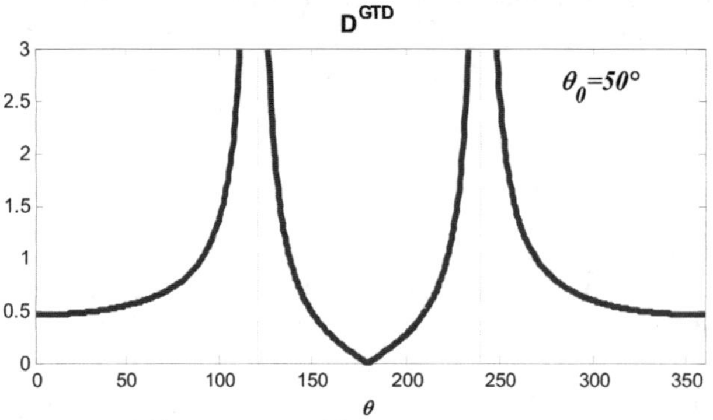

Figure 2: diffraction coefficient for Neumann boundary condition and incidence angle φ_0=50°

The situation would be essentially the same when the obstacle is an infinite wedge. The only difference between the wedge and halfplane would manifest itself only in the angular dependence of the edge-wave diffraction pattern, i.e. in the dependence of the diffraction coefficient D on the wedge angle Φ. Let us now consider a wedge of angle Φ (see Figure 3) with the polar coordinates (r, θ), this wedge occupies the region $\{(r, \theta): 2\pi - \Phi \leq \theta \leq 2\pi\}$.

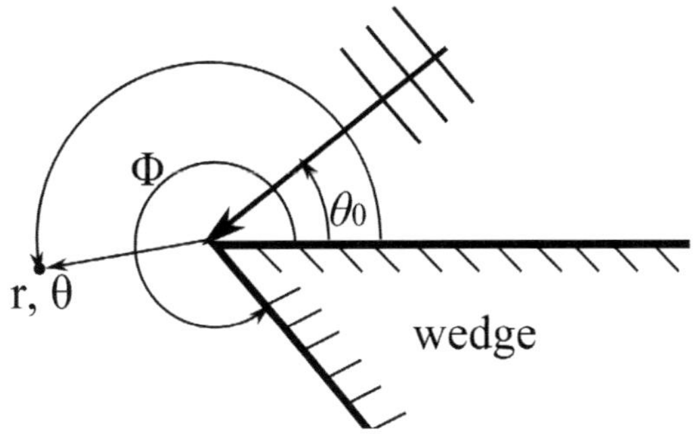

Figure 3: Scattering of a plane wave by a wedge with the wedge angle Φ.

The scattered field from the wedge has the same form as that for the halfplane (eqn.(1)) with diffracted field written as

$$U_{\Phi}^{\mathrm{GTD}} = \frac{e^{ikr}}{\sqrt{kr}} D_{\Phi}^{\mathrm{GTD}}(\theta,\theta_0), \tag{5}$$

with the diffraction coefficient for a wedge of angle Φ [5] defined by

$$\begin{aligned} D_{\Phi}^{\mathrm{GTD}}(\theta,\theta_0) &= -\frac{e^{i\pi/4}}{2\sqrt{2\pi}}\frac{\pi}{\Phi}\left[H(\theta+\pi,\theta_0,\Phi) - H(\theta-\pi,\theta_0,\Phi)\right], \\ H(\alpha,\beta,\Phi) &= \left[\operatorname{ctg}\frac{\pi}{2\Phi}(\alpha-\beta) + \operatorname{ctg}\frac{\pi}{2\Phi}(\alpha+\beta)\right]. \end{aligned} \tag{6}$$

This diffraction coefficient has also two poles (corresponding to zeros of ctg's arguments) at $\theta = \pi \pm \theta_0$, i.e. on the light-shadow boundaries of the incident and reflected waves.

2.2. Uniform asymptotic theory (UAT) for scattered waves

Several uniform theories derived from GTD exist. One of the two most studied methods, the so-called uniform asymptotic theory of diffraction (UAT), involves the application of Fresnel integral near the light-shadow boundaries [5, 6] in order to smooth the abrupt field shift through the boundaries. Let us take the example of halfplane and there are two light-shadow boundaries: $\theta = \pi + \theta_0$ for the incident wave and $\theta = \pi - \theta_0$ for the reflected wave. The uniform solution U^{UAT} of eqn.(1) is written as

$$U^{\mathrm{UAT}} = U^{\mathrm{GA}} F\{\sqrt{k(s_e - s_{i,r})}\}, \tag{7}$$

where F is the Fresnel integral. For a plane incident wave its argument can be given as follows:

$$\sqrt{k(s_e - s_i)} = \sqrt{2kr}\cos\frac{\theta-\theta_0}{2}, \quad \sqrt{k(s_e - s_r)} = \sqrt{2kr}\cos\frac{\theta+\theta_0}{2}, \tag{8}$$

where s_i, s_e and s_r are respectively the eikonals of incident wave, the edge wave and the reflected wave. In the case of wedge, the eqn.(7) becomes:

$$U^{\mathrm{UAT}} = U^{\mathrm{GA}} F\{\sqrt{k(s_e - s_{i,r})}\} + \frac{e^{ikr}}{\sqrt{r}} \sum_{n=0}^{\infty} (i/k)^{n+1/2} C_n(\theta,\theta_0), \tag{8}$$

The expansion associated to C_n corresponds to the differences in the boundary conditions between the halfplane and the wedge, and the zero order of expansion C_n is given as follow [9]:

Journal of Physics: Conference Series **353** (2012) 012009 doi:10.1088/1742-6596/353/1/012009

$$C_0(\theta,\theta_0)=\left[D_\Phi^{\rm GTD}(\theta,\theta_0)+\left(2\sqrt{2\pi}\right)^{-1}\sec\frac{\theta-\theta_0}{2}\pm\left(2\sqrt{2\pi}\right)^{-1}\sec\frac{\theta+\theta_0}{2}\right], \tag{9}$$

where $D_\Phi^{\rm GTD}(\theta,\theta_0)$ is given by eqn. (6). Such a zero-th order approximation defines the primary term of UAT.

Another method proposes a modification of the diffraction coefficient and consists in suppressing the coefficient poles by multiplying it with a transition function having zeros at the poles. This procedure known as the uniform geometrical theory of diffraction (UTD) was described in [8].

2.3. Asymptotic solution for scattered waves from a finite impedance obstacle

To deal with the solid targets immersed in a fluid medium, we have to take into account non perfectly rigid boundary condition at the target surface. Thus the real acoustic impedances of the propagation medium and the obstacle have to be taken into account

$$Z=\rho c \tag{10}$$

with ρ the density and c the acoustic velocity in the corresponding medium. In our case, the acoustic wave propagates in a liquid sodium and is reflected on a steel target; we can here define the admittance of such fluid-solid interface as

$$\beta=Z_f/Z_s, \tag{11}$$

where Z_f and Z_s represent the acoustic impedance of sodium (fluid) and steel (solid) respectively. The boundary condition of the wave equation on this fluid-solid interface can be given by

$$\frac{\partial U(\mathbf{x})}{\partial \mathbf{x}}-\mathrm{i}k\beta U(\mathbf{x})=0, \tag{12}$$

where $U(\mathbf{x})$ is the acoustic potential field and $\mathbf{x}$ denote the chosen Cartesian coordinate system. The mathematical formulations for the diffraction of an acoustic plane wave by a finite impedance wedge have been given by Willams [10] and Pierce *et al.* [11]. In our case, we consider that the interface sodium/steel is a nearly rigid interface, since Z_f (200°C) = 2.23 × 10^6 kg/ (m^2.s) and for a typical steel alloy ρ = 7700 kg/m^3, with a longitudinal velocity c_l of 6000 m/s, Z_s = 46.2× 10^6 kg/ (m^2s), thus, β=0.048 ≪ 1. The diffraction coefficient for a nearly rigid wedge [11] is given by

$$D_\Phi^*(\theta,\theta_0,\beta)=D_\Phi^{\rm GTD}\times[1+S_\Phi(\theta,\theta_0)\cdot\beta]. \tag{13}$$

This is a modification of the GTD diffraction coefficient given previously by multiplying a term containing a function S_Φ (θ, θ_0) and the admittance β of this interface. This asymptotic solution for the scattered-field is an extension of GTD for nearly-rigid wedge. In order to find the function S_Φ (θ, θ_0), let us define firstly the function

$$M_\nu(\theta)=\frac{\cos(\nu\pi)-\cos(\nu\theta)}{\nu\sin(\nu\pi)}, \tag{14}$$

with $\nu=\pi/\Phi$ and the function

$$\begin{aligned}Q_\Phi(\theta)=&-\nu\sin(\nu\pi)\sum_{n=1}^{1/2(p-1)}\frac{1}{\sin\left[\nu(\theta+2n\pi)\right]\sin\left\{\nu\left[\theta+\pi(2n-1)\right]\right\}}\\&-\sum_{m=0}^{q-1}\frac{\sin(\theta-2m\Phi)+\sin\left[\theta-\beta(2m+1)\right]}{\sin(\theta-2m\Phi)\sin\left[\theta-\beta(2m+1)\right]},\end{aligned} \tag{15}$$

in which the wedge angle Φ should take the form $p\pi/2q$ (i.e. for instance for a right-angled wedge $\Phi=3\pi/2$, p = 3 and q = 1). Finally the function S_Φ (θ, θ_0) is given as follows:

$$S_\Phi(\theta,\theta_0)=2\left[M_\nu(\theta+\theta_0)+M_\nu(\theta-\theta_0)\right]^{-1}-Q_\Phi(-\theta)-Q_\Phi(-\theta_0). \tag{16}$$

With the modified diffraction coefficient $D_\Phi^*(\theta,\theta_0,\beta)$, the non-uniform solution of scattered field from a nearly rigid wedge can be written as

$$U_{\Phi}^{\text{Nearly-rigid}} = U^{\text{Inc}} - \frac{1-\sin\theta_0/\beta}{1+\sin\theta_0/\beta} U^{\text{Ref}} + \frac{e^{ikr}}{\sqrt{r}} D_{\Phi}^{*}. \tag{17}$$

The second term of (17) refers to the reflected field, where the reflection coefficient R is

$$R = -\frac{1-\sin\theta_0/\beta}{1+\sin\theta_0/\beta}. \tag{18}$$

In the extreme case where the obstacle is perfectly rigid $Z_s \to \infty$, so $\beta \to 0$ we have $R = 1$ or where it is a perfect acoustic absorbent surface $Z_s \to 0$, $\beta \to \infty$, we get $R = -1$. In our case, the module and the phase of reflection coefficient on the sodium/steel interface are given below with $\varphi = \pi - \theta_0$.

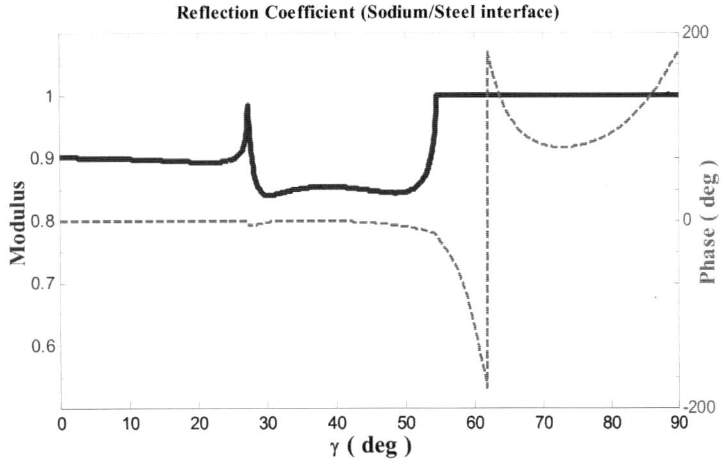

Figure 4: Reflection coefficient on the interface sodium/steel versus the complementary angle of incidence ($\gamma = \pi/2 - \theta_0$): modulus (solid curve) and phase (dash curve).

3. Kirchhoff approximation

The geometrical theory of diffraction provides short-wave asymptotic solutions accurate to the given orders of k for some typical model problems. Unfortunately, many model problems of practical interest have neither rigorous solutions to extract short-wave asymptotics nor appropriate asymptotics. Under these circumstances, one has to resort to approximate methods. A widely used method employed for large flaw size compared to the wavelength is the Kirchhoff approximation (KA) [3]. For any geometry, however complicated, the solution of KA is formulated as an integral of the field over the illuminated side of the reflector.

Consider how the KA method formulates a solution to the scattering field from a halfplane (see Figure 1). Let us introduce the associated Cartesian coordinate system $\mathbf{x}$ (x_1, x_2), so that we have

$$x_1 = r\cos\theta \quad \text{and} \quad x_2 = r\sin\theta. \tag{19}$$

Let us consider this acoustic problem with the acoustic potential field $U(\mathbf{x})$ satisfying the Helmholtz equation. Fundamental of this method is the use of Green's function G to obtain, by superposition of elementary fields, an expression in the form of integral equation for a given boundary S $(x_1, x_2{=}0)$

$$U^{\text{Scat}}(\mathbf{x}) = \int_S \left[G(\mathbf{x},\mathbf{x}') \frac{\partial U(\mathbf{x}')}{\partial \mathbf{n}} - U(\mathbf{x}') \frac{\partial G(\mathbf{x},\mathbf{x}')}{\partial \mathbf{n}} \right] ds(\mathbf{x}'). \tag{20}$$

Here, $U(\mathbf{x})$ refers to the field on the surface S, $\partial/\partial\mathbf{n}$ implies differentiation along the inward-directed normal to S, $\mathbf{x}$' denote one point on the surface S and the Green function G ($\mathbf{x}$, $\mathbf{x}$') for the two-dimensional problem studied here takes the form :

$$G(\mathbf{x},\mathbf{x}') = (i/4) H_0^{(1)}(k|\mathbf{x}-\mathbf{x}'|) \tag{21}$$

with $H_0^{(1)}$ the Hankel function of first kind.

3.1. Approximation solutions for a perfect boundary condition

The Kirchhoff approximation is based on the assumption that for $\lambda = (2\pi/k) \ll L$ (L is a typical size of the reflector), i.e. for $kL \gg 1$, one can use the approximation of geometrical acoustics for the total field $U(\mathbf{x}')$ near the surface. In the shadow region, we can set

$$U(x') = \frac{\partial U(x')}{\partial \mathbf{n}} = 0. \tag{22}$$

When evaluating $U(\mathbf{x})$ and $\partial U(\mathbf{x})/\partial \mathbf{n}$ in the illuminated region, the total field is equal to the sum of incident and specularly reflected fields for the Neumann condition or to a difference of these fields for the Dirichlet condition. Therefore, on S:

$$U(\mathbf{x}') = 2U^{\mathrm{Inc}}(\mathbf{x}') \text{ and } \frac{\partial U(\mathbf{x}')}{\partial \mathbf{n}} = 0 \quad \text{for the Neumann condition ;} \tag{23-a}$$

$$U(\mathbf{x}')=0 \text{ and } \frac{\partial U(\mathbf{x}')}{\partial \mathbf{n}} = 2\frac{\partial U^{\mathrm{Inc}}(\mathbf{x}')}{\partial \mathbf{n}} \quad \text{for the Dirichlet condition.} \tag{23-b}$$

Thus, in the KA, the scattered field from a perfectly rigid surface (Neumann condition) can be written as

$$U^{\mathrm{KA}}(x) = -2\int_0^{+\infty} U^{\mathrm{Inc}}(\mathbf{x}')\frac{\partial G(\mathbf{x},\mathbf{x}')}{\partial \mathbf{n}} ds(\mathbf{x}'), \tag{24}$$

where s is the surface element along the illuminated surface of halfplane. In the wedge case, this integration must be carried out on each wedge face when the two faces are illuminated by the incidence wave.

3.2. Approximated solutions for an impedance boundary condition

The eqn.(20) can also be calculated for a given arbitrary boundary condition whose corresponding refection coefficient is R. In this condition, the assumption (22) can remain but the approximations given by eqn. (23) should be rewritten [12] as follows:

$$U(\mathbf{x}') = [1+R]U^{\mathrm{Inc}}(\mathbf{x}'), \quad \frac{\partial U(\mathbf{x}')}{\partial \mathbf{n}} = [1-R]\frac{\partial U^{\mathrm{Inc}}(\mathbf{x}')}{\partial \mathbf{n}}. \tag{25}$$

R = 1 corresponds to the Neumann condition and R = -1 gives the Dirichlet condition. Therefore the Kirchhoff integral for a given impedance interface is given by

$$U^{*\mathrm{KA}}(\mathbf{x}) = \int_S \left[G(\mathbf{x},\mathbf{x}')\frac{\partial U^{\mathrm{Inc}}(\mathbf{x}')}{\partial \mathbf{n}}(1-R) - U^{\mathrm{Inc}}(\mathbf{x}')(1+R)\frac{\partial G(\mathbf{x},\mathbf{x}')}{\partial \mathbf{n}} \right] ds(\mathbf{x}'). \tag{26}$$

4. Refinement of the Kirchhoff approximation

The Kirchhoff approximation has some limitations; the most important one is the incorrect prediction of the diffraction wave amplitudes. To overcome this limitation we are going to correct the Kirchhoff approximation by employing GTD diffraction coefficient. As we can see the GTD diffraction coefficient can be computed in an efficient manner using algorithm given by eqn.(4), this refinement of the Kirchhoff approximation should be quite fast.

In order to correct the Kirchhoff approximation, we should identify different parts inside the Kirchhoff integral (eqn.(24) for a Neumann boundary condition for example). Using the stationary phase method, we find that the integral (24) involves two critical points, a stationary point corresponding to the geometrical field U^{GA} and the lower limit contribution where $\mathbf{x}'=0$ corresponding to the diffraction field. This diffraction field contribution has the same form as the U^{GTD} with a different diffraction coefficient:

$$U^{\mathrm{KA(Diff)}} = \frac{e^{ikr}}{\sqrt{kr}} D^{\mathrm{KA}}(\theta,\theta_0). \tag{27}$$

Using [5] to find the asymptotic contribution of integration domain boundary, the Kirchhoff diffraction coefficient, for a Neumann boundary condition (plus sign) and a Dirichlet boundary condition (minus sign), turns to be

$$D^{\mathrm{KA}}(\theta,\theta_0) = -\frac{e^{i\pi/4}}{2\sqrt{2\pi}}\left(\tan\frac{\theta-\theta_0}{2} \pm \tan\frac{\theta+\theta_0}{2}\right). \tag{28}$$

This means that outside the penumbrae areas, the non-uniform asymptotics of the Kirchhoff integral are

$$U^{\mathrm{KA}}_{\text{non-uniform}}(\mathbf{x}) = U^{\mathrm{GA}}(\mathbf{x}) + U^{\mathrm{KA(Diff)}}(\mathbf{x}). \tag{29}$$

Thus the KA integral has been decomposed in two parts. The refinement of the Kirchhoff approximation is to correct the diffraction field amplitudes by employing the GTD

$$\begin{aligned} U^{\mathrm{RKA}}(\mathbf{x}) &= U^{\mathrm{KA}}(\mathbf{x}) + U^{\mathrm{GTD}}(\mathbf{x}) - U^{\mathrm{KA(Diff)}}(\mathbf{x}) \\ &= U^{\mathrm{KA}}(\mathbf{x}) + \frac{e^{ikr}}{\sqrt{kr}}\left(D^{\mathrm{GTD}}(\mathbf{x}) - D^{\mathrm{KA}}(\mathbf{x})\right). \end{aligned} \tag{30}$$

Finally the refinement of the Kirchhoff (RKA) consists in correcting, thanks to GTD, the KA contribution corresponding to the field scattered by the edge. This correction leads to add a corrective term to the KA field which is the difference of wave amplitudes diffracted by the edge given by GTD and KA. The diffraction coefficients for KA diffraction contribution and GTD have the same singularities at $\theta = \pi + \theta_0$ and $\theta = \pi - \theta_0$ (see Figure 5); when we make the difference of the two coefficients, their singularities cancel each other.

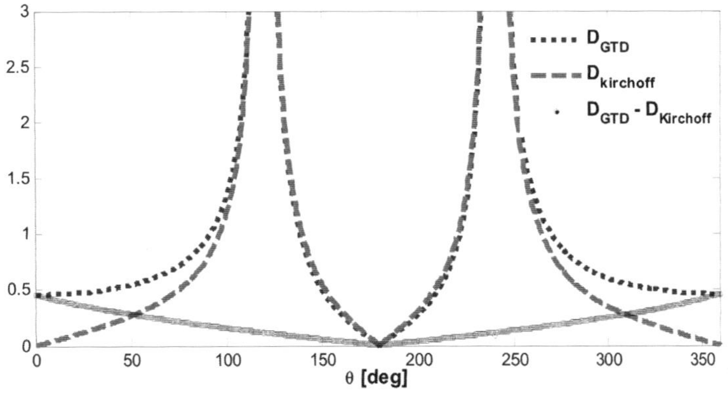

Figure 5: Diffraction coefficient for GTD and KA for θ_0=50°.

5. Models comparisons and discussion

5.1. Perfectly rigid halfplane

The scattering of a plane wave by a halfplane is a canonical problem and it has an exact solution which allows us to compare with the GTD non-uniform eqn.(1) and uniform solutions eqn.(7). Those results are represented by their radiation pattern (containing the maximum power) and shown in Figure 6 and Figure 7. The incidence angle θ_0 is taken at 50°; the observation points are located around the edge for two distances from the edge r = λ and r = 5λ where λ is the wave length. According to the incidence θ_0 the light-shadow boundaries are θ_1= 130° and θ_2=230° on which we find the singularities of the non-uniform GTD solution (black dash-dot curve). However the uniform solutions (green dash curve) coincide quite well with the exact solution (red solid curve).

Journal of Physics: Conference Series **353** (2012) 012009 doi:10.1088/1742-6596/353/1/012009

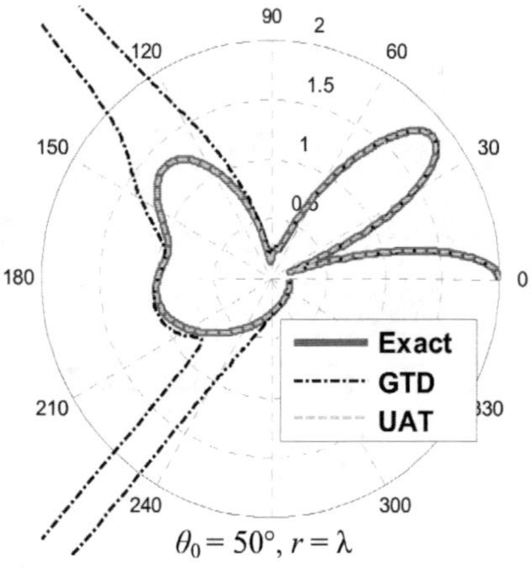

Figure 6: Radiation pattern comparison of the scattered field by a perfectly rigid halfplane given by exact solution, GTD and UAT.

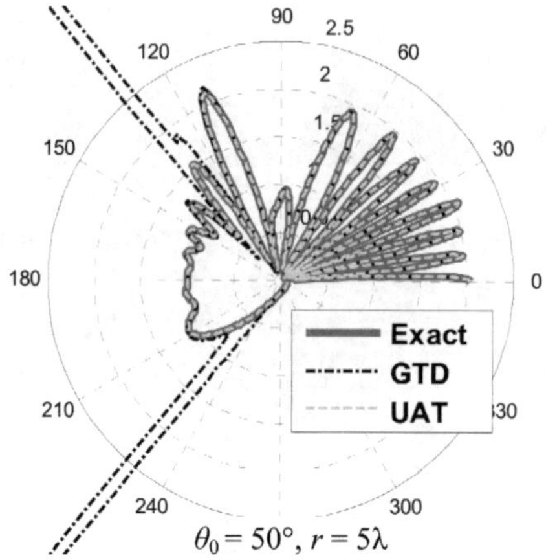

Figure 7: Radiation pattern comparison of the scattered field by a perfectly rigid halfplane given by exact solution, GTD and UAT.

Applying Kirchhoff approximation to the perfectly rigid halfplane in the same configuration as in Figure 6 and Figure 7, we obtain the results in Figure 8 and Figure 9. Kirchhoff approximation provides a qualitatively correct description of the scattered field. Errors can be found near the boundaries and in the shadow region where the edge diffraction wave dominate. When the observation is done far from the boundaries ($r = 5\lambda$) and it is usually the case in non destructive evaluation, the KA field coincide quite well with the exact solution (Figure 9).

In a two dimensional space, the distribution of KA's normalized errors around the halfplane can be represented in Figure 10. The errors located near the edge are caused by the approximations we made in eqn.(23) on the boundary condition. The errors away from the edge are indeed due to the incorrect prediction of the diffracted wave amplitudes.

To correct the errors produced by the Kirchhoff approximation, we apply the refinement of the Kirchhoff approximation by employing GTD diffraction coefficients. The configurations of Figure 8 and Figure 9 have been recalculated using the refined Kirchhoff approximation. The results are given in Figure 11 and Figure 12. The results after the refinement coincide quite well with the exact solutions. The errors are greatly reduced Figure 13 (from 20% max to 1.5% max).

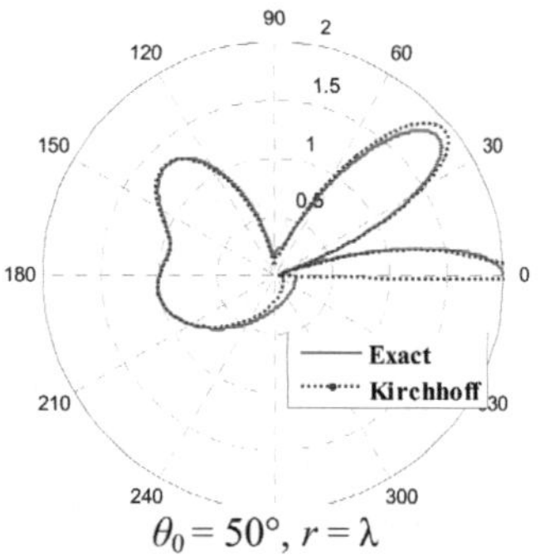

Figure 8: Radiation pattern comparison of scattered field from a rigid halfplane given by exact solution and Kirchhoff approximation.

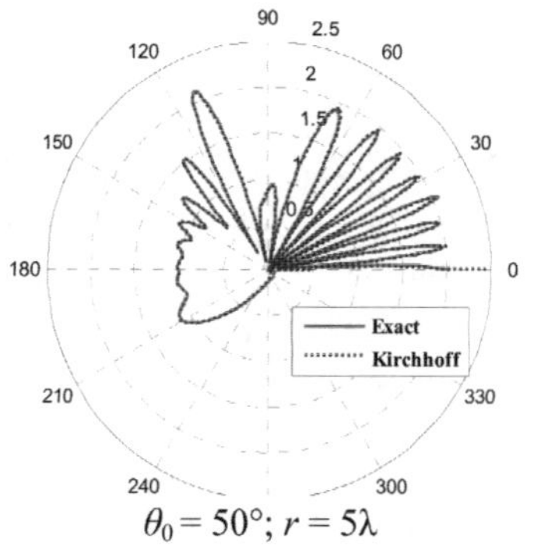

Figure 9: Radiation pattern comparison of scattered field from a rigid halfplane given by exact solution and Kirchhoff approximation.

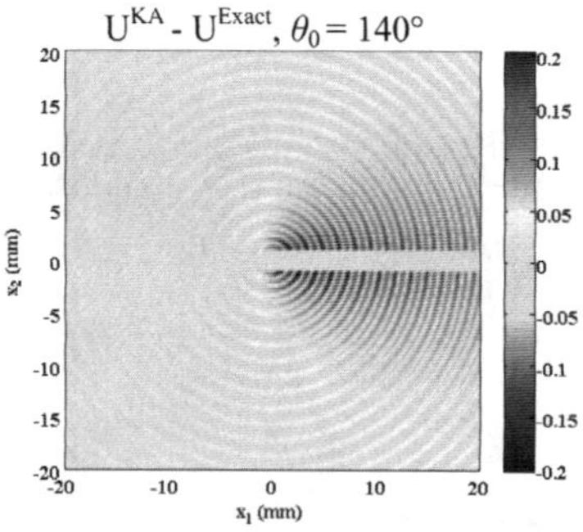

Figure 10: Distribution of KA errors to the exact solution with the $\theta_0 = 50°$.

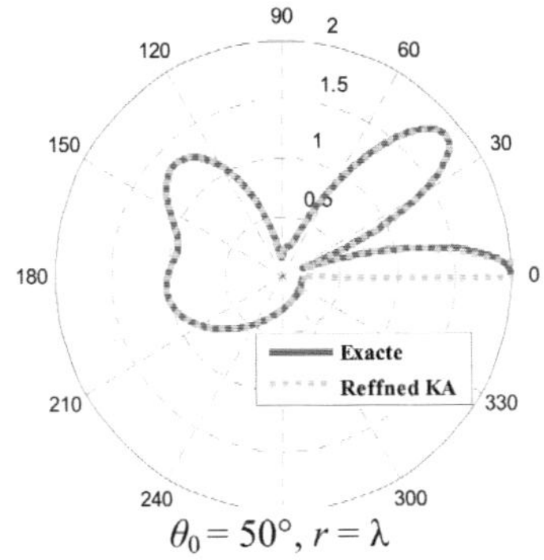

Figure 11: Radiation pattern comparison of scattered field from a rigid halfplane given by exact solution and refined KA.

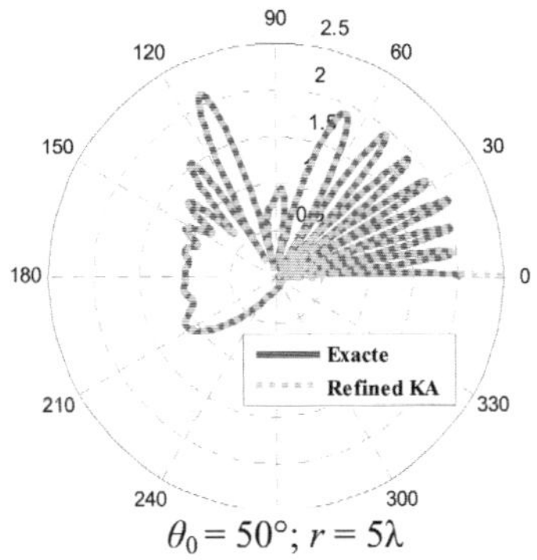

Figure 12: Radiation pattern comparison of scattered field from a rigid halfplane given by exact solution and refined KA

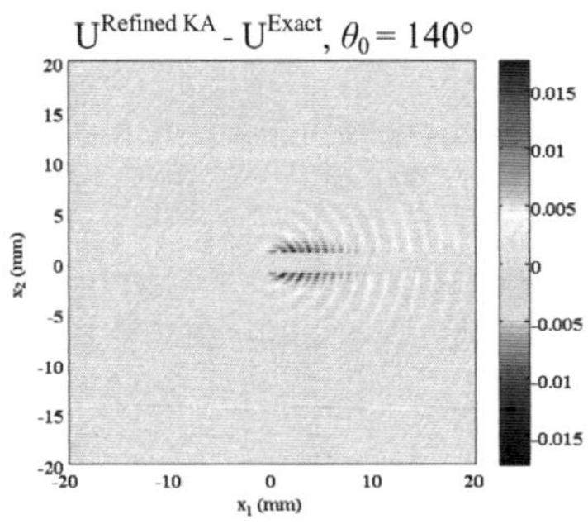

Figure 13: Distribution of refined KA's errors to the exact solution with $\theta_0 = 50°$.

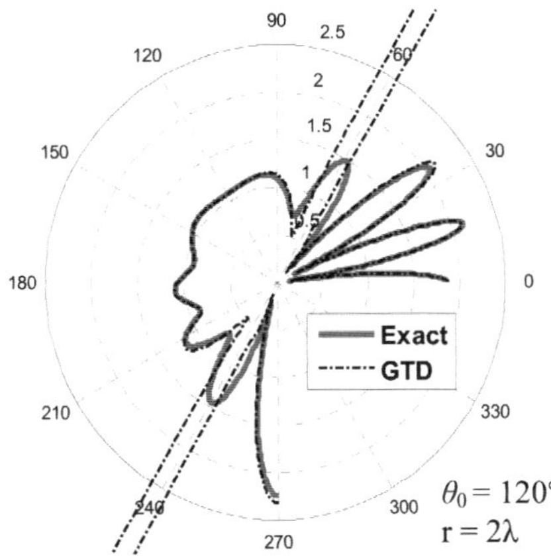

Figure 14: Radiation pattern comparison of the scattered field from a perfectly rigid right-angle wedge given by GTD and exact solution.

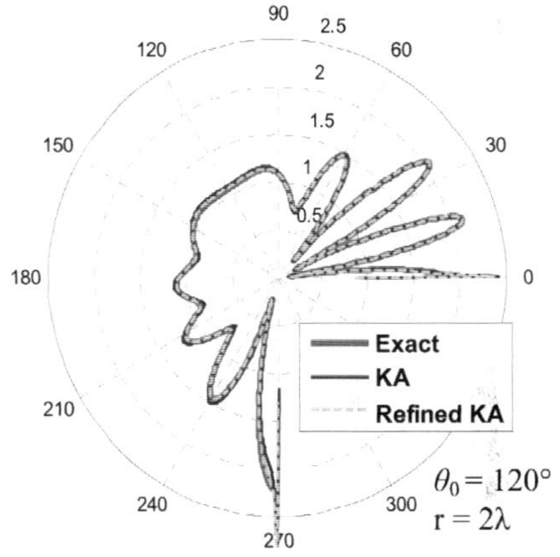

Figure 15: Radiation pattern Comparison of the scattered field from a perfectly rigid right-angle wedge given by KA and the refined KA.

5.2. Perfectly rigid wedge

The GTD extension for a wedge (6) has been applied to a rigid right-angled wedge (Figure 14). For an incidence of 120°, there is no shadow region outside the wedge and the incident wave is reflected on both surfaces in the direction of 60° and 240°. Two singularities are found in the these two specular directions, but outside the singularity zones, the GTD model gives a quite good prediction. The Kirchhoff approximation can be also used to calculate the scattered field by a wedge and if two surfaces are illuminated, the integral (24) should be calculated on both surfaces. The refinement procedure of Kirchhoff approximation can be extended for a rigid wedge. The radiation pattern of KA and refined KA are illustrated in Figure 15 and we see the refined KA leads to a slight improvement.

5.3. Nearly rigid wedge

The scattered fields from a nearly rigid wedge have been calculated for two different incidences: θ_0 = 50° (before the abrupt phase shift of the reflection coefficient, see Figure 16) and θ_0 = 20° (after the abrupt phase shift, see Figure 17). The result, given by eqn.(17) (GTD $_{nearly\text{-}rigid}$: red solid curve), is a non-uniform asymptotic solution (GTD extension) and considered as an almost exact solution outside the penumbrae areas. Then we compare this GTD result (outside the penumbrae areas) to those given by the Kirchhoff approximation with the Neumann boundary condition [eqn. (24)] (KA_{rigid}: green dot curve). For the incidence $\theta_0 = 50°$, the amplitude of KA_{rigid} is little high than that of $GTD_{nearly\text{-}rigid}$ which can be explained by the modulus of the reflection coefficient $|R(\theta_0 = 50°)| \approx 0.85$ in the nearly rigid case and $|R(\theta_0 = 50°)| = 1$ in the perfectly rigid case. At this incidence the pattern lobes of the two results have the same orientations, since there is no phase-shift yet, and for the incidence $\theta_0 = 20°$, there's no obvious amplitude difference, but the pattern lobes have different orientations which is due to the phase shift of R. We add in the Figure 16 and Figure **17** the results given by eqn.(26) (KA_{impe}: black dash curve): the Kirchhoff approximation solution for a finite impedance boundary condition. We find a quite good coincidence between $GTD_{nearly\text{-}rigid}$ and KA_{impe} outside the penumbrae areas.

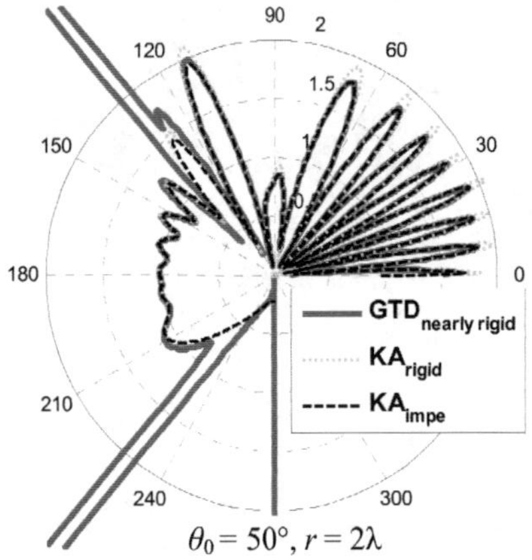

$\theta_0 = 50°, r = 2\lambda$

Figure 16: Radiation pattern comparison of the scattered field from a nearly-rigid wedge

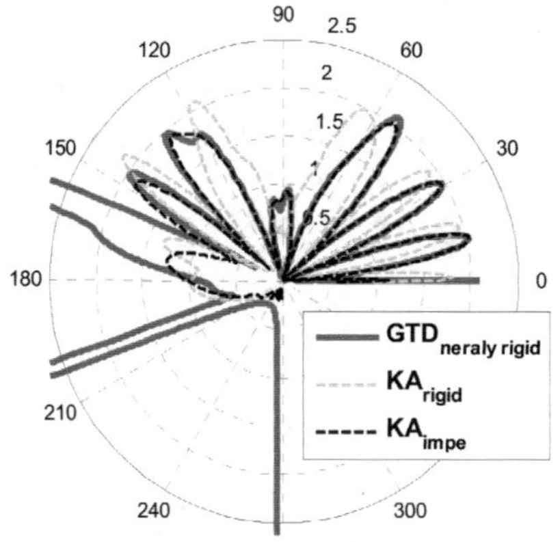

$\theta_0 = 50°, r = 2\lambda$

Figure 17: Radiation pattern comparison of the scattered field from a nearly-rigid wedge

6. Conclusion

Two classic wave scattering models: Geometrical Theory of Diffraction (GTD) and the Kirchhoff approximation have been studied for a rigid halfplane model and a rigid wedge. The results have been compared with exact solutions. The asymptotic GTD formalism can be computed in an efficient manner and gives a perfectly adequate description of geometrical regions but is inapplicable inside transition regions. A more sophisticated uniform GTD is required to complete the description. The Green's-function-based KA formalism's results are uniform with respect to the observation point. KA correctly describes the geometrical field but leads to qualitatively correct description of diffraction wave but with incorrect amplitudes. In order to eliminate the deficiencies of GTD and KA and combine their advantages, the refinement of KA has been developed which consists in correcting, thanks to GTD, the KA contribution corresponding to the field scattered by the edge. This correction leads to add a corrective term to the KA field which is the difference of wave amplitudes diffracted by the edge given by GTD and KA. The refined KA gives accurate results compared to the exact solutions and with its simple formalism it can deal with obstacles of a complicated shape.

As to the wedge with finite impedance which is representative of the solid targets immersed in fluid which are inspected by telemetry, a non-uniform asymptotic model has been implemented. It is a modification of the classic GTD model by account of the real admittance of the fluid-solid interface. This model gives accurate description of the scattered field from a finite impedance surface outside the transition regions. The reflection coefficient of the fluid-solid interface can be taken into the KA integral and this KA extended to impedance wedges yields scattered fields from a nearly-rigid wedge coinciding with the modified GTD fields.

By coupling the impedance Kirchhoff model to the wave propagation model in inhomogeneous medium, it will be possible to simulate the complete telemetry technique and predict the possibility to locate or detect objects contained in the reactor main vessel, which will help the development of visualization tools under sodium for the new generation nuclear reactor.

References

[1] Lu B. Darmon M. Potel C. 2010 *Proc. of the 10th French Acoustic Congress.*
[2] Keller J.B. 1962 *J. Opt. Soc. Am.* **4** 116-130.
[3] Meecham W.C. 1956 *J. Ration. Mech. Anal.* **5** 323–333.
[4] Ufimtsev P. Ya. 2007 Fundamentals of the Physical Theory of Diffraction, Wiley & Sons, Inc., Hoboken, New Jersey.
[5] Borovikov V.A. 1994 *Uniform stationary phase method*, IEE, London.
[6] Ahluwalia D.S. 1970 *SIAM J. Appl. Math.,* **18**, 287-301.
[7] Lewis R.M. Boersma J. 1969 *J. Math. Phy.* **10**, 2291-2305.
[8] Kouyoumjian R.G. and Pathak P.H. 1974 *Proc. IEEE*, **62**, 1448-1461.
[9] Borovikov V.A. Kinber B. Ye. 1994 *Geometrical theory of diffraction*, IEE, London.
[10] Williams W.E. 1959 *Proc. R. Soc.* London Ser. A**252**, 376-393.
[11] Pierce A.D. Hadden Jr. W.J. 1978 *J. Acoust. Soc. Am.* **63**, 17-27.
[12] Bass F.G. and Fuks I.M. 1973 *Wave scattering from statistically rough surfaces*, Pergamon Press, Oxford.

Journal of Physics: Conference Series **353** (2012) 012010 doi:10.1088/1742-6596/353/1/012010

Scattering of the field of a multi-element phased array by human ribs

P Gélat[1,2,†], G ter Haar[3] and N Saffari[2]

[1] National Physical Laboratory, Hampton Road, Teddington TW11 0LW, United Kingdom.

[2] Department of Mechanical Engineering, University College London, Torrington Place, London WC1E 7JE, United Kingdom.

[3] Therapeutic Ultrasound Group, Physics Department, Institute of Cancer Research, Sutton SM2 5NG, United Kingdom.

[†] E-mail: Pierre.Gelat@npl.co.uk

Abstract. The efficacy of high intensity focused ultrasound (HIFU) for the non-invasive treatment of cancer has been demonstrated for a range of different cancers including those of the liver, kidney, prostate and breast. As a non-invasive focused therapy, HIFU offers considerable advantages over other techniques such as chemotherapy and surgical resection, in terms of invasiveness and risk of harmful side effects. Despite its advantages, however, there are a number of significant challenges currently hindering its widespread clinical application. One of these challenges is the need to transmit sufficient energy through the ribcage to induce tissue necrosis at the required foci whilst minimising the formation of side lobes. Multi-element random arrays are currently showing great promise in overcoming the limitations of single-element transducers. Nevertheless, successfully treating a patient for liver tumours requires a thorough understanding of the way in which the ultrasonic pressure field from a HIFU array is scattered by the ribcage. A mesh of quadratic pressure patches was generated using CT scan data for ribs nine to twelve on the right side. A boundary element approach based on a Generalised Minimal Residual (GMRES) implementation of the Burton-Miller formulation was used, in conjunction with phase conjugation techniques to focus the field of a 256-element random HIFU array past the ribs at both intercostal and transcostal treatment locations. This method has the advantage of accounting for full effects of scattering and diffraction in three dimensions under continuous wave excitation.

1. Background

The efficacy of high-intensity focused ultrasound (HIFU) has been demonstrated for a range of cancers [1], with over 38000 prostate cancer treatments and over 50000 treatments for abdominal tumours, osteosarcoma, uterine fibroids and thyroid worldwide [2]. HIFU offers considerable advantages over techniques such as chemotherapy and surgical resection, in terms of its non-invasiveness and low risk of harmful side effects. There nevertheless are a number of significant challenges currently hindering its more widespread clinical application. In the cases of liver, kidney and pancreatic cancers, one of these challenges is the need to transmit sufficient energy through the rib cage to induce tissue necrosis at the required location whilst minimising the formation of side lobes. In addition to this, a common side effect of focusing ultrasound in regions located behind the rib cage is the overheating of bone and surrounding tissue, which can lead to skin burns [3], [5].

Multi-element random spherical arrays are showing promise in overcoming the limitations of single-element transducers, such as side lobe formation, lack of electronic steering and lack of beam shaping capabilities [5], [6], [7], [8], [9]. Successfully treating hepatic, renal and pancreatic cancers nevertheless requires a thorough understanding of the way in which the ultrasonic pressure field generated by a HIFU array is scattered by the rib cage. Hence, an important part of the treatment planning procedure is likely to involve a model capable of predicting the pressure distribution both on the ribs and in the surrounding tissue.

The theoretical and experimental feasibility and efficacy of transcostal HIFU (THIFU) has been extensively investigated. Given the considerable computational challenges involved at Megahertz frequencies for large domain dimensions, many models have relied on simplified shadowing techniques [10], [11]. Although these techniques predict features of wave propagation during THIFU treatments, they fail to address the actual scattering and diffraction mechanisms involved in complex 3D structures. As such, they are of limited value to treatment planning applications. An experimental method based on the decomposition of the time reversal operator (DORT) goes some way towards addressing challenges in treatment planning [12]. It was implemented on human ribs immersed in water and shows promise in sparing the ribs from the incident ultrasonic field. A 3D model capable of reproducing these results theoretically would considerably further the understanding of HIFU fields in the presence of scatterers and would constitute a valuable tool for optimisation of phased array design and performance.

Finite difference time domain (FDTD) methods of modelling HIFU fields scattered by ribs have been implemented in 2D, in conjunction with the time reversal method [8]. These techniques have the advantage of being relatively straightforward to realise for purely acoustic problems, can be generalised to 3D and can be readily implemented on a computer cluster. They nevertheless possess a number of disadvantages. These include having to mesh the entire computational domain, dealing with artefacts due to domain truncation and relying on pulsed wave solutions. There is clearly a requirement to implement a three-dimensional continuous wave model of a multi-element HIFU array irradiating a scatterer of arbitrary shape. Exterior boundary element (BE) methods have been successfully applied to a range of acoustic scattering problems. They rely on the discretisation of the Helmholtz integral equation. BE methods have the advantage of being a frequency domain technique and are therefore suitable for simulating the quasi-continuous wave conditions encountered in HIFU. Provided the medium surrounding the scatterer can be treated as homogeneous, it is not necessary to mesh any portion of the exterior domain. Also, it is not necessary to truncate the exterior domain as in FDTD methods.

The discretisation of the Helmholtz integral equation leads to a fully populated asymmetrical system of linear equations. The conventional Helmholtz boundary integral equation (BIE) formulation for exterior wave problems yields non-unique solutions at the eigenfrequencies associated with the (interior) Dirichlet problem. When the coefficient matrix resulting from discretisation is inverted at frequencies approaching these (fictitious) eigenfrequencies, the problem becomes ill-conditioned. Burton and Miller proposed a method that guarantees uniqueness of the solution [13], which involves a formulation employing a linear combination of the conventional BIE and its derivative with respect to the outward normal. Provided the coupling coefficient has a non-zero imaginary part, this method provides unique solutions at all frequencies. Computational limitations have traditionally hampered BE methods from becoming widespread. Inverting a large, fully populated asymmetrical matrix on a stand-alone computer is impractical when considering the memory allocation necessary to store all elements simultaneously. This challenge was addressed by Ochmann *et al* by using an iterative generalised minimum residual (GMRES) based boundary element solver for acoustic scattering [14]. Shen and Liu investigated a range of scattering problems using an adaptive fast multipole BE method, also using an iterative GMRES scheme [15]. The GMRES method is amenable to cluster computing, as it performs matrix-vector multiplications at each iteration. Thus, employing the GMRES scheme offers the opportunity to significantly reduce run times.

In this paper, a feasibility study is described in which a BE approach was used to model the scattering of the field of a multi-element HIFU phased array by human ribs. The array was modelled using a linear superposition of plane circular piston sources. The ribs were assumed to be perfectly rigid. The surrounding medium was assumed to be homogeneous and inviscid. After validation on spherical scatterers, a mesh of quadratic pressure patches was generated using CT scan data for ribs nine to twelve on the right side of an adult male. A GMRES implementation of the Burton-Miller formulation based on the method proposed in [16] was used, in conjunction with phase conjugation techniques [17] to focus the field of a 256-element random HIFU array 3 cm deep behind the rib cage at locations accessible either inter- or trans-costally. The calculations were distributed over a number of cores on a dedicated computer cluster at University College London Centre for Medical Image Computing.

2. Theory

Consider an exterior domain V_{ext} bounded by a surface Λ and by a closed smooth surface S. Let Λ be at a sufficiently large distance from the acoustic sources and from the surface S. Let the boundary condition on Λ satisfy Sommerfeld's acoustic radiation condition. The propagation of time-harmonic acoustic waves in a homogeneous isotropic inviscid medium is described by the Helmholtz equation:

$$\nabla^2 p(\vec{r}) + k^2 p(\vec{r}) = 0,\ \forall \vec{r} \in V_{ext} \tag{1}$$

where p is the acoustic pressure, $k = \omega / c$ is the acoustic wave number where ω is the angular frequency, c the sound speed in the external medium and $\vec{r}$ is the position vector. A source term may be included in the right hand side of (1) (incident pressure wave), but it is convenient to split the total pressure $p(\vec{r})$ as the sum of the incident pressure $p_i(\vec{r})$ and the pressure scattered by the surface S, $p_s(\vec{r})$, $\forall \vec{r} \in V_{ext}$.

For scattering problems, the integral representation of the solution to (1) is given by the Helmholtz integral equation:

$$\int_S \left[p(\vec{r}_q) \frac{\partial \mathrm{G}(\vec{r} \mid \vec{r}_q)}{\partial n_q} - \frac{\partial p(\vec{r}_q)}{\partial n_q} \mathrm{G}(\vec{r} \mid \vec{r}_q) \right] dS = \frac{p(\vec{r})}{2} - p_i(\vec{r}),\quad \forall \vec{r} \in S \tag{2}$$

where the Green's function in a three-dimensional space is given by:

$$\mathrm{G}(\vec{r} \mid \vec{r}_q) = \frac{1}{4\pi |\vec{r} - \vec{r}_q|} e^{-ik|\vec{r} - \vec{r}_q|} \tag{3}$$

and $p(\vec{r}_q)$ is the pressure on the surface S at point $\vec{r}_q$ and $\vec{n}_q$ is the outward normal.

The problem as described by the integral equation (2) suffers from non-uniqueness at frequencies of excitation approaching an eigenvalue of one of the (fictitious) modes of the cavity inside the scatterer. In cases where the dimensions of the scatterer are large compared with the wavelength in the propagating medium, such as in HIFU applications, this is likely to occur. The matrix formed by discretizing equation (2) then becomes close to singular. The method which appears to offer the best compromise in terms of application of the Helmholtz integral equation to exterior acoustic problems involving scatterers of arbitrary shape remains the Burton-Miller formulation [18], which solves for a linear combination of equation (2) and its derivative with respect to the outward normal vector on S at $\vec{r}$.

By discretising a linear combination of equation (2) and its normal derivative, for all position vectors $\vec{r}$ on S corresponding to each node on the mesh of the surface, a linear system of equations may be generated.

$$[H]\{p\}-[G]\left\{\frac{\partial p}{\partial n}\right\}+\{p_i\}+\alpha\left\{\frac{\partial p_i}{\partial n}\right\}=\{0\} \tag{4}$$

For a rigid boundary, the normal derivative vector is $\{0\}$.

By knowledge of the incident pressure field and its normal derivative on S, together with $[H]$ and $[G]$, $\{p\}$ may be obtained through matrix inversion if the problem size is small enough. For larger problems, where storing all the elements of $[H]$ and $[G]$ simultaneously becomes challenging, the GMRES method may be used. The pressure $p(\vec{r})$ at any location in the exterior volume V_{ext} may be obtained at the post-processing stage.

The HIFU transducer modelled as part of the underlying work was assumed to be spherically shaped and to contain a central aperture (used to insert an ultrasonic diagnostic treatment head). The HIFU array was assumed to be populated with N = 256 plane circular elements mounted onto its surface. Guidelines regarding the dimensions involved were obtained from the literature and from prior *in-vivo* applications [5], [11], [19]. The elements were each of a = 3 mm radius. Larger values for a will result in an increased likelihood of scattering from the ribs whilst smaller values would be difficult to manufacture and may result in insufficient acoustic power generation to induce tissue necrosis. A radius of curvature of D = 18 cm was used, to ensure the applicability to deep-seated tumours. The diameter of the central aperture was chosen to be 4 cm, allowing for an ultrasound imaging transducer to be housed. The outer diameter of the HIFU transducer was chosen as 16 cm. The elements were randomly spatially distributed on the surface of the array. A frontal view of the array is shown in figure 1. The incident pressure field p_i in equation (4), i.e. the field produced by the HIFU array in absence of scatterers, was modelled as a superposition of plane circular piston sources, assumed to be rigidly vibrating in an infinite baffle.

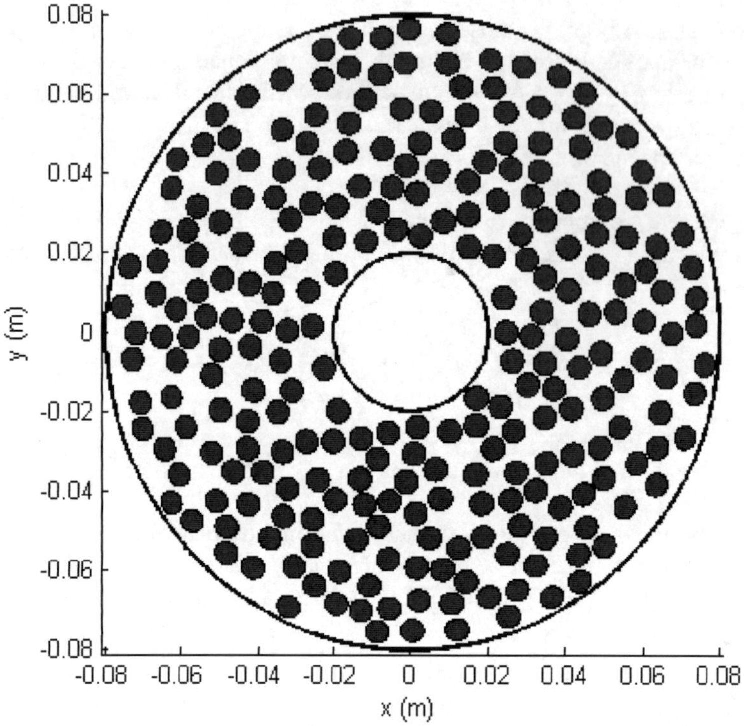

Figure 1. Frontal view of HIFU random phased array configuration. 6 mm element diameter, 4 cm diameter central aperture, 16 cm array diameter, 18 cm focal length, 1 MHz frequency of operation.

3. Results

The GMRES Burton-Miller boundary element approach was implemented on a dedicated computer cluster. The routines employed to generate the boundary element matrices were obtained from the PAFEC (Program for Automatic Finite Element Calculations) VibroAcoustics software with permission from PACSYS Ltd. Validation on simple scatterers was carried our prior to simulations on human ribs. The results displayed here are for a 1 MHz unit amplitude plane wave scattered by a 0.5 cm radius rigid sphere. These were compared against the analytical solution provided in [20] in the shadow zone between 1.1 cm and 5.0 cm (see figure 2).

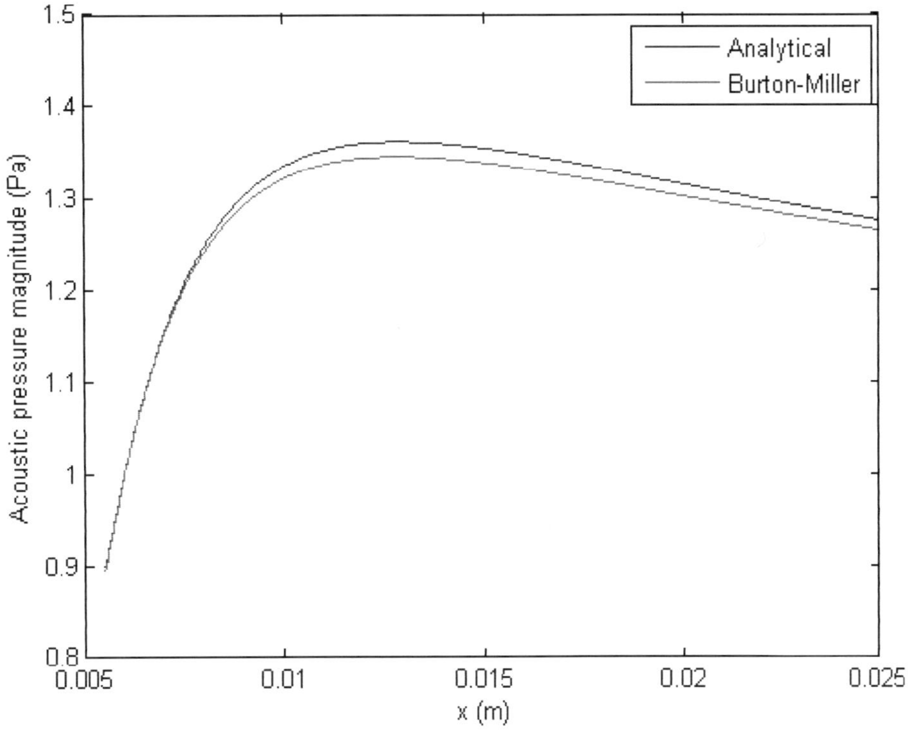

Figure 2. GMRES implementation of Burton-Miller BE formulation on a spherical scatterer of 0.5 cm radius. Incident field: unit amplitude 1 MHz plane wave travelling in positive x direction. Comparison against anaytical solution in shadow zone.

The scatterer was meshed using isoparametric eight-noded quadratic patches ensuring at least three elements per wavelength for a wavespeed of 1500 m s^{-1}. The density of the medium was assumed to be 1000 kg m^{-3}. The coupling coefficient in the Burton-Miller formulation was chosen as $\alpha = 0.1i/k$, as this resulted in quicker convergence of the solution than using i/k, as recommended in [21]. In all cases of using the GMRES scheme, the initial trial pressure vector was chosen as the incident field at nodal positions on the surface of the scatterer. 40 iterations were generally sufficient for the residual norm to fall below 1% of its initial value. Figure 2 shows agreement within 1.2% of the analytical solution.

Using CT scan data obtained from [22] for ribs nine to twelve on the right side of an adult male, a surface mesh was obtained using isoparametric eight-noded quadratic elements, ensuring at least three elements per wavelength for a wavespeed of 1500 m s^{-1}. The density of the medium was assumed to be 1000 kg m^{-3}. CATIA v5 Advanced Meshing Tools was used to generate the mesh, which contained 250370 nodes, corresponding to 83454 surface patches.

Two treatment locations positioned at an intercostal and transcostal space approximately 3 cm deep into the rib cage were selected. It was assumed that the HIFU array was positioned so that its geometric focus coincided with the treatment location. These arrangements are displayed in figures 3-5. To focus the beam into the rib cage, the phase conjugation method was used, which involved two stages.

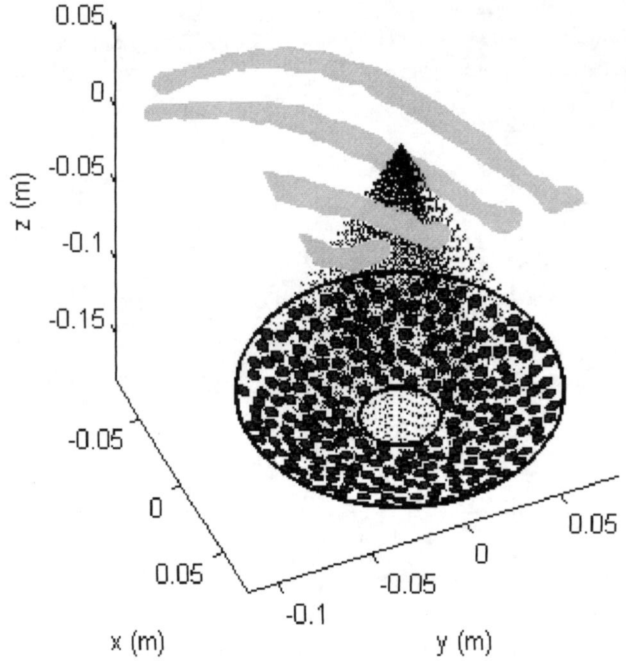

Figure 3. Intercostal treatment between ribs 10 and 11 on right side, approximately 3 cm deep into rib cage (view 1).

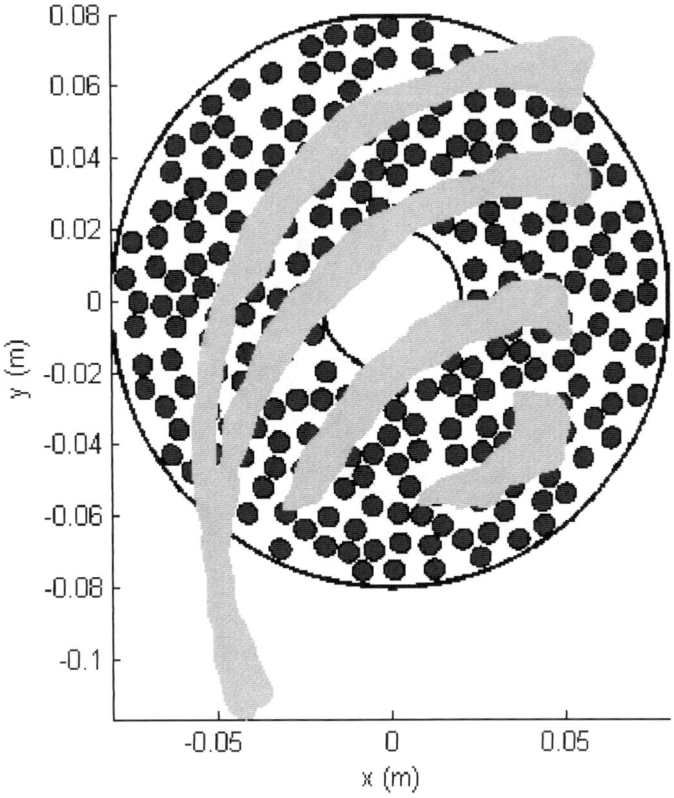

Figure 4. Intercostal treatment between ribs 10 and 11 on right side, approximately 3 cm deep into rib cage (view 2).

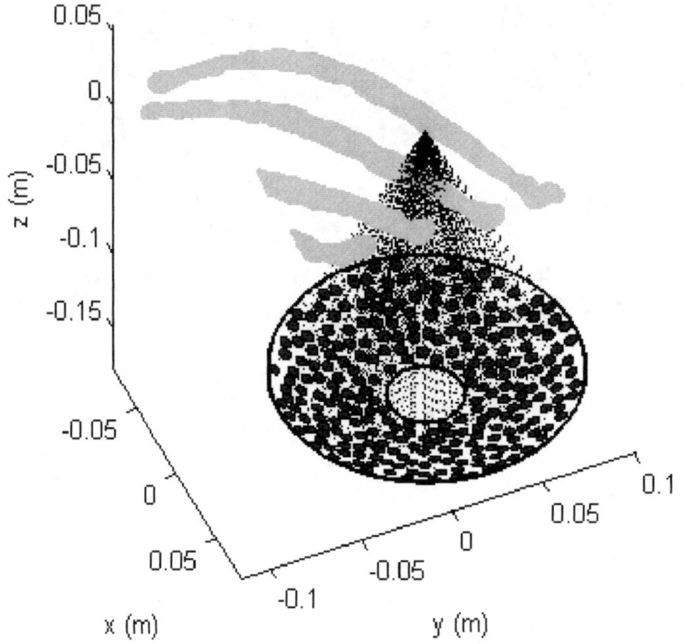

Figure 5. Transcostal treatment approximately 3 cm from rib 10 into rib cage, on right side.

In the first stage, a calculation was carried out in which the incident pressure field was that of a unit source strength point source positioned at the treatment location. The scattered pressure on the rib cage was then computed employing the GMRES implementation of the Burton-Miller formulation and the acoustic pressures at field locations corresponding to points on the surface of the HIFU phased array elements was calculated at a post-processing stage. The total pressure was then averaged over the surface of each element, and the complex conjugate of each average was calculated. These quantities were then normalised so as to produce 256 source velocities generating a peak acoustic pressure of 10 MPa at the geometric focus in absence of the scatterers.

In the second stage, the incident field was computed from the linear combination of the pressures resulting from the 256 source velocities obtained from the point source calculation. The GMRES implementation of the Burton-Miller BE formulation was then used to obtain the scattered pressure on the ribs. Pressure values within a 10 cm^3 cubic volume, centred around the treatment region field locations, were then obtained at the post-processing stage.

Figure 6 shows the total pressure field for the intercostal treatment in successive planes parallel to the *x-y* plane, of the phased array scattered by the ribs after applying the phase conjugation technique and scaling the source velocities so as to produce 10 MPa at the geometric focus in absence of the scatterers. Although there is a slight decrease in the peak pressure, the methodology appears to work well based these results. This particular configuration of ribs features intercostal spacing of around 2.5 cm and rib widths of around 1.5 cm, which allows the majority of the ultrasonic energy to be delivered in the vicinity of the geometric focus. Figure 7 shows the corresponding acoustic pressure distribution

on the surface of the ribs showing maximum pressures over 2.2 MPa, which could potentially be a cause for concern in terms of potential bone and tissue heating.
http://www.npl.co.uk/acoustics/ultrasound/figure-6

Figure 6. Total acoustic pressure field of 1 MHz random phased HIFU array for intercostal treatment location approximately 3 cm deep into rib cage between ribs 10 and 11 on right side. Source velocity distribution obtained using phase conjugation based on point source calculation in Fig.8 (colour bar in MPa).

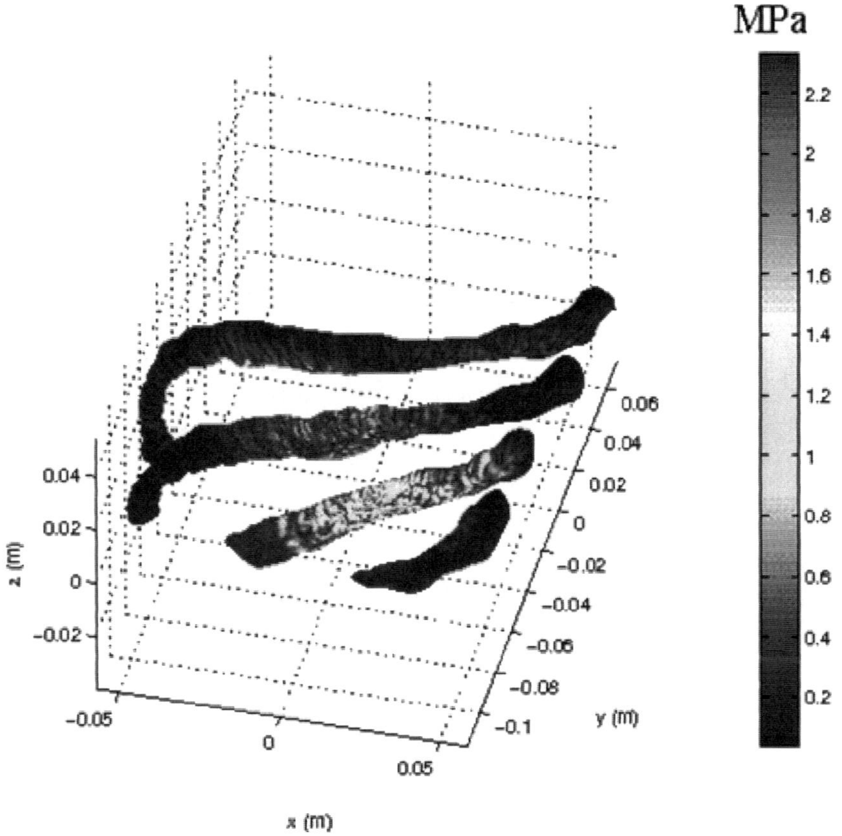

Figure 7. Total acoustic pressure on surface of ribs 9 to 12 on right side resulting from field of 1 MHz random phased HIFU array for intercostal treatment location between ribs 10 and 11 on right side approximately 3 cm deep into rib cage.

Figure 8 shows the total pressure field for the transcostal location in successive planes parallel to the *x-y* plane, of the phased array scattered by the ribs after applying the phase conjugation technique and scaling the source velocities so as to produce 10 MPa at the geometric focus in absence of the scatterers. Again, despite a slight reduction in the peak pressure at the focus, the method appears to work well, owing to the large intercostal spacing when compared with the rib cross-section. Figure 9 shows the corresponding acoustic pressure distribution on the surface of the ribs showing maximum pressures just over 1.8 MPa.

http://www.npl.co.uk/acoustics/ultrasound/figure-8

Figure 8. Total acoustic pressure field of 1 MHz random phased HIFU array for transcostal treatment location approximately 3 cm deep into rib cage behind rib 10 on right side. Source velocity distribution obtained using phase conjugation based on point source calculation in Fig.13 (colour bar in MPa).

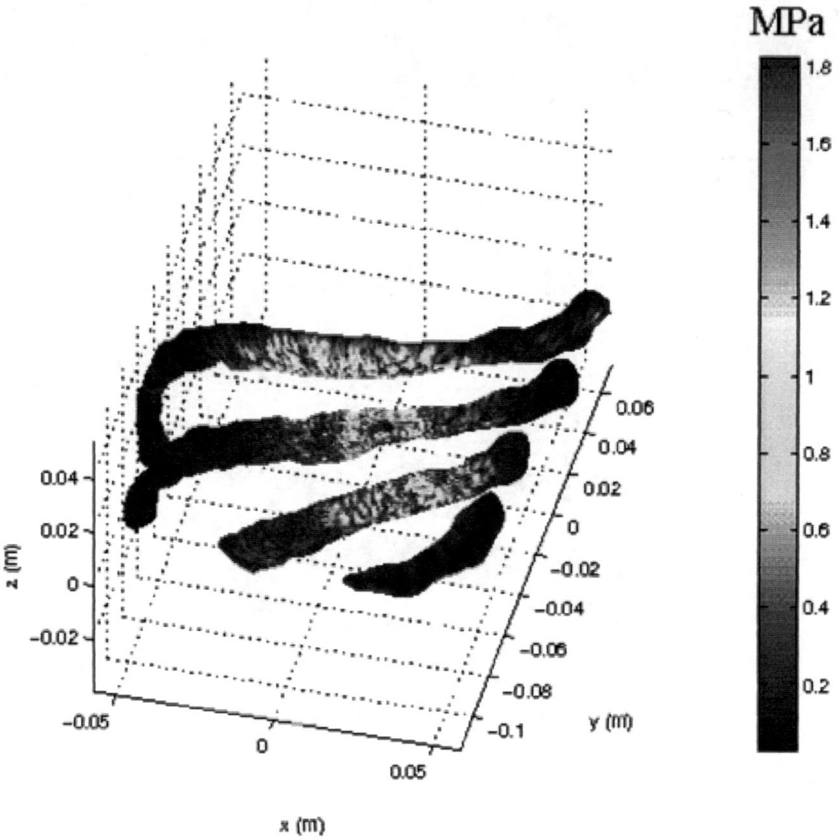

Figure 9. Total acoustic pressure on surface of ribs 9 to 12 on right side resulting from field of 1 MHz random phased HIFU array for transcostal treatment location approximately 3 cm deep into rib cage behind rib 10 on right side.

4. Conclusions

An approach based on a GMRES implementation of the Burton-Miller BE formulation was developed to model the scattering of the field of a multi-element random HIFU phased array by 3D objects of an arbitrary geometry under continuous wave excitation. The BE code was run on a dedicated computer cluster and the problem distributed over a number of cores. After validation on spherical scatterers, a mesh of quadratic pressure patches was generated using CT scan data for ribs nine to twelve on the right side of an adult male. Intercostal and transcostal treatments were simulated respectively between ribs 10 and 11, and behind rib 10, at locations approximately 3 cm deep into the rib cage. The phase conjugation method was used to focus the beam at the required treatment location. Acoustic pressure maps were obtained in 10 cm^3 cubic volumes centred around each treatment region. For the HIFU array and rib topology investigated, the method successfully focused the beam at the required locations, and no effects of splitting at the focus were observed. Nevertheless, maximum acoustic pressures predicted at the ribs were of the order of a quarter of those at the focus, which may need to be addressed from the point of view of treatment planning. To further validate the model, calculations are required on periodic structures consisting of rectangular cuboids, with rounded spherical edges, where it is hoped that effects of focus splitting consistent with those reported in [11].

Acknowledgements
The authors would like to acknowledge the support of the National Measurement Systems Directorate of the UK Department of Business, Innovation and Skills and EPSRC EP/F025750/1. The authors are grateful to PACSYS Ltd. for supplying stand-alone executables of the Program for Automatic Finite Element Calculations boundary element routines as well as for useful discussions. The authors would equally like to express their gratitude to Ian Rivens, Adam Shaw, Tristan Clark, Erik-Jan Rijkhorst, Simon Woodford, Grigory Vilenskiy, David Sinden and Svetlana Bobkova for useful discussions.

References
[1] Crum L, Bailey M, Hwang J H, Khokhlova V and Sapozhnikov 2010 O Therapeutic ultrasound: Recent trends and future perspectives *Physics Procedia* **3** 25
[2] Communication with Gail ter Haar, Therapeutic Ultrasound Group, Physics Department, Institute of Cancer Research, Sutton, UK. March 2011
[3] Wu F, Zhi-Biao W, Wen-Zhi C, Hui Z, Jin B, Jian-Zhong Z, Ke-Quan L, Cheng-Bing J, Fang-Lin X, Hai-Bing S 2004 Extracorporeal high intensity focused ultrasound ablation in the treatment of patients with large hepatocellular carcinoma *Ann. Surg. Oncol.* **11** 1061
[4] Li J-L, X-Zh Liu, Zhang D, Gong X-F 2007 Influence of ribs on nonlinear sound field of therapeutic ultrasound. *Ultrasound Med. Biol.* **33** 1413
[5] Gavrilov L R, Hand J W 2000 A theoretical assessment of the relative performance of spherical phased arrays for ultrasound surgery *IEEE Trans. Ultrason. Ferroelectr. Freq. Control* **47** 125
[6] Pernot M, Aubry J F, Tanter M, Thomas J L, Fink M 2003 High power transcranial beam steering for ultrasonic brain therapy *Phys. Med. Biol.* **48** 2577
[7] Tanter M, Pernot M, Aubry J F, Montaldo G, Marquet F, Fink M 2007 Compensating for bone interfaces and respiratory motion in high intensity focused ultrasound *Int. J. Hyperthermia* **23** 141.
[8] Aubry J-F, Pernot M, Marquet F, Tanter M, Fink M 2008 Transcostal high-intensity-focused ultrasound: *ex vivo* adaptive focusing feasibility study *Phys. Med. Biol.* **53** 2937
[9] Hand J W, Shaw A, Sadhoo N, Dickinson R J, Gavrilov L R 2009 Initial testing of a prototype phased array device for delivery of high intensity focused ultrasound (HIFU) *Phys. Med. Biol.* **54** 5675
[10] Botros Y Y, Ebbini E S, Volakis J L 1998 Two-step hybrid virtual array-ray (VAR) technique for focusing through the ribcage *IEEE Trans. Ultrason. Ferroelectr. Freq. Control* **45** 989
[11] Bobkova S, Gavrilov L, Khokhlova, Shaw A and Hand J. Focusing of high-intensity ultrasound through the rib cage using a therapeutic random phased array 2010 *Ultrasound Med. Biol.* **36** 888
[12] Cochard E, Prada C, Aubry J F, Fink M. Ultrasonic focusing through the ribs using the DORT method 2009 *Med. Phys.* **36** 3495
[13] Burton A J and Miller G F 1971 The application of integral equation methods to the numerical solution of some exterior boundary-value problems *Proc. R. Soc. London Ser.* **A323** 201
[14] Ochmann M, Homm A, Makarov S and Semenov S 2003 An iterative GMRES-based boundary element solver for acoustic scattering *Eng. Anal. Bound. Elem.* **27** 717
[15] Shen L and Liu Y J 2007 An adaptive fast multipole boundary element method for three-dimensional acoustic wave problems based on the Burton-Miller formulation *Comput. Mech.* **40** 461
[16] Liu Y and Rizzo F J 1992 A weakly singular form of the hypersingular boundary integral equation applied to 3-D acoustic wave problems *Comput. Method Appl. M.* **96** 271

[17] Zel'dovich B Y, Pilipetsky N F, and Shkunov V V 1985 Principles of Phase Conjugation. Berlin. CO: Springer-Verlag

[18] Chien C C, Rjiyah H and Atluri S N 1990 An effective methodfor solving the hypersingular integral equations in 3-D acoustics *J. Acoust. Soc. Amer.* **88** 343

[19] Visioli A G, Rivens I H, ter Haar G R, Horwich A, Huddart R A, Moskovic E, Padhani A, Glees J 1999 Preliminary results of a phase I dose escalation clinical trial using focused ultrasuond in the treatment of of localised tumours *Eur. J. Ultrasound* **9** 11

[20] Morse P M and Ingard K U, Theoretical Acoustics 1968 New York. CO: McGraw-Hill 419

[21] Kress R 1985 Minimizing the condition number of boundary integral operators in acoustic and electromagnetic scattering *Q. J. Mech. Appl. Math.* **38** 323

[22] Christ A, Kainz W, Hahn E G, Honegger K, Zefferer M, Neufeld E, Rascher W, Janka R, Bautz W, Chen J, Kiefer B, Schmitt P, Hollenbach H-P, Shen J, Oberle M, Szczerba D, Kam A, Guag J W and Kuster N 2010 The Virtual Family—development of surface-based anatomical models of two adults and two children for dosimetric simulations *Phys. Med. Biol.* **55** N23

Scattering of obliquely incident guided waves by a stiffener bonded to a plate

L Taupin[1], A Lhémery[1], V Baronian[1] and A-S Bonnet-BenDhia[2]
[1] CEA, LIST, F-91191 Gif-sur-Yvette, France
[2] POEMS, CNRS-INRIA-ENSTA UMR 2706, ENSTA ParisTech, 32 boulevard Victor, F-75015 Paris, France

E-mail: alain.lhemery@cea.fr

Abstract. Nondestructive testing of aerospace structures often requires their immobilization or disassembly. Structural health monitoring (SHM) can overcome these problems; the use of guided elastic waves (GW) in SHM is of great interest, because they propagate long distance in the structure thickness. Structures being stiffened, optimally positioning sensors implies to determine the number of stiffeners a wave can go through while remaining detectable. Here, the diffraction of GW by a stiffener bonded to a plate is considered. Elastic and geometric invariances along stiffener axis lead to 2D computations involving the three components of wave particle displacement, whatever its incidence on stiffener. A hybrid model is developed combining the semi-analytical finite element method for GW propagation and a finite element method (FE) for the stiffener diffraction. Optimal hybridization is obtained thanks to the development of transparent boundaries of the FE domain. Such boundaries have been obtained for GW normally incident onto a scattering feature, thanks to Fraser's bi-orthogonality relation, which unfortunately does not hold for oblique incidence. A numerical approach is developed to numerically approximate it, then, to derive boundary conditions in the wanted form. Their use minimizes the size of the FE domain and avoids any artificial reflection. They provide a mean for projecting diffracted fields on modes reflected on or transmitted through the stiffener; corresponding coefficients are obtained as functions of the direction of incidence.

1. Introduction

Structural Health Monitoring (SHM) by guided waves (GW) is studied as an alternative or complementary mean for the non-destructive examination (NDE) of aeronautical structures, since classical NDE methods often require partial immobilization and disassembly [1]. Generally speaking, SHM allows both permanent monitoring and scheduled tests to be achieved. Many aircrafts parts are thin plate-like structures made of composite materials or Aluminium alloy, which are stiffened for reinforcement. SHM by GW is particularly promising since GW propagate long distance in plates while probing their whole thickness. A stiffener or a set of stiffeners of course scatter GW.

Present work aims at quantitatively predicting scattering phenomena arising when a GW is incident on a stiffener. This question arises when designing a SHM method: optimally positioning sensors on the structure requires the knowledge of reflection upon and transmission through a stiffener of GW. Once corresponding coefficients are known for the various propagating modes, it becomes easier to optimally position sensors needed to achieve full coverage in the examination of the structure with sufficient signal amplitude for further processing and interpretation.

This works is part of a more general attempt to develop simulation tools capable of addressing typical SHM configurations for optimizing them at a lower cost and more efficiently than with the sole experimentation. In this work, plate-like structures can be made either of homogeneous and isotropic material (Aluminium alloy) or of multiple layers of viscoelastic anisotropic material (composite). A study of GW propagation in these latter structures with application to SHM was described in [2].

It is also part of a general development at CEA of a tool for simulating NDT methods based on GW [3] to be included in the CIVA-platform. Inclusion in this industrial software implies that tools must be generic and compatible with intensive use. The overall approach consists in dealing with the whole testing configuration as a series of connected sub-problems: GW radiation and reception by transducers, propagation in uniform portions where waves propagate as GW in an unbounded guide, scattering by whatever scatters (a defect, a junction, a stringer, etc.). It fundamentally relies on a description of GW as modal series in uniform portions; then, complex phenomena involving local non uniformity of the configuration as far as GW are concerned are interpreted as factors affecting modal amplitudes and phases. Scattering phenomena are treated by various approaches [3], a most promising one consisting in developing a hybrid model linking modal descriptions in uniform guides with a local numerical method (*e.g.*, FE) to deal with non-uniformities. Note that complex scattering behaviour of GW does not generally admit accurate and fast asymptotic solutions since typical scatterer dimensions are of the same order as guide dimension and as the wavelength. A crucial ingredient in the development of the hybrid model was the obtaining of exact transparent boundary conditions adapted to the modal description of waves in uniform guides [3, 4]. This allows restricting the FE computation zone to a minimum.

Typical stiffener geometry and elastic properties are however invariant along one direction; as it will be shown, one can benefit of this symmetry for deriving an appropriate hybrid modelling approach minimizing computation effort. Therefore, the problem can be solved in a 2D computation. The existing hybrid method is developed for GW propagating along a single guiding direction. In the case considered herein, an arbitrary direction of propagation must be taken into account to predict wave behaviour between several sensors or actuators arbitrarily positioned on the stiffened plate.

The paper is therefore organized as follows. At first, the theoretical derivation of a hybrid modelling approach to solve the problem in hands is described; this section includes a brief reminder of the originally proposed hybrid approach restricted to normal incidence. In the course of the derivation, a modified Semi-Analytical Finite Element method is proposed to compute in one run all possible reflected and transmitted modes than can be generated by the scattering process. Then, simple numerical examples are given to check and proof the accuracy of the numerical scheme. Lastly, a study using it is carried out to evaluate how scattering by a realistic stiffener varies with incident direction depending on the incident mode considered. Implications to practical design of SHM configurations are briefly tackled.

The subject of our paper was recently considered independently in [5]; the aims of both studies are identical and the methods themselves are quite similar. The difference is the way transparent conditions are obtained for hybridizing modal series and FE scheme: in [5] this relies on the use of absorbing boundary conditions. Advantage and drawbacks of both methods will be briefly discussed.

2. Hybrid SAFE/FE method for a GW at oblique incidence

2.1. Governing equations for the symmetry of the configuration in hand

The structure considered is a plate with a straight infinitely long scattering feature (called "the scatterer" in what follows) along the *y*-axis (see figure 1). The whole 3D structure is entirely described by its 2D trace in the (*x-z*) plane. It is assumed that the same symmetry holds for elastic properties.

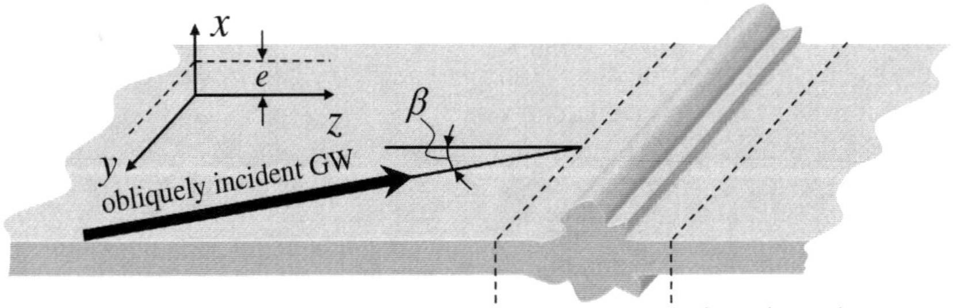

Figure 1. Schematic view of the configuration of interest.

The field radiated by an arbitrary time-dependent source in a plate can always be decomposed as a sum of monochromatic guided waves propagating in all the directions in the plate, by means of adapted space and time Fourier transforms. Let us consider a source radiating in the stiffened plate under study. Thanks to the same decomposition, predicting the field radiated in the presence of the stiffener can be made by solving the problem of a monochromatic GW making an arbitrary angle of incidence relatively to the straight scatterer, followed by the required inverse Fourier transforms to get the space and time-dependent solution. Our aim is therefore to find the elementary solution for each possible incident mode existing at a given frequency parameterized by the angle of incidence β its wavevector makes with the normal to the scatterer. Thanks to the symmetry of the problem in hands, modes reflected on or transmitted through the scatterer resulting from the interaction of a given incident mode, have all in common that the y-component of their wavevector $\mathbf{k}$ is equal to that of the incident mode considered. The conservation law at the basis of this result is nothing but an expression of the Snell-Descartes' law. Let us call this component γ defined as

$$\gamma = -k \sin \beta . \tag{1}$$

For the symmetry considered and by taking into account this conservation law for wavevectors, the dependence of any elastodynamic quantity upon the y-coordinate can be factorized as

$$\mathbf{u}(x, y, z) = \mathbf{u}^{\gamma}(x, z) e^{-j\gamma y} . \tag{2}$$

Operators can be defined by accounting for the symmetry: for instance, the divergence of a vector writes

$$div(\mathbf{u}) = \frac{\partial u_x}{\partial x} + \frac{\partial u_y}{\partial y} + \frac{\partial u_z}{\partial z} = \left(\frac{\partial u_x^{\gamma}}{\partial x} - i\gamma u_y^{\gamma} + \frac{\partial u_z^{\gamma}}{\partial z} \right) e^{-i\gamma y} = div_{\gamma}(\mathbf{u}^{\gamma}) e^{-i\gamma y} . \tag{3}$$

Eventually, the general 3D problem to be solved can be mathematically dealt with a sum over γ of solutions of simpler 2D problems in the (x, z) plane (vector variables having three components still) at fixed γ. The γ-dependent equation of wave motion in the volume V of the part considered, of free boundary ∂V is given by:

$$\begin{cases} div_{\gamma}(\boldsymbol{\sigma}^{\gamma}(\mathbf{u}^{\gamma})) = -\omega^2 \rho \mathbf{u}^{\gamma} & \text{in } V \\ \boldsymbol{\sigma}^{\gamma}(\mathbf{u}^{\gamma}).\mathbf{n} = 0 & \text{on } \partial V \end{cases} \tag{4}$$

where ω denotes the pulsation and ρ the mass density.

2.2. *SAFE method at fixed γ*

The Semi-Analytical Finite Element (SAFE) method is becoming a standard to efficiently compute guided modes in a variety of guide geometries made of arbitrary material. Classically, the SAFE method allows computing, at a given frequency, a series of guided modes (propagative, inhomogeneous and evanescent) assuming one given guiding direction. The FE computation is limited to a solution of the wave equation in a plane perpendicular to the guiding direction considered, the dependency upon this direction of the solution being handled analytically. The SAFE formulation leads to the prediction of wavevectors and associated particle displacements as being the eigenvalues and eigenvectors of the SAFE system of equations.

Here, we can derive a specific SAFE formulation to compute all the modes having the same γ. The unknown eigenvalues are no longer the wave numbers, k, but their projection k_z on the z-axis. Classical matrices of SAFE formulation (recalled in [2]) are slightly modified; the system now becomes

$$(\mathbf{A} - k_z\mathbf{D})\mathbf{Q} = 0 \tag{5}$$

with,

$$\mathbf{Q} = \begin{bmatrix} \mathbf{d} \\ k_z\mathbf{d} \end{bmatrix}, \tag{6}$$

$$\mathbf{A} = \begin{bmatrix} 0 & \gamma^2\mathbf{K}_{22} - j\gamma\mathbf{K}_{12} + \mathbf{K}_{11} - \omega^2\mathbf{M} \\ \gamma^2\mathbf{K}_{22} - j\gamma\mathbf{K}_{12} + \mathbf{K}_{11} - \omega^2\mathbf{M} & (-j\mathbf{K}_{13} + \gamma\mathbf{K}_{23}) \end{bmatrix}, \tag{7}$$

$$\mathbf{D} = \begin{bmatrix} \gamma^2\mathbf{K}_{22} - j\gamma\mathbf{K}_{12} + \mathbf{K}_{11} - \omega^2\mathbf{M} & 0 \\ 0 & -\mathbf{K}_{33} \end{bmatrix}. \tag{8}$$

ω denotes the angular frequency and **d** the nodal displacement, $\mathbf{K}_{mn}$ are the stiffness matrices (in the FE sense of the word) and **M** the mass matrix. Stiffness and mass matrices are defined in [2].

An example of solution is shown on figure 2 where an incident mode gives rise to two reflected modes. The figure actually displays the slowness curves of wave modes existing below the first cut-off frequency in a plate made of unidirectional (transversely isotropic) composite material (see [2] for the stiffness constants used). The symmetry axis of the composite makes an angle of $\pi/4$ with the y-direction. The value of γ of the incident wave mode is such that *i)* there is no mode conversion in the reflection into the fastest mode (in red) *ii)* but the two other modes contribute to the reflected field.

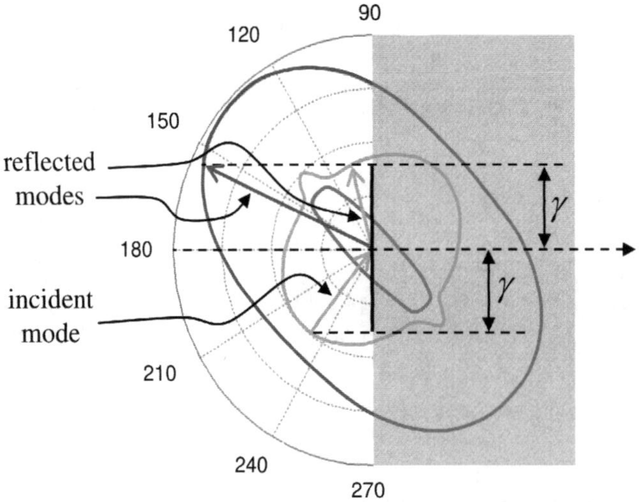

Figure 2. Slowness curves of guided modes in a unidirectional composite plate illustrating Snell's law as it applies to reflected modes computed for a given γ in the modified SAFE-γ method.

2.3. Transparent boundaries: short review and brief recall on previous work for normal incidence

In a hybrid model such as the one we want to derive, a boundary is defined between two physical portions of the volume, where two different methods apply. At this boundary, the two methods exchange data that represent physical quantities of interest for the overall computation. Suppose now this boundary is at a position in the volume where the physical quantities studied are continuous; in the present case, consider an artificial boundary arbitrarily positioned in the propagation medium where particle displacement and stress associated with the propagating waves are continuous. The way the two methods exchange data must ensure that no artificial reflection takes place at it.

In the literature, transparent boundaries have been developed for elastodynamics. Methods based on Perfectly Matched Layers (PML) originally developed for electromagnetic waves [6] constitute a class of computationally efficient methods available for various equations. However, in the case of elastic guided waves, the method may be unstable in the presence of inverse modes, but problems occur since the sign of the normal component of group velocity and phase velocity are different [7, 8]. A classical way of obtaining locally transparent boundaries for wave equations consists in defining absorbing zones [5, 9] to artificially decrease unwanted contributions. The size of the absorbing zone that must be meshed by finite elements is proportional to the wavelength in the direction of absorption, this leading to costly computation at typical frequencies involved in GW testing configurations. In acoustic problems, the Dirichlet-to-Neumann (DtN) method consists in a boundary condition where the normal derivative of the pressure is expressed as a (non-local) function of the pressure on the same boundary. The natural extension to elasticity would be a boundary condition with a normal stress expressed as a function of the displacement. This is unfortunately not tractable. Instead, hybrid variables **X** and **Y** (mixing components of both the displacement and the normal stress) have been used by Baronian *et al.* [4] to develop the YtX method adapted to guided modes. This last method will be extended in this study to the case of an obliquely incident GW on a scatterer invariant in one direction. We briefly recall the underlying principle of this recently developed method before deriving what is needed for its extension.

Two mixed variables **X** and **Y** are first introduced, defined as

$$\mathbf{X}=\begin{pmatrix}\mathbf{t}_S\\ u_z\end{pmatrix}=\begin{pmatrix}\sigma_{xz}\\ \sigma_{yz}\\ u_z\end{pmatrix},\quad \mathbf{Y}=\begin{pmatrix}\mathbf{u}_S\\ t_z\end{pmatrix}=\begin{pmatrix}u_x\\ u_y\\ -\sigma_{zz}\end{pmatrix}, \tag{9}$$

As GW displacement and stress can be expressed as modal series, both variables can similarly be decomposed as:

$$\begin{pmatrix}\mathbf{X}\\ \mathbf{Y}\end{pmatrix}=\sum_n A_n^+(z)\begin{pmatrix}\mathcal{X}_n(x)\\ \mathcal{Y}_n(x)\end{pmatrix}. \tag{10}$$

Let $(\cdot|\cdot)_S$ denotes the following product

$$\left(\mathbf{Y}\,|\,\mathbf{X}\right)_S=\int_S(\mathbf{t}_S\cdot\mathbf{u}_S+t_z u_z)\,dS\,, \tag{11}$$

where S is the section of the guide. Fraser's relation [10] expresses the bi-othogonality of two modes m and n through the bi-linear relation given by equation (11) as

$$\left(\mathcal{Y}_n\,|\,\mathcal{X}_m\right)_S=\delta_{nm}\,. \tag{12}$$

Let us now consider the following problem depicted on figure 3. We want to define a transparent boundary between a homogeneous guide for which a modal decomposition is known, called the exterior domain, and a scattering feature, called the interior domain. The transparent boundary condition to be built must relate the two variables **X** and **Y** in both the two domains while ensuring continuity of these fields across it.

Journal of Physics: Conference Series **353** (2012) 012011 doi:10.1088/1742-6596/353/1/012011

The so-called YtX operator must be obtained to relate $\mathbf{X}^{\text{int}}$ to $\mathbf{Y}^{\text{int}}$ of the internal field at the boundary Σ between the scatterer and the guide. It is built in three steps, as shown on figure 3.

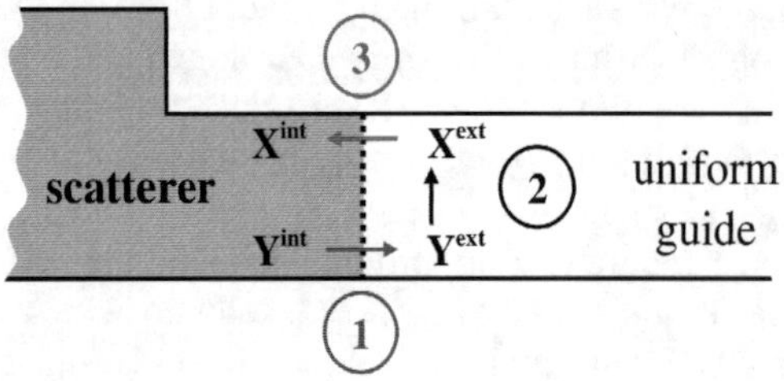

Figure 3. The artificial boundary between the uniform guide (external medium) and the scatterer (internal medium) where transparency is wanted, is obtained in three steps.

In steps 1 and 3, continuity relations for **X** and **Y** fields across the boundary separating the interior domain and the exterior domain are used. These relations are:

$$\begin{cases} \mathbf{Y}^{\text{ext}} = \mathbf{Y}^{\text{int}} & \text{on } \Sigma, \\ \mathbf{X}^{\text{int}} = \mathbf{X}^{\text{ext}} & \text{on } \Sigma. \end{cases} \tag{14}$$

In step 2, using Fraser's relation, it is possible to decompose the exterior field onto the modal solution. The amplitude of the field for the various modes is simply obtained as

$$A_n^+ = \left(\mathbf{Y}_{\Sigma^+}^{\text{ext}} | \mathcal{X}_n\right). \tag{15}$$

Noting that in equation (10), the same coefficient A_n^+ appears in the modal decomposition for the two mixed variables **X** and **Y** and using equation (15) for the projection of the external field on modes, a relation between $\mathbf{X}^{\text{ext}}$ and $\mathbf{Y}^{\text{ext}}$ is readily obtained, given by

$$\mathbf{X}_{\Sigma}^{\text{ext}} = \sum_{n\in\mathbb{N}} \left(\mathbf{Y}_{\Sigma^+}^{\text{ext}} | \mathcal{X}_n\right)_S \mathcal{X}_n. \tag{16}$$

By combining equation (16) with the continuity relation in equations (14), the YtX operator denoted by **T** coupling $\mathbf{X}^{\text{int}}$ and $\mathbf{Y}^{\text{int}}$ can be expressed as

$$\mathbf{X}_{\Sigma^+}^{\text{int}} = \sum_{n\in\mathbb{N}} \left(\mathbf{Y}_{\Sigma^+}^{\text{int}} | \mathcal{X}_n\right)_S \mathcal{X}_n = \mathbf{T}(\mathbf{Y}_{\Sigma^+}^{\text{int}}), \tag{17}$$

with

$$\mathbf{T} = \begin{pmatrix} \sum\limits_{n\in\mathbb{N}} \left(* | \mathcal{T}_S^n\right)_S \mathcal{T}_S^n & \sum\limits_{n\in\mathbb{N}} \left(* | \mathcal{U}_z^n\right)_S \mathcal{T}_S^n \\ \sum\limits_{n\in\mathbb{N}} \left(* | \mathcal{T}_S^m\right)_S \mathcal{U}_z^n & \sum\limits_{n\in\mathbb{N}} \left(* | \mathcal{U}_z^m\right)_S \mathcal{U}_z^n \end{pmatrix}, \tag{18}$$

where $\mathcal{U}_z^n$ denotes the axial displacement of mode n, and $\mathcal{T}_S^n$ denotes the following vector:

$$\mathcal{T}_S^n = \begin{pmatrix} \sigma_{xz}^n \\ \sigma_{yz}^n \end{pmatrix}. \tag{19}$$

2.4. Transparent boundaries for an obliquely incident wave

This sub-section describes how to derive transparent boundary conditions by generalizing the YtX method to deal with obliquely incident guided waves. The formulae given hereafter concern the two boundaries $\Sigma^{\pm}$ between the scatterer and the portions of the uniform guide, as shown on figure 4. The appropriate field quantities are again the mixed variables which now depend on γ to account for the symmetry described in the first section. These variables are defined as follows:

$$\mathbf{X}^{\gamma} = \begin{pmatrix} \mathbf{t}_S^{\gamma} \\ u_z^{\gamma} \end{pmatrix} = \begin{pmatrix} \sigma_{xz}^{\gamma} \\ \sigma_{yz}^{\gamma} \\ u_z^{\gamma} \end{pmatrix} \text{ and } \mathbf{Y}^{\gamma} = \begin{pmatrix} \mathbf{u}_S^{\gamma} \\ t_z^{\gamma} \end{pmatrix} = \begin{pmatrix} u_x^{\gamma} \\ u_y^{\gamma} \\ -\sigma_{zz}^{\gamma} \end{pmatrix}. \tag{20}$$

where the stress and the displacement are calculated by the modified SAFE method at fixed γ, and notations with γ are defined in relations (2) et (3).

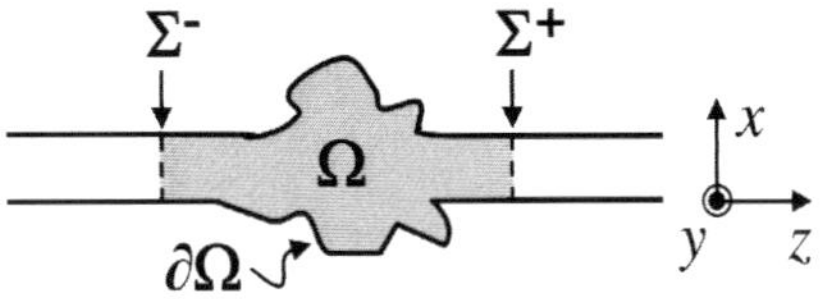

Figure 4. FE domain for diffraction problem.

The originally developed YtX method relied on Fraser's relation. For obliquely incidence (as for anisotropic materials), the Fraser's relation is no longer verified so that modes cannot be projected straightforwardly on each others. To overcome this difficulty, a new way of projecting modes must be developed. At first, a Gramm matrix $\mathcal{O}$ is defined, whose components are given by:

$$\mathcal{O}_{nm} = \left(\mathcal{Y}_n^{\gamma} | \mathcal{X}_m^{\gamma}\right)_S \tag{21}$$

where $(\cdot|\cdot)_S$ denotes the same bilinear relation as that defined by equation (11) for γ dependent variables. For normal incidence, Fraser's relation implies that $\mathcal{O}$ is a diagonal matrix, defined in equation (12). But for oblique incidence, the matrix is no longer diagonal.

The YtX operator, to be defined at fixed γ, is derived similarly in three steps. Since the overall procedure to get the new transparent conditions is basically the same as that already shown in the case of normal incidence, we restrict the description of the procedure for oblique incidence to what differs in the two cases. Moreover, as this was done in the previous subsection, the formulae are detailed for the case of the right boundary Σ^+ where waves leave the interior domain, assuming that the incident waves arise from the left. Boundary conditions on Σ^- will be deduced from their expression on Σ^+ at the end of this subsection.

The external field $\mathbf{Y}^{\gamma^{\text{ext}}}$ is projected on a mode m of the modal series $\mathcal{X}_m^{\gamma}$ using the matrix $\mathcal{O}$, defined by equation (21). One gets

$$\left(\mathbf{Y}_{\Sigma^+}^{\gamma^{\text{ext}}} | \mathcal{X}_m^{\gamma}\right)_S = \sum_{n\in\mathbb{N}} A_n^+ \left(\mathcal{Y}_n^{\gamma} | \mathcal{X}_m^{\gamma}\right)_S = \sum_{n\in\mathbb{N}} \left(\mathcal{O}^T\right)_{mn} A_n^+ . \tag{22}$$

The inversion of the system given by equation (22) leads to a new expression of the modal amplitude of the field on Σ^+, written as

$$A_n^+ = \sum_{m\in\mathbb{N}} \left[\left(\mathcal{O}^T\right)^{-1}\right]_{nm} \left(\mathbf{Y}_{\Sigma^+}^{\gamma^{\text{ext}}} | \mathcal{X}_m^{\gamma}\right)_S . \tag{23}$$

The Gramm matrix being neither unitary nor diagonal, its numerical inversion requires a truncation of the modal series.

Now, the modal decomposition of the external field $\mathbf{X}^{\gamma^{\text{ext}}}$ can be rewritten as follows:

$$
\begin{aligned}
\mathbf{X}_{\Sigma^+}^{\gamma^{ext}} &= \sum_{n\in\mathbb{N}} A_n^+ \mathcal{X}_n^\gamma \\
&= \sum_{n\in\mathbb{N}} \mathcal{X}_n^\gamma \sum_{m\in\mathbb{N}} \left[\left(\mathcal{O}^T\right)^{-1}\right]_{nm} \left(\mathbf{Y}_{\Sigma^+}^{\gamma^{ext}} | \mathcal{X}_m^\gamma\right)_S \\
&= \sum_{n\in\mathbb{N}} \sum_{m\in\mathbb{N}} \mathcal{X}_n^\gamma \left[\left(\mathcal{O}^T\right)^{-1}\right]_{nm} \left(\mathbf{Y}_{\Sigma^+}^{\gamma^{ext}} | \mathcal{X}_m^\gamma\right)_S .
\end{aligned} \tag{24}
$$

Finally, combining equation (24) with continuity relations for the interior and exterior fields at the boundary, the YtX operator $\mathbf{T}^{\gamma+}$ for obliquely incident guided waves on Σ^+ is expressed as.

$$
\mathbf{X}_{\Sigma^+}^{\gamma^{int}} = \mathbf{T}^{\gamma+}(\mathbf{Y}_{\Sigma^+}^{\gamma^{int}}), \tag{25}
$$

$$
\mathbf{T}^{\gamma+} = \begin{pmatrix}
\sum_{n\in\mathbb{N}}\sum_{m\in\mathbb{N}} \mathcal{T}_{S,n}^\gamma \left[\left(\mathcal{O}^T\right)^{-1}\right]_{nm} \left(\mathcal{T}_{S,m}^\gamma | *\right)_S & \sum_{n\in\mathbb{N}}\sum_{m\in\mathbb{N}} \mathcal{T}_{S,n}^\gamma \left[\left(\mathcal{O}^T\right)^{-1}\right]_{nm} \left(\mathcal{U}_{z,m}^\gamma | *\right)_S \\
\sum_{n\in\mathbb{N}}\sum_{m\in\mathbb{N}} \mathcal{U}_{z,n}^\gamma \left[\left(\mathcal{O}^T\right)^{-1}\right]_{nm} \left(\mathcal{T}_{S,n}^\gamma | *\right)_S & \sum_{n\in\mathbb{N}}\sum_{m\in\mathbb{N}} \mathcal{U}_{z,n}^\gamma \left[\left(\mathcal{O}^T\right)^{-1}\right]_{nm} \left(\mathcal{U}_{z,m}^\gamma | *\right)_S
\end{pmatrix}, \tag{26}
$$

where $\mathcal{U}_{z,m}^\gamma$ denotes the axial displacement for the mode m, and $\mathcal{T}_{S,m}^\gamma$ denotes the following vector:

$$
\mathcal{T}_{S,m}^\gamma = \begin{pmatrix} \sigma_{xz,m}^\gamma \\ \sigma_{yz,m}^\gamma \end{pmatrix}. \tag{27}
$$

For the diffraction problem, YtX operator is used to model the outgoing conditions on the boundaries of the FE domain. The construction of the YtX operator $\mathbf{T}^{\gamma-}$ on the left boundary Σ^- is similar and involves the modal decomposition of the field on the left-going modes. For example, if the plate is isotropic, one has $\mathbf{X}_{-n} = -\mathbf{X}_n$ and $\mathbf{Y}_{-n} = \mathbf{Y}_n$ so that $\mathbf{T}^{\gamma-} = -\mathbf{T}^{\gamma+}$, right-going modes being again used in its expression.

We can now write the variational formulation for the FE computation at fixed γ of the diffraction by a straight scatterer of volume Ω, with $\Sigma^{+/-}$ transparent boundary conditions and with $\partial\Omega - \Sigma^{+/-}$ modelled as free boundaries. This formulation is given by the following system of equations

$$
\begin{cases}
div_\gamma(\boldsymbol{\sigma}^\gamma(\mathbf{u})) = -\omega^2\rho\mathbf{u}^\gamma & \text{in } \Omega, \\
\boldsymbol{\sigma}^\gamma(\mathbf{u}^\gamma).\mathbf{n} = 0 & \text{on } \partial\Omega - \Sigma^{+/-}, \\
\mathbf{X}_{\Sigma^{+/-}}^{\gamma^{int}} = \mathbf{T}^{\gamma+/-}(\mathbf{Y}_{\Sigma^{+/-}}^{\gamma^{int}}) & \text{on } \Sigma^{+/-}.
\end{cases} \tag{28}
$$

3. Validations of the hybrid method

The hybrid method has been implemented in the Melina code (FE code developed at IRMAR and POEMS [11]). In this section, elementary tests are computed to check the accuracy of the proposed hybrid method. A first validation test consists in considering a scattering zone to be computed by FE which has the same geometric and elastic characteristics as the guiding plate; expected results must show that the propagation computed by FE in this zone remains identical to that in the plate as computed by the SAFE-γ (no mode-conversion, no reflection, no energy lost). The second test consists in checking energy conservation in the presence of a non-trivial scattering feature.

The various examples of simulated results given in this section and in the next one have been computed by considering 50 modes (propagative, evanescent and inhomogeneous) in the series taken into account for obtaining transparent boundaries as it was explained en section 2. It must be noted that, presently, the influence of this number on accuracy has not been fully studied. The number considered, as it will be shown, leads to accurate transparent conditions, but is probably higher than really needed. Optimizing this number could be studied for minimizing computation times.

3.1. "Scattering" by a zone geometrically and elastically identical to the plate

This first numerical experiment is useful for proving the efficiency and quality of transparent boundaries: the FE computation must leave unperturbed the incident wave so that one expects both null-reflection and total transmission without mode-conversion. A 5-mm-thick plate made of Aluminium alloy is considered and the FE zone is 10-mm-long, as shown by Figure 5.

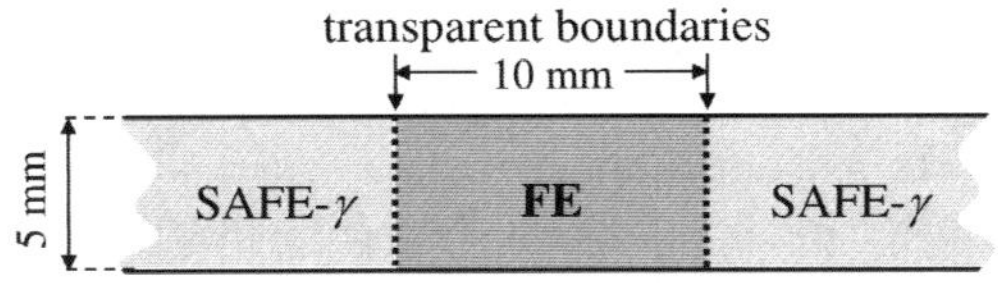

Figure 5. FE computation for an unperturbed zone (identical to the plate) for first validations.

For an excitation frequency of 500 kHz, there are five propagative modes in the unperturbed plate: three Lamb modes A0, S0 and A1 and two shear horizontal modes SH0 and SH1. Each mode is successively taken to be an incident wave for various values of γ. For each mode, a quadratic error for the total displacement field at the input boundary Σ^- and the output boundary Σ^+ is computed between the FE solution that includes transparent boundaries and the modified-SAFE solution (identical at the two boundaries). All the values of quadratic errors are less that 0.03% at both boundaries.

Further tests not discussed herein demonstrated that transparency is accurately obtained even for incident angles of 89.5°. As it may be anticipated, the 90° case is singular but it has no actual physical meaning as far as the SHM applications are concerned.

3.2. Energy conservation in the presence of a scatterer

Let us consider now the case of a stiffened isotropic plate. The geometry considered is shown on Figure 6. The plate in Aluminium alloy is 1.6-mm-thick.

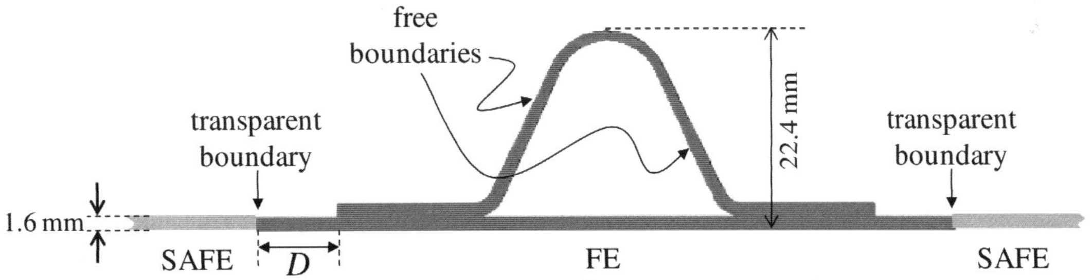

Figure 6. Configuration for stiffened plate computation.

In the present case, reflection and transmission phenomena are not trivial. The test consists in checking that the numerical method ensures energy conservation. For this, the sum of squared modulus of reflection and transmission coefficients (each coefficient being normalized by the square root of the ratio between its power flow and that of the incident mode, measured in the direction normal to the straight feature) shall equal 1, according to

$$E = E_{reflected} + E_{transmitted} = \sum_i \left|R_{ij}\right|^2 + \left|T_{ij}\right|^2 = 1,$$

where R_{ij} (respectively T_{ij}) denotes the reflection (respectively, transmission) coefficient of the i-th mode for the j-th incident mode.

In this example, we also check the influence of the distance of the transparent boundaries to the stiffener. Boundaries can be considered as being transparent if transmission and reflection coefficients do not depend on that distance. Distances *D* considered are 7.5 (I), 10 (II) and 20 (III) mm. Previous studies in the case of normal incidence shown that the boundary position can be all the closer to the scatterer while ensuring transparency since the number of modes accounted for in the series is high.

Whatever the distance *D* considered, energy conservation is obtained with a total energy equal to the incident energy within an error less than 0.01 %. This was checked for all the possible incident modes and for various incident angles.

4. Example of parametric study for a realistic stiffened plate

The same stiffened plate with $D = 10$ mm is now used in several parametric studies given to exemplify the interest of the method for practical applications.

At first, transmission and reflection coefficients for a S0 incident mode are shown on figure 7 at a frequency of 300 kHz. The behaviour of these coefficients is rather complex to describe; globally, transmission tends to decrease (and reflection to increase) for increasing incident angle. At about 40°, both the reflection and transmission coefficients of the S0 incident mode are of low value whereas other coefficients corresponding to mode conversion are higher.

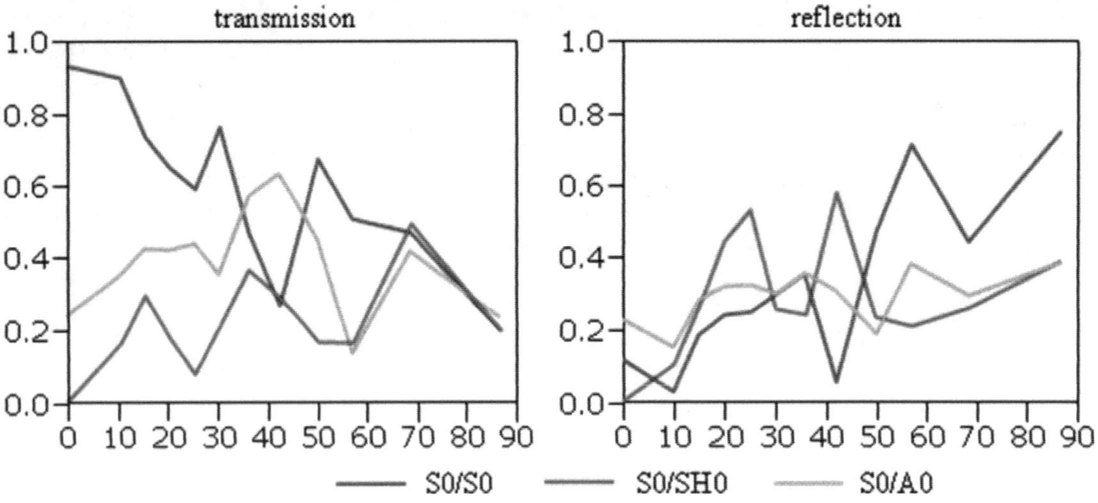

Figure 7. Transmission (left) and reflection (right) coefficients for a S0 incident mode at 300 kHz *vs*. incident angle (°).

Now, at 500 kHz, still below the first cut-off frequency so that there are only three propagative modes A0, S0 and SH0, we compare transmission coefficients for two different incident modes, S0 and A0, as shown by figure 8.

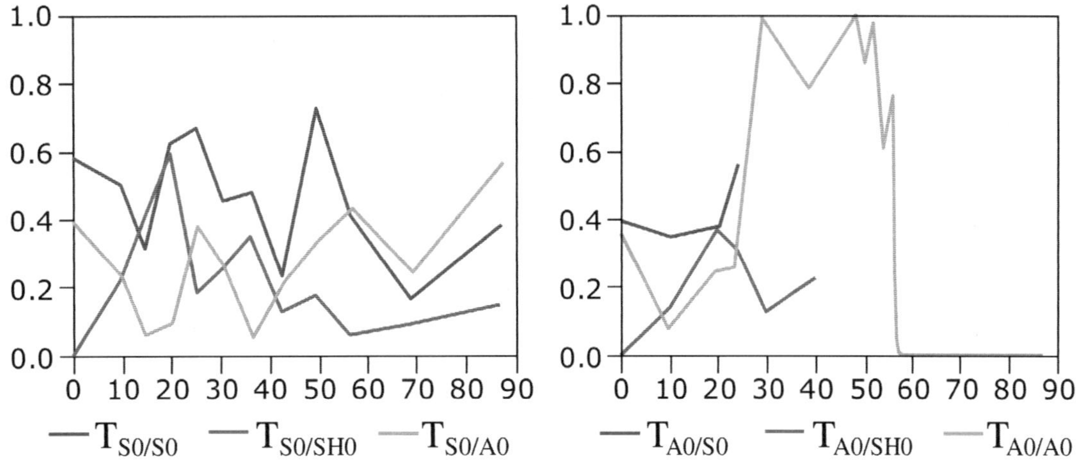

Figure 8. Transmission coefficients at 500 kHz *vs.* incident angle. Left: S0 incident mode; right: A0 incident mode.

The behaviour of the various transmission coefficients very much differs depending on the incident mode. As it may have been anticipated, in the case of an A0 incident mode, two critical angles are observed, since the A0 mode is the slowest of the three modes. Interestingly, it is also observed that for incident angles higher than 59°, even A0 is not transmitted anymore through the stiffener.

We can also compare figures 7 and 8 to study the influence of the frequency on the transmission coefficients for a S0 incident mode. The global behaviour tends to become more complex as the frequency increases.

At last, to illustrate total transmission and total reflection phenomena observed in the case of a A0 incident mode depending on the incident angle considered, the following figure display the various components of the total field inside the FE zone for two different incident angles. For a 50° of incidence, total transmission is observed while for a 70° incident angle, total reflection occurs.

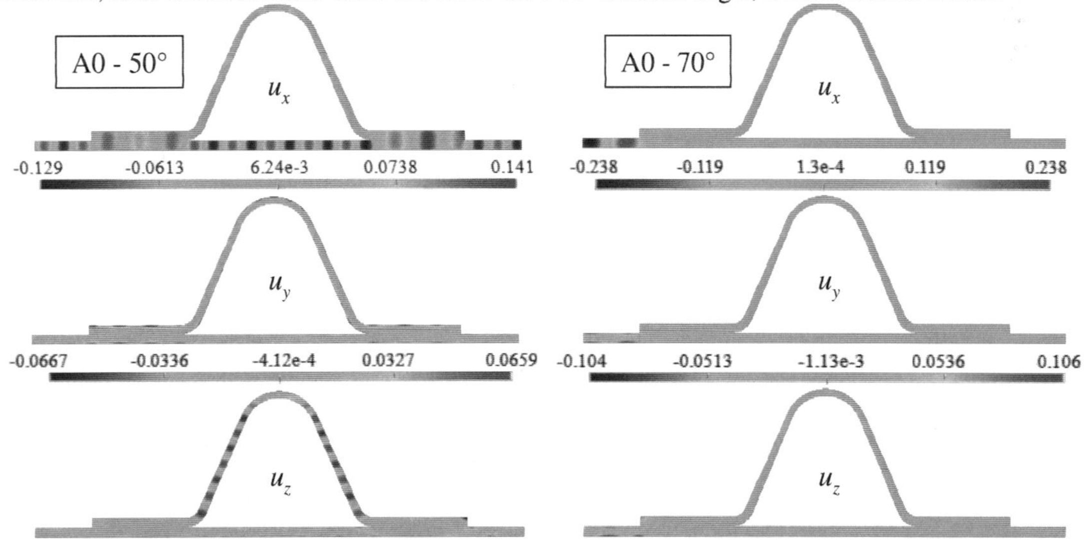

Figure 9. Components of the displacement for A0-incident mode at 500 kHz. Left: total transmission at 50° of incidence; right: total reflection at 70° of incidence.

These first results of scattering coefficients for obliquely incident guided waves demonstrate how complex are the scattering phenomena that take place in a stiffened plate and how these phenomena are highly dependent on the angle of incidence, on the working frequency and on the type of incident mode considered. To quantitatively address such a complexity, adapted simulation tools are obviously required.

5. Conclusion

A hybrid method has been proposed to simulate the scattering of elastic guided waves in a plate by an obliquely incident wave onto a straight feature. The model hybridizes a semi analytical finite element method developed for computing all the guided modes in the unperturbed plate that have the same wavevector component along the axis of the scattering feature and a finite element method for computing the scattering phenomena. For the latter method, a classical finite element method is used and complemented by specific transparent boundary conditions. These conditions are derived in details in the paper; they are expressed through the decomposition of the fields on modes existing in the unperturbed plate. They have been shown to be very accurately transparent and can be very close to the scatterer, all the closer since more inhomogeneous and evanescent modes are taken into account in the modal series used.

The method has been validated by checking the quality of transparency of the specifically developed boundaries and its ability to satisfy global energy conservation. The method has also been used in a dedicated parametric study of the scattering of guided waves by a realistic stiffener, expressed in terms of variation of reflection and transmission coefficients with the incident angle, the type of incident mode etc. These latter results are typical of what is needed for optimizing the number and position of transducers in SHM applications.

The overall computations being modal, the interpretation of simulated results in terms of mode reflection, transmission, conversion etc. is straightforward. Moreover, the FE zone being reduced to a minimum thanks to the specific boundaries developed, results in typical parametric studies needed for demonstrating the performance of a method can be efficiently computed in an industrial context.

The hybrid method has been implemented at first to deal with isotropic materials as the examples given show. Its implementation to deal with multilayered anisotropic case was under test at the time of this conference; it may be noticed that its application to composite plate and stiffener has been considered since [12].

Our purpose was the development of a simulation tool to predict these phenomena. In practice, the full understanding of the complex wave behavior can be obtained using this tool to study in details how the elastic energy propagates in the stiffener depending on the given set of configuration parameters; results obtained suggest that an optimal SHM configuration could be designed but would be specific to a given material and a given geometry of the stiffener.

Acknowledgments

Laura Taupin's grant for her PhD studies is partly supported by EADS France – Innovation Works which is gratefully acknowledged. The authors would like to thank Benoît Petitjean (EADS–IW) and Colin Chambeyron (ENSTA) for their help in the course of this work.

References

[1] Petitjean B, Barut S and Simonet D 2007 Impact damage detection on aeronautical composite parts: issues and challenges, *in Proc. 6th Int. Workshop on SHM (Stanford, 2007)* 1127

[2] Taupin L, Lhémery A and Inquiété G 2011 *J. Phys.: Conf. Ser.* **269** 012002

[3] Lhémery A, Baronian V, Jezzine K and Bonnet-BenDhia A-S 2010 Simulation of inspections of elastic waveguides of arbitrary section containing arbitrary local discontinuities or defects *Review of Progress in QNDE* **29**, ed D O Thompson and D E Chimenti (Melville: AIP Conf. Proc. **1211**) 145

[4] Baronian V, Bonnet-BenDhia A-S and Lunéville E 2010 *J. Comp. Appl. Math.* **234** 1945

[5] Wilcox P D, Velichko A, Drinkwater B W, Croxford A J and Todd M D 2010 *J. Acoust. Soc. Am.* **128** 2715
[6] Bérenger J P 1997 *IEEE Trans. Antennas and Propagation* **45** 466
[7] Skelton E A, Adams S D M and Craster R V 2007 *Wave Motion* **44** 573
[8] Bécache E, Bonnet-Ben Dhia A-S, Legendre G 2004 *SIAM J. Numer. Anal.* **42** 409
[9] Castaings M and Lowe M 2008 *J. Acoust. Soc. Am.* **123** 696
[10] Fraser W B 1976 *J. Acoust. Soc. Am.* **59** 215
[11] Martin D, on line documentation of MELINA, http://anum-maths.univ-rennes1.fr/melina/
[12] Taupin L, Lhémery A, Baronian V Bonnet-Ben Dhia A-S and Petitjean B 2012 Hybrid SAFE / FE model for the scattering of guided waves in a stiffened multilayered anisotropic plate to appear in *Review of Progress in QNDE* **31**, ed D O Thompson and D E Chimenti (Melville: AIP Conf. Proc.)

Lamb wave interactions through dispersion 2D filters

L Martinez[1], N Wilkie-Chancellier[1], E Caplain[1], B Sarens[2] and C Glorieux[2]

[1] Université de Cergy Pontoise, ENS Cachan, UMR CNRS 8029, Laboratoire Systèmes et Applications des Techniques de l'Information et de l'Energie (SATIE), 5 mail Gay-Lussac, F 9500 Cergy-Pontoise, France
[2] Katholieke Universiteit Leuven, Laboratorium voor Akoestiek en Thermische Fysica (LATF), Celestijnenlaan 200D, B-3001 Leuven, Belgium

E-mail: loic.martinez@u-cergy.fr , nicolas.wilkie-chancellier@u-cergy.fr

Abstract. Acoustic surface waves are widely used to sense and map the properties of the propagation media. In order to characterise local space-time waves, methods such as Gabor analysis are powerful. Nevertheless, knowing which wave is observed, extracting its full bandwidth contribution from the others and to map it in the signal domain is also of great interest. In the Fourier domain, the acoustic energy of a wave is concentrated along the wave-number frequency (k-ω) dispersion curve, a way to extract one wave from others is to filter the signals by mean of k-ω band-pass area that keeps only the selected surface wave. The objective of the present paper is to propose 2D Finite Impulse Response (FIR) filters based on an arbitrary area shape designed to extract selected waves. FIR filtering is based on convolving the impulse response of the filter with the signals. Impulse responses derived from using k-ω elliptical areas (E-FIR) are presented. The E-FIR filters are successfully tested on three experimental space-time signals corresponding to the propagation of Lamb waves measured by standard transducers on a cylindrical shell, by laser Doppler on a plate and generated by a circular pulse and observed by shearography on a rectangular plate.

1. Introduction

Acoustic surface waves are widely used to sense and map the properties of propagation media. For decreasing both the acquisition time of such signals along the region of interest (ROI) and reach useful space-time resolutions, impulse generation for wideband acquisitions are required. This is at the cost of increasing the post-signal processing complexity [1-4]. Methods such as 2D and 3D Gabor analysis are powerful for localizing and identifying both transient and surface waves, as they need no prior knowledge upon the involved signals [5-8]. Nevertheless, knowing which wave is observed, extracting the full bandwidth contribution of one surface wave from the others and to map it in the signal domain is also of great interest [9-11]. In the Fourier domain, the acoustic energy is concentrated along the wave number-frequency (k-ω) dispersion curves. A way to extract one wave from others is to filter the signals by mean of a (k-ω) band-pass area that keeps only the selected surface wave. The aim of the present paper is to propose 2D Finite Impulse Response (FIR) filters based on an elliptical area (k- ω) shape to perform the identification of surface waves.

2. Context

Laser detection methods allow the investigation of ultrasonic transient phenomena in both space and time dimensions. Used along a two dimensional surface, laser ultrasonic leads to three dimensional (3D) space–time signal collections. The advantage of collecting impulse response transient signals is to acquire a wide frequency range from one experiment. However, post-processing depends on what goal is aimed at. For extracting stationary propagation aspects, 3D Fourier analysis or classical high resolution signal processing methods can be used to identify the wave propagation information such as dispersion curves [1-4]. However these methods are not adapted for identifying where and when the waves are generated (Figure 1).

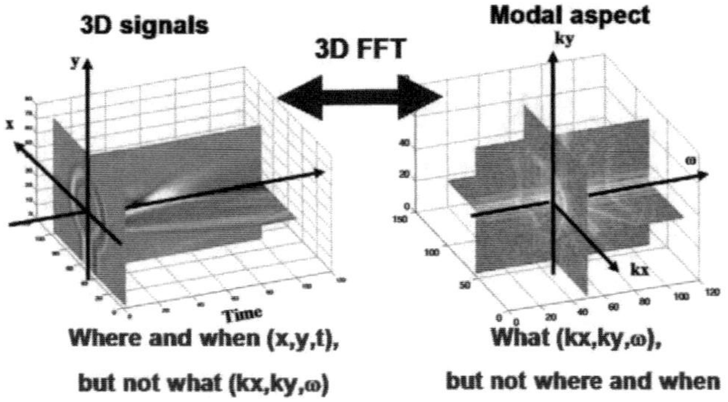

Figure 1. Advantages and drawbacks of space-time or modal representation.

Localizing wave packets through time was introduced in 1947 by Gabor [5]. It gave the principle of the time–frequency analysis by short-time Fourier transform (STFT), with the help of a sliding gaussian window [6-9]. The early applications dealt with audio signals and some efficient compression methods rose from the understanding of their frequencies components time shape. The 2D Gabor transform were introduced in acoustics [9] in order to localize ultrasonic phonons in space–time domains and their conversions on a cylindrical shell. The 3D Gabor analysis [10-11] is both an extension to three dimensions and an improvement of the early 2D (Figure 2).

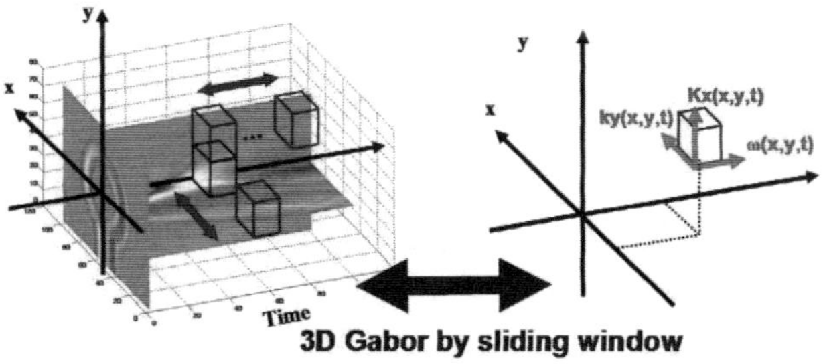

Figure 2. 3D Gabor transform principle.

Having the dispersion curves from Fourier analysis and the transient aspects from Gabor analysis, both methods are complementary and enable ones to identify waves properties from experimental signals. One of the goals of the 2D Finite Impulse Response (FIR) filters presented here is to go to the reverse direction: knowing dispersion curves, how to extract selected surface acoustic waves from wide-band data sets containing several waves? This kind of wideband wave selection can be done by zero-filling unwanted surface waves from their Fourier space, and recover the filtered signals by reverse Fourier transform. Straightforward for 1D signals, this method lacks of simplicity for implementing 2D trajectories in Fourier domain and controlling Gibbs oscillations in signal domain. In the following proposed method, wave selection is done in the signal domain by convolution with the impulse response of the filter: the advantage is that for elliptical Fourier pattern, such response is an analytical expression of the filter parameters.

The filter impulse response design takes advantage from a method derived by Shepp and Logan and used in Magnetic Resonance Imaging [12-14]. Shepp and Logan method generates elliptical patterns in image domain, by using analytical expression in Fourier domains (Figure 3). Compared to MRI domain, Fourier and signal domains are permuted, as for the proposed application, elliptical patterns are required in Fourier domain, not in signal domain or image as in MRI. Elliptical areas transfer function Finite Impulse Response filters are investigated (E-FIR).

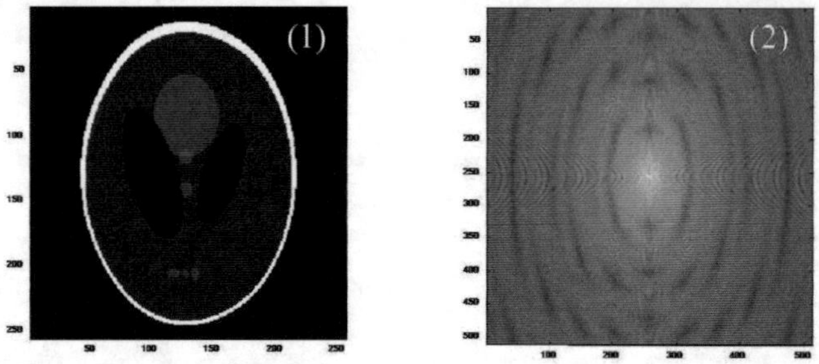

Figure 3. The Fourier reconstruction of a head section ([12]).

3. Filter design

The method lays on the convolution of the FIR impulse response with the data (Figure 4).

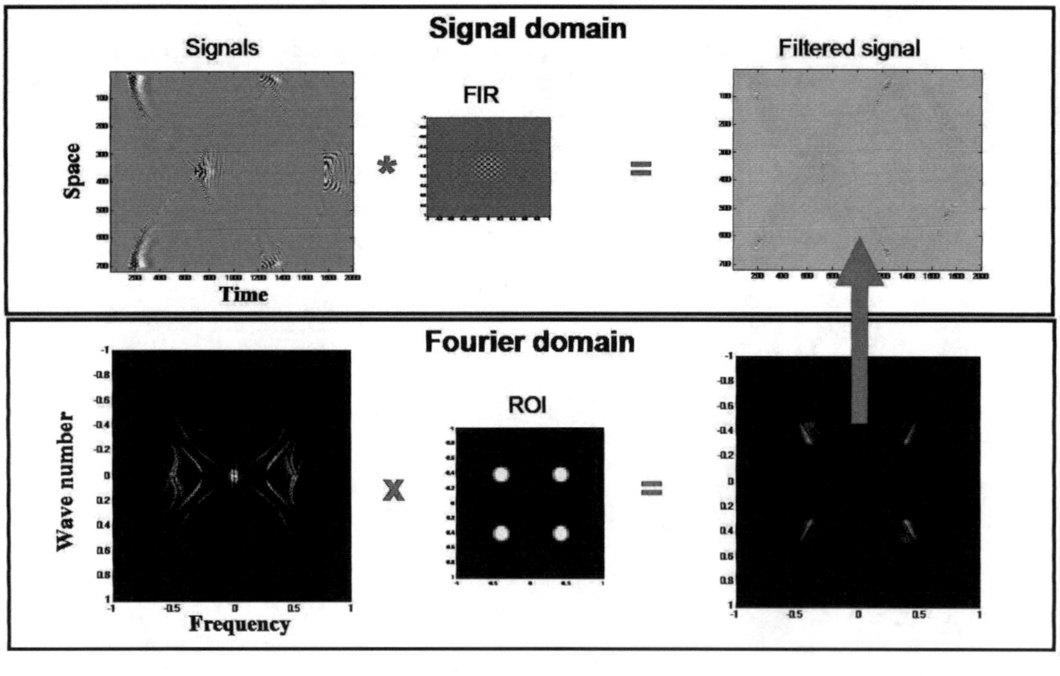

Figure 4. 2D FIR filter scheme.

The filter pattern is a frequency (k-ω) Region Of Interest (ROI) with an elliptical shape, chosen for its versatily and its known analytical inverse Fourier transform. The main axis of the elliptical area is oriented along the selected surface wave dispersion curve (Figure 5). As surface waves dispersion curves (k-ω) can be split in linear segments, the elliptical function parameters are well suited to keep surface waves of interest and eliminate the others.

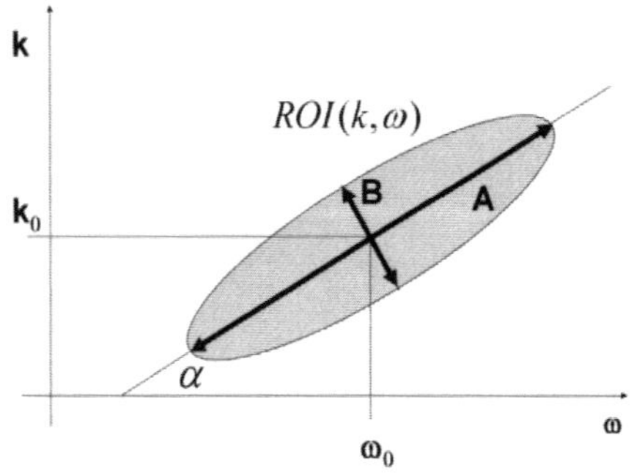

Figure 5. Filter transfert function and parameters.

The Impulse Response (IR) of such a filter is:

$$IR(x,t) = \rho e^{-i(K_0 x + \omega_0 t)} \frac{A J_1\left(B\sqrt{(u' AB^{-1})^2 + v'^2} \right)}{\sqrt{(u' AB^{-1})^2 + v'^2}}$$

where : $u' = x\cos\alpha + t\sin\alpha$ and $v' = -x\sin\alpha + t\cos\alpha$.

In this equation, J_1 is the first order Bessel function of the first kind, (K_0,ω_0) is the center of the ellipse, ρ the intensity, α its orientation and A and B are the lengths of the horizontal and vertical axes. In order to recover real valued filtered signals or select the direction of propagation, a great care must be taken into account to mirror or not this pattern in the three other quadrants of the $(k\text{-}\omega)$ spaces. The finite length of the filter is carefully chosen to optimize resolution in the space-time and in the frequency domain.

4. Experimental validation : Fluid filled cylindrical shell

4.1. Experimental set-up.

The propagation of surface waves is investigated on an aluminium cylindrical shell of external radius a=20.6mm and of thickness e=0.62mm (internal radius b=19.98mm). The characteristic ratio internal radius to external radius is b/a=0.97. This shell is immersed in water.

A pulse of 0.1µs long with 200V of amplitude is sent to a broadband transducer of central frequency 2.25MHz. In the enlightened zone of the shell, the plane bulk wave generates Lamb waves that propagate with continuous energy losses (Figure 6). The space-time signals collection s(θ,t) is collected for 720 angular positions equally spaced between 0 and 360 degrees. Each time signal is 2000 points long, for a 100µs time FOV.

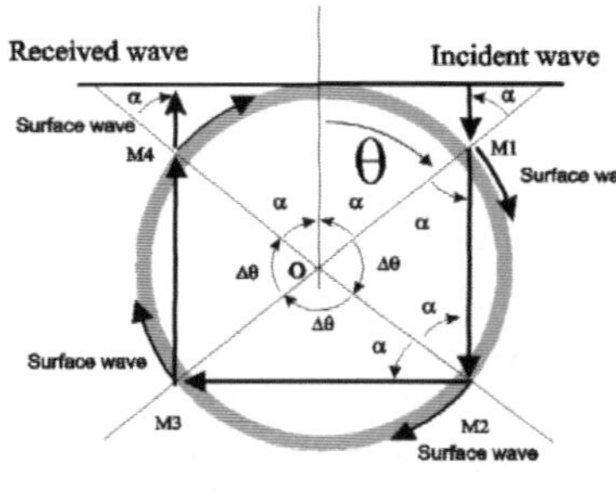

Figure 6. Surface wave coupling with internal waves.

4.2. Experimental results.
Wideband signals collected around the cylindrical shell are presented in Figure 7.

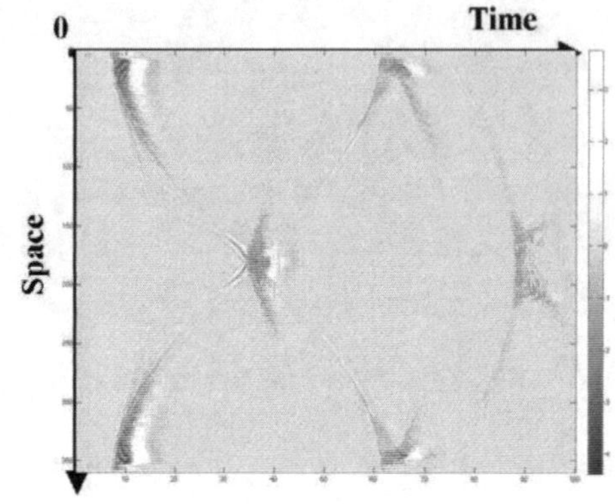

Figure 7. Experimental signals collected around the cylindrical shell.

Using the Elliptical FIR filters, the S_0 mode low frequency and high frequency dispersion curves are extracted (Figure 8). The waves are fully isolated from the other waves seen on Fig. 8 (4). As shown in Gabor analysis [9], the S_0 mode is clearly generated again at points M2 and M3 from acoustic bulk waves transmitted inside the shell. As the echoes M1, M2 are resolved, a Fourier analysis of each echo can give access to the mode dispersion curve and attenuation.

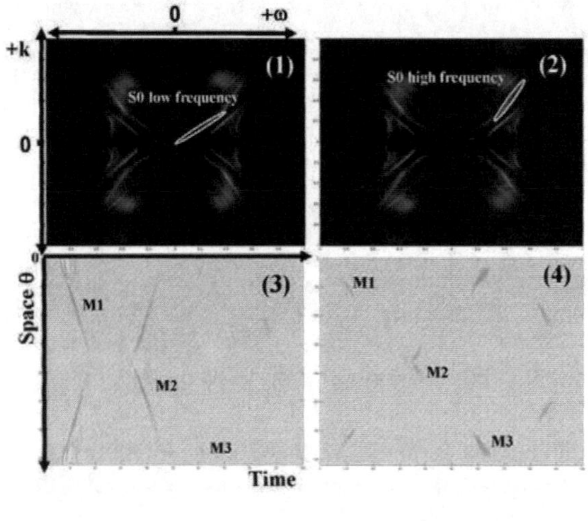

Figure 8. Filtered signals for S_0 Lamb mode in (3) and (4) using filter function defined respectively in (1) and (2).

The Lamb mode S_1 is also studied near its frequency cutoff in Figure 9. In this region, the S_1 dispersion curve implies two branches, one with positive group velocity, the other one with a negative group velocity. The two branches are identified in Figure 9. The S_1 branch with a negative group velocity (M2, M4) has a strong energy contribution in the shadow side of the shell, whereas the S_1 branch with a positive group velocity (M1, M3) has a strong energy contribution in the insonified zone. Due to the small elliptic areas used in the filter design, the FIR filter used for Figure 9 is close to Gabor analysis [9] and point out the generation zones with a better understanding than the corresponding Gabor analysis. To the question "where are S_1 positive or negative group velocity branches generated?", the Figure 9 answers with a good space-time resolution.

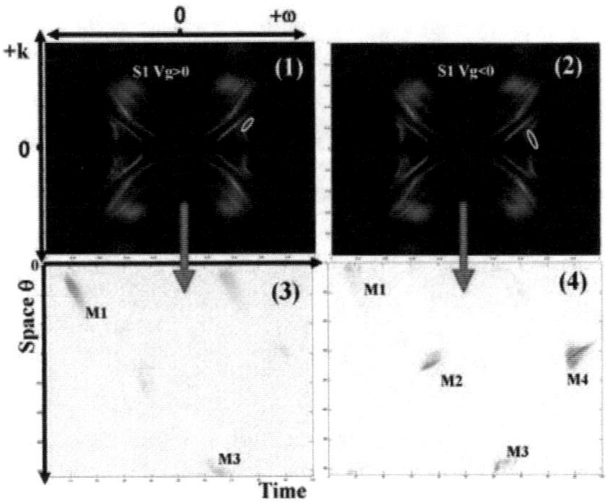

Figure 9. S_1 mode with positive group velocity in (3) and negative group velociy in (4) using filter function defined respectively in (1) and (2).

5. Experimental validation : Lamb wave on a plate

5.1. Experimental set-up.
The propagation of the surface waves is investigated on a plane plate (length L=60 mm and thickness e=2mm). The aluminium plate is immersed in a water tank. A pulse of 0.1μs long with 200V of amplitude is sent to a broadband transducer (2.25 MHz). In the enlightened zone of the plate, the plane bulk wave generates Lamb waves that propagate along the plate (Figure 10). A Polytec laser vibrometer is used for the vibration measurement. The incidence angle used is 13°.

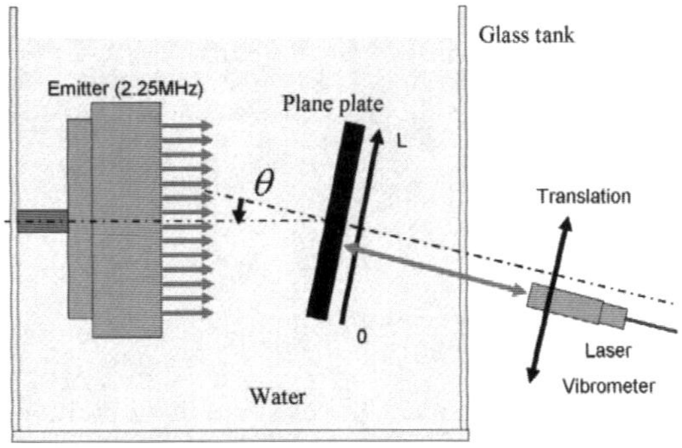

Figure 10. Experimental set-up

5.2. Experimental results.
The complete signal is presented in Figure 11. Surface waves are generated mainly by the edges of the plate (x=0 and x=L).

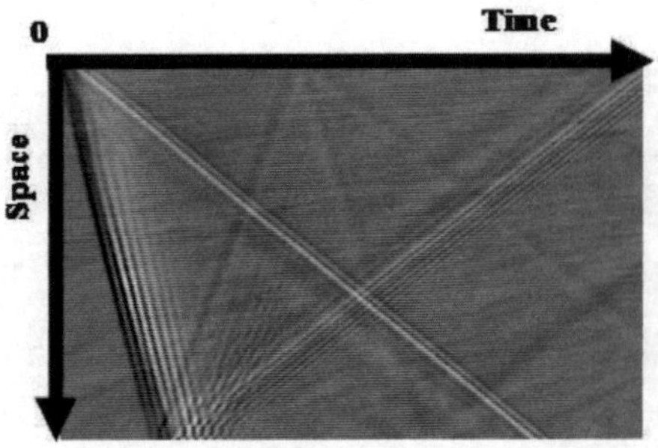

Figure 11. Space-time signals

Fluid born A-wave [4, 15] and A_0 mode are extracted from the complete signal using the FIR method (Figure 12). The mode conversion sequence is recovered : the A-wave is generated at both plate borders while the A_0 mode is generated at one border. At time t_1: the A_0 mode is reflected at the border and part of its energy is converted in A-wave. The FIR filter has a good resolution, even with the low frequency noise included in the signals.

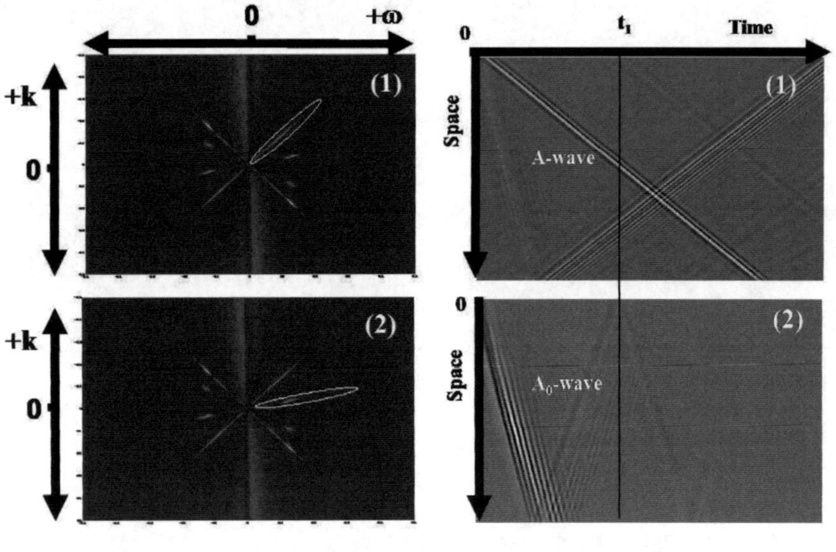

Figure 12. Extracted surface waves : A-wave (1) and A_0 wave (2)

6. Experimental validation : Surface acoustic waves from laser induced circular source

In the previous experiments, the signals were bi-dimensional, and scanned by classical transducers or laser vibrometers. The signal to noise ratio was increased by the use of time-averaging for each position. In the following experiments, the signals are acquired by grating interferometer measurement [16]. This allows the fast acquisition of the 2D images through time.

In the following experimental results, high power laser beam was diffracted to produce a circular acoustic source in thermo-elastic mode (10ns laser pulse). The resulting signals S(x,y,t) are 3D and reach a high spatial resolution, at the cost of a lower signal to noise ratio due to the lack averaging and optical artefacts (Figure 13).

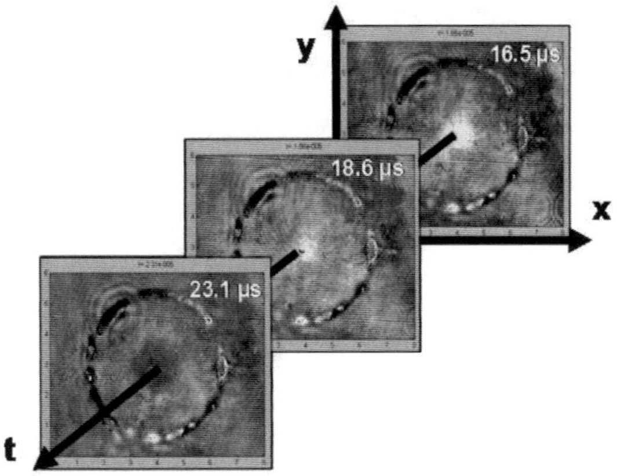

Figure 13. Space-time signals (6x8mm scanned area)

However, in the frequency domain, the artefact are localized close to zero frequency, whereas the observed surface waves (kx-ky-ω) dispersion curves are localized along trajectories far from the zero frequency origin (Figure 14).

In order to extract propagating surface waves from stationary patterns, the E-FIR has been designed following Fig. 14. The filter was applied on to successive pass: one along x-t and one along y-t dimensions. The resulting filtered images clearly exhibits the two waves propagating respectively towards and outwards the source centre, by nearly eliminating the non oscillating image background.

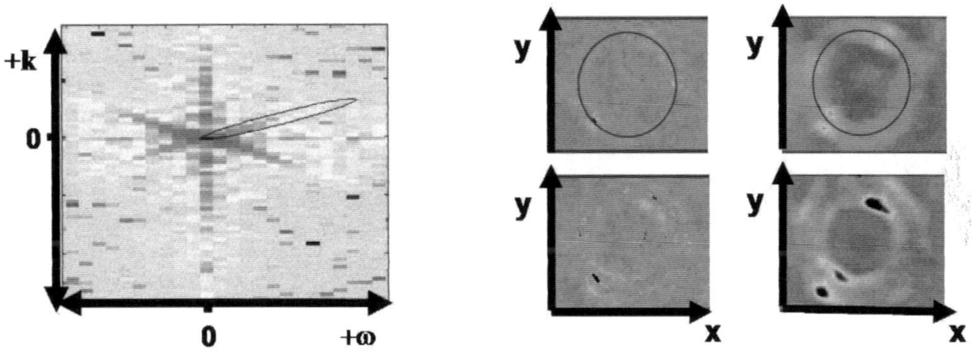

Figure 14. Filtered signals

7. Conclusion

The 2D FIR filters have been successfully employed for extracting individual surface waves present in transient space time signals. Whereas classical wave localization often lays on choosing one frequency or one wave number, the proposed E-FIR filters give the opportunity to extract the wave impulse response of a target for each wave. By using wide k-bandwidth and narrow ω-bandwidth E-FIR filter, Gabor analysis is also reached. The method is straightforward to extend to 3D signals obtained from surface scans with laser vibrometers.

8. References

[1] Alleyne J D and Cawley P 1991 A two-dimensional Fourier transform method for the measurement of propagating multimode signals *J. Acoust. Soc. Am.* **89(3)** 1159-1168

[2] Vollmann J and Dual J 1997 High-resolution analysis of the complex wave spectrum in a cylindrical shell containing a viscoelastic medium. Part I. Theory and numerical results *J. Acoust. Soc. Am.* **102** 896-908

[3] Vollmann J and Dual J 1997 High-resolution analysis of the complex wave spectrum in a cylindrical shell containing a viscoelastic medium. Part II. Experimental results versus theory *J. Acoust. Soc. Am.* **102** 909-920

[4] Martinez L 1998 New surface wave analysis methods. A-wave propagation on a curved plate *PhD Thesis, Le Havre University, France*

[5] Gabor D 1946 Theory of communication, *Proc. Inst. Electric. Eng.*, **93 (26)** 429–457

[6] Claasen T A C M and Mecklenbraüker W F G 1980, The Wigner distribution. A tool for time-frequency analysis, Part I: Continuous-time signals *Phillips journal of research* **35 (3)** 217-250

[7] Claasen T A C M and Mecklenbraüker W F G 1980, The Wigner distribution. A tool for time-frequency analysis, Part II: Discrete-time signals *Phillips journal of research* **35 (3)** 276-300

[8] Claasen T A C M and Mecklenbraüker W F G 1980 The Wigner distribution. A tool for time-frequency analysis, Part III: Relations with other time frequency Continuous-time signal representations *Phillips journal of research* **35 (6)** 373-389

[9] Martinez L, Morvan B and Izbicki J L 2004 Space-time-wave number-frequency X(x,t,k,f) analysis of SAW generation on fluid filled cylindrical shells *Ultrasonics International* **42 (1)** 383-389

[10] Martinez L, Goossens J, Glorieux C, Wilkie-Chancellier N, Ould Ehssein C and Serfaty S 2006 3D Gabor analysis of transient waves propagating along an AT cut quartz disk *Ultrasonics International* **43** 1173-1177

[11] Goossens J 2007 Elastic characterization of heterogeneous materials and structures by laser ultrasonic methods *PhD Thesis, KU Leuven, Belgium*

[12] Shepp L A and Logan B F 1974 The Fourier reconstruction of a head section *IEEE Transactions on Nuclear Science* **21** 21-43

[13] Shepp L A 1980 Computerized tomography and nuclear magnetic resonance *J Comput Assist Tomogr* **4** 94-107

[14] Koay C G, Sarlls J E and Zarslan E O 2007 Three-Dimensional Analytical Magnetic Resonance Imaging Phantom in the Fourier Domain *Magnetic Resonance in Medicine* **58** 430–436

[15] Martinez L, Duclos J and Tinel A 1996 Propagation of A-waves in a plane plate. An experimental study *Acoustics Letters* **20 (6)** 111-120

[16] Sarens B 2009 Full-Field optical methods for imaging ultrasonic waves and the detection of defects *PhD thesis, KU Leuven, Belgium*

10th Anglo-French Physical Acoustics Conference (AFPAC 2011) IOP Publishing
Journal of Physics: Conference Series **353** (2012) 012013 doi:10.1088/1742-6596/353/1/012013

The application of low frequency longitudinal guided wave mode for the inspection of multi-hole steel floral pipes

Z H Liu, X D Xie, B Wu, Y H Li and C F He

College of Mechanical Engineering and Applied Electronics Technology, Beijing University of Technology, Beijing 100124, China

E-mail: liuzenghua@bjut.edu.cn

Abstract. Shed-pipe grouting technology, an effective advanced supporting method, is often used in the excavation of soft strata. Steel floral pipes are one of the key load-carrying components of shed-pipe grouting supporting structures. Guided waves are a very attractive methodology to inspect multi-hole steel floral pipes as they offer long range inspection capability, mode and frequency tuning, and cost effectiveness. In this contribution, preliminary experiments are described for the inspection of steel floral pipes using a low frequency longitudinal guided wave mode, L(0,2). The relation between the number of grouting holes and the peak-to-peak amplitude of the first end-reflected signal was obtained. The effect of the grouting holes in steel floral pipes on the propagation velocity of the L(0,2) mode at 30 kHz was analyzed. Experimental results indicate that the typical grouting holes in steel floral pipe have no significant effect on the propagation of this mode. As a result, low frequency longitudinal guided wave modes have potential for the non-destructive long range inspection of multi-hole steel floral pipes. Furthermore, the propagation velocity of the investigated L(0,2) mode at 30 kHz decreases linearly with the increase of the number of grouting holes in a steel floral pipe. It is also noticeable that the effect of the grouting holes cumulates along with the increase in the number of grouting holes and subsequent increase in reflection times of longitudinal guided waves in the steel floral pipe. The application potential of the low frequency longitudinal guided wave technique for the inspection of embedded steel floral pipes is discussed.

1. Introduction

With the acceleration of development in west China, civil construction activities for the transportation infrastructure such as rail and highway have increased significantly. Because of the mountainous terrain of this area, underground operations are extensively used in the civil construction. Shed-pipe grouting technology, an effective advanced supporting method, is often used in the excavation of soft strata. As a key load-carrying component, steel floral pipes provide support for shed-pipe grouting supporting structures by tying the separated strata to form a whole. In general, the length of these pipes varies from several meters to tens of meters. Therefore, it is important to provide a nondestructive testing method to evaluate the construction quality, including embedment quality and length measurement of steel floral pipes, and verify the safety of complete advanced supporting structures. Due to its advantages [1], which include long range inspection capability, mode and frequency tuning, and cost effectiveness, guided wave technique is suitable for the inspection of various structures including pipes, tubes [2-6]. As an emerging nondestructive testing method, the technique allows for the inspection of structures with limited access from a single transducer position. Therefore, it is suitable for the inspection of steel floral pipes embedded in soft rock strata.

In order to measure the embedment length and evaluate support conditions of steel floral pipes, it is necessary to obtain end-reflected echoes. Low frequency guided wave modes with long wavelength and low sensitivity to grouting holes can be one option to achieve this aim. In this study, the experiments are conducted to investigate the propagation ability of a low frequency longitudinal guided wave mode in multi-hole steel floral pipes and the effect of grouting holes on the propagation velocity of this guided wave mode. The application possibility and potential of longitudinal guided wave technique for the inspection of steel floral pipes are also discussed.

2. Geometry of multi-hole steel floral pipes

Steel floral pipes acquire their name from the peach-like distribution of small grouting holes along their length. Figure 1 illustrates the configuration of a steel floral pipe and its grouting hole distribution. The grouting holes are distributed along four axial lines of the steel floral pipe. Here, $2l$ is the distance of adjacent holes on the same axial line, and d is the diameter of grouting hole and between two neighbouring grouting holes on adjacent axial lines, the axial distance is l. Thus if the cylinder was unfurled the flat surface would look like a triangular lattice.

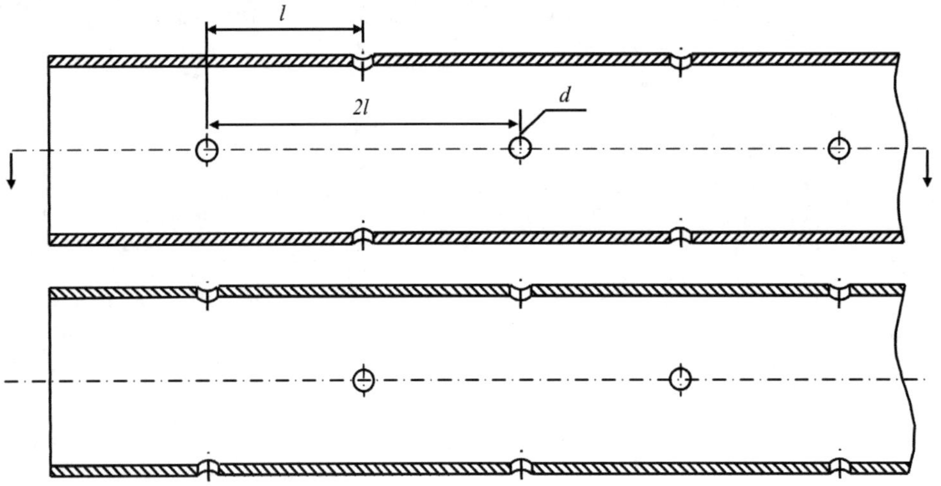

Figure 1. Configuration of steel floral pipe and its grouting hole distribution

In engineering construction, after a steel floral pipe has been inserted into the rock stratum, cement slurry, a mixture of cement and water, is pumped in under pressure from the end of steel floral pipe and forced out through the grouting holes. The cement slurry mixes with the surrounding soil and seeps into the rocks to strengthen the whole stratum. The surrounding stratum can achieve a higher strength after the cement solidification.

3. Longitudinal guided wave modes in steel pipe

Figure 2 shows the group velocity dispersion curves of longitudinal guided wave modes for a steel pipe generated using the Disperse software [7]. Here, the steel pipe external diameter is 89 mm and the wall thickness is 5 mm. The material properties used for the steel pipe are as follows: longitudinal and shear wave velocities: 5960 m/s and 3260 m/s; respectively; density 7932 kg/m^3.

In general, the grouting holes can be considered as defects to be inspected using guided waves. Here, in order to evaluate the construction quality of the steel floral pipe and surrounding stratum, the interaction between the grouting holes and guided waves should be decreased to a minimum, resulting in low attenuation of the guided wave mode. Guided waves are generated at one of pipe end, propagate along the whole pipe and are reflected from the other pipe end. The end-reflected signals will be received by the receivers. Measuring the amplitude and time-of-flight of the reflection signal, the safety of the whole advanced supporting structure can be evaluated. Low frequency longitudinal guided wave modes have long wavelength and are not sensitive to these grouting holes with low diffraction loss. Therefore, low frequency longitudinal guided wave modes have a certain potential to achieve this aim.

As shown in figure 2, the chosen excitation frequency of 30 kHz is close to the cutoff frequency, approximately 20 kHz, of the L(0,2) mode. At 30 kHz, the wavelength of the L(0,2) mode is about 189 mm and the mode shows relative low dispersion from the theoretical prediction. Figure 3 displays the radial and axial displacement of the L(0,2) mode at 30 kHz. The axial displacement is large compared to its radial displacement. Due to these propagation characteristics, the longitudinal guided wave mode can be excited efficiently using transducers generating an axial force in the pipe. Therefore, in this paper, the low frequency longitudinal guided wave mode, L(0,2) at 30 kHz, is chosen for the long range inspection of steel floral pipes.

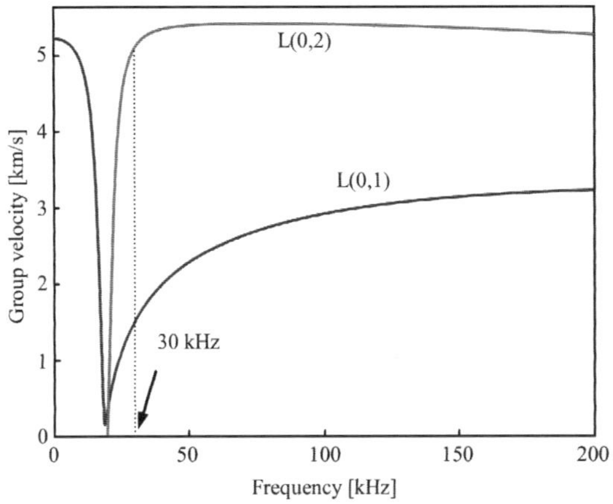

Figure 2. Group velocity dispersion curves of longitudinal guided wave modes for a steel pipe

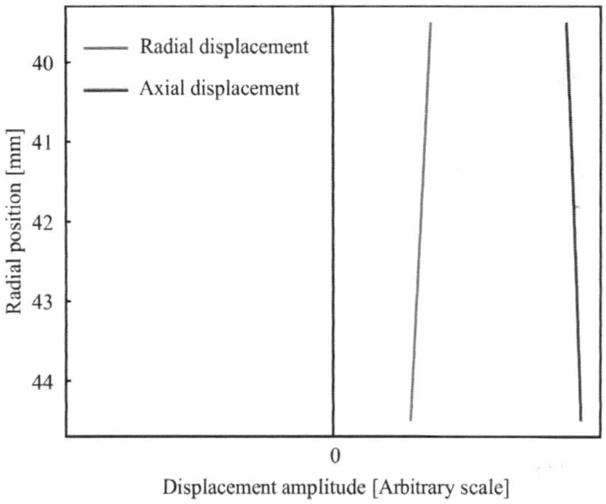

Figure 3. Radial and axial displacement distributions of L(0,2) mode at 30 kHz

4. Experimental investigation

4.1. Experimental setup

The propagation ability of low frequency longitudinal guided wave modes was investigated experimentally. Furthermore, the effect of grouting holes on propagation velocity of low frequency longitudinal guided wave modes was analyzed.

Figure 4 shows the experimental setup for low frequency longitudinal guided wave mode inspection in steel floral pipes. It consists of a power amplifier Ultra 2020, computer, digital oscilloscope DPO 4054, function generator 33120A, the investigated 4.2 m long steel pipe sample and the developed PZT array. The dimensions of the steel pipe are as stated above, external diameter 89 mm and wall thickness 5 mm. For the efficient excitation and reception of low frequency longitudinal guided wave modes at low frequency, a new type of PZT array with longitudinal vibration mode was developed. For this type of PZT element, the working frequency decreases with its length. The length of each PZT element 75 mm and its width and thickness are 4 mm and 1 mm, respectively. 24 similar PZT elements were bonded equally around the circumference of one of the pipe ends as shown in figure 5. They were connected in parallel to form an array.

A 10-cycle sinusoidal tone burst modulated by a Hanning window is generated by the function generator and amplified using the power amplifier as the excitation signal. The PZT array is excited by this amplified excitation signal through switcher. Longitudinal guided waves are generated and propagate along the pipe. When these signals are reflected from the grouting holes or the far end of the pipe, the same PZT array acts as a receiver for the reflected signals. The signals are averaged (512 averages) and recorded on the oscilloscope and transferred to the computer via an Ethernet connection.

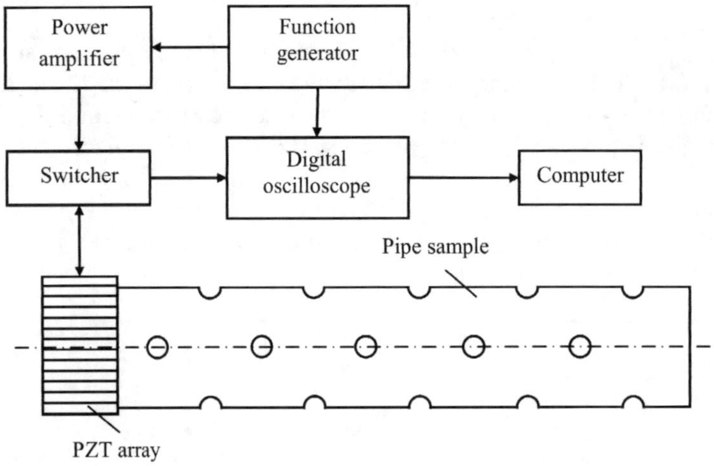

Figure 4. Experimental setup for low frequency longitudinal guided wave mode inspection in steel floral pipes

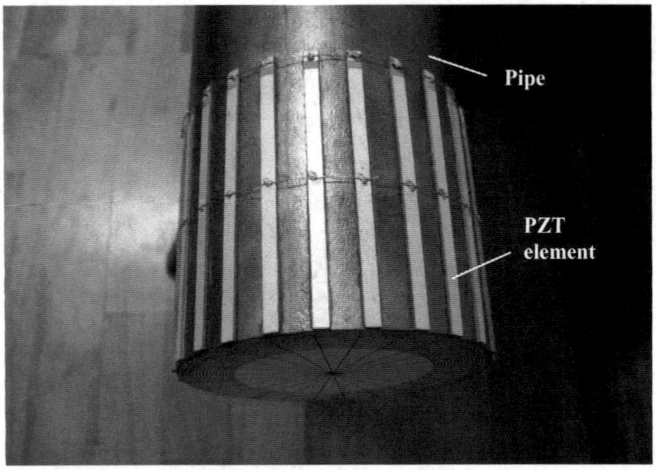

Figure 5. PZT array distributed equally around the circumference of pipe end

4.2. Experimental results and analysis

For the investigated steel floral pipe the axial distance of adjacent grouting holes on the same axial line, $2l$, was 150 mm and the diameter of grouting holes was $d = 10$ mm. These dimensions are often used in engineering practice. For the experiments, the grouting holes were drilled one by one into the steel pipe, starting 1 m away from the PZT array. The holes are numbered sequentially by their distance from the transducer. The maximum number of the grouting holes is 60 and at last all of these grouting holes will be distributed in a 2.25 m long section in the steel pipe. The position of the drilled grouting holes is illustrated in figure 1. Longitudinal guided wave signals were recorded after the drilling of each grouting hole.

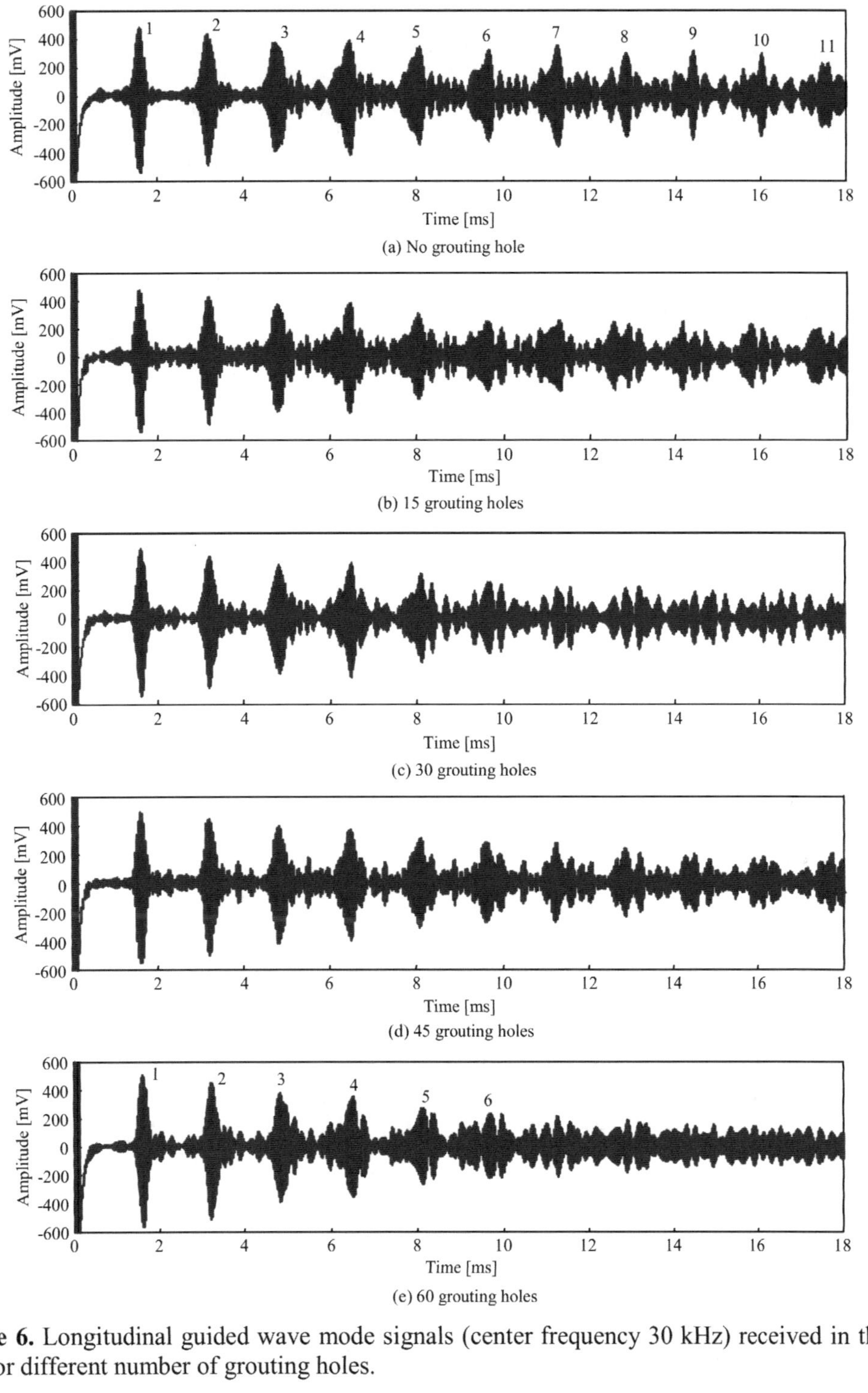

Figure 6. Longitudinal guided wave mode signals (center frequency 30 kHz) received in the steel pipe for different number of grouting holes.

Figure 6 shows the longitudinal guided wave mode signals (center frequency 30 kHz) received from the steel pipe for different numbers of grouting holes. As shown in figure 6(a), in the time range from 0-18 ms 11 end-reflected echoes of the L(0,2) mode can be seen when the pipe has no grouting holes. Therefore, the mode can propagate more than 90 meters in the pipe without significant loss of energy. With the increase of the number of grouting holes in the steel pipe, additional, smaller reflected pulses in these signals can be observed between the end reflection echoes (figures 6(b-d)). When the number of grouting holes reaches 60, six end-reflected echoes can still be identified in figure 6(e). This shows that the L(0,2) mode at 30 kHz can propagate a distance of more than 50 m in the steel pipe with 60 grouting holes (diameter 10 mm). Figure 7 shows the relation between the number of grouting holes and the peak-to-peak amplitude of the first longitudinal guided wave mode signal received from the end reflection in the steel pipe. The effect of the number of grouting holes on the amplitude of the first end-reflected echo is almost imperceptible for the typical geometry investigated here. The wavelength of the mode is significantly larger than the hole diameter. It can be concluded that the grouting holes have relative limited effect on the propagation of low frequency longitudinal guided wave modes. In figure 7, some data fluctuations, partially due to inconsistent operation of experimental instruments, can be seen as the collection of data lasted a long time.

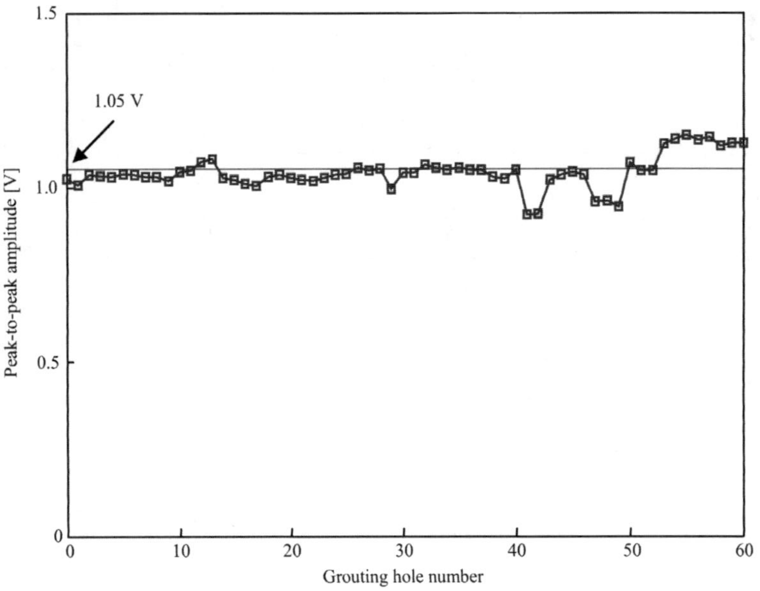

Figure 7. Relation between grouting hole number and peak-to-peak amplitude of the first end-reflected longitudinal guided wave mode.

However, with the increase in the number of grouting holes and subsequent increase in reflection times of longitudinal guided wave mode signals in the steel floral pipe, the effect of the grouting holes cumulates. When the grouting holes increase from 0 to 60, only the first six end-reflected echoes can be identified in the time traces. The other end-reflected echoes become gradually indistinguishable from the complex scattered signals at the grouting holes and cannot be identified from their waveform shape.

The effect of grouting holes on the propagation velocity of the low frequency longitudinal guided wave mode, L(0,2), at 30 kHz, was investigated. Figure 8 shows the zoomed waveforms of the first end-reflected echoes for different numbers of grouting holes from figure 6. It can be seen that the arrival time of the first end-reflected signals gradually increases with the increase of the number of grouting holes. This implies that the group velocity decreases when the number of grouting holes increases in the steel floral pipe.

Figure 9 illustrates the relation between the number of grouting holes and the group velocity of the L(0,2) mode with a center frequency of 30 kHz. From this figure, it is found that the group velocity of this mode decreases almost linearly with the number of grouting holes. Using a linear least square fit, the slope coefficient can be confirmed to be -0.43 (m/s)/(per hole).

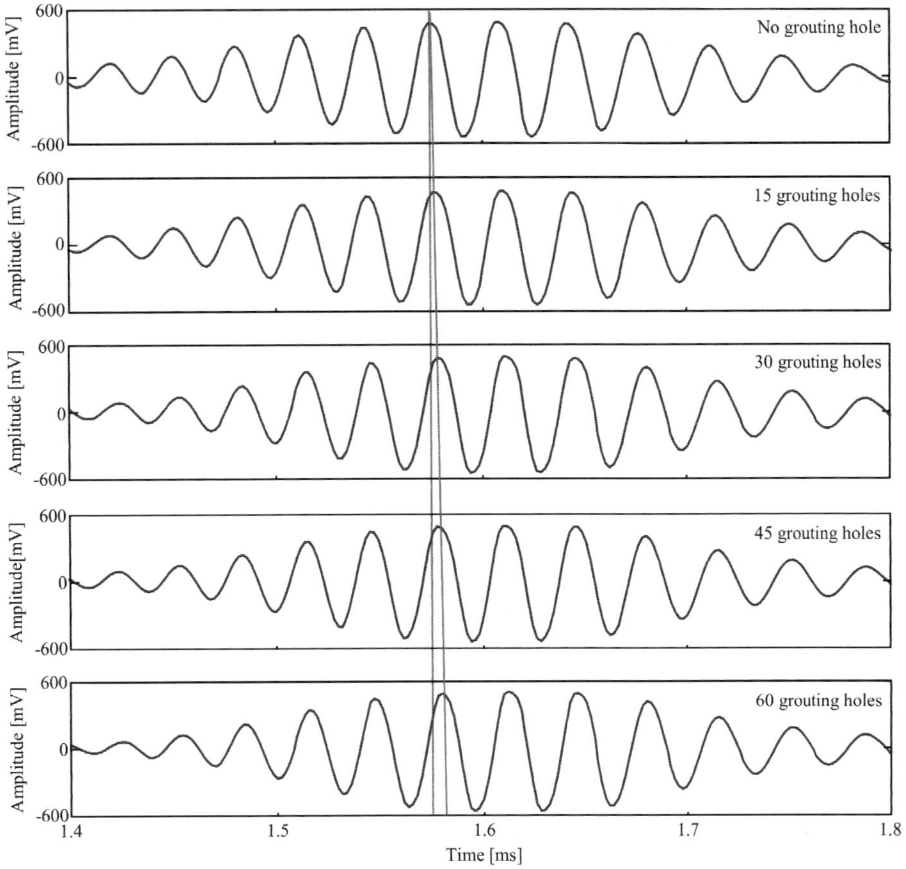

Figure 8. Zoom on the first end-reflected pulse from figure 6.

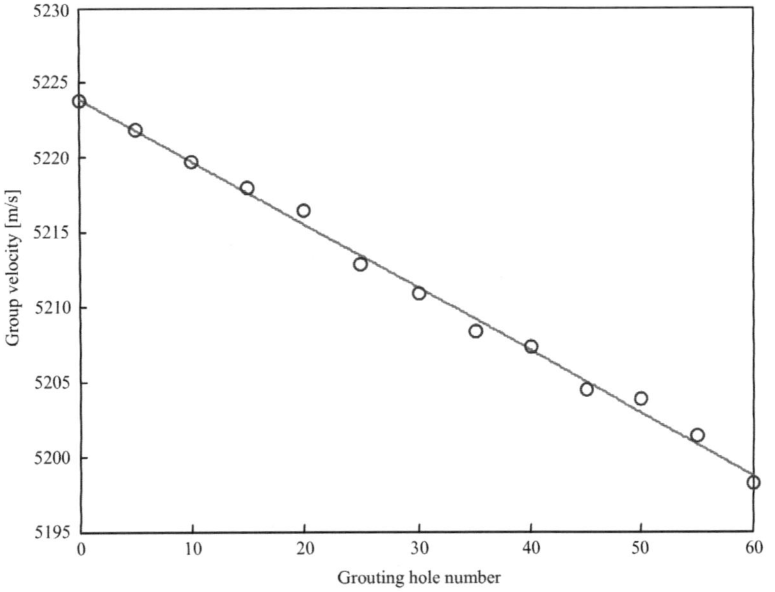

Figure 9. Relation between the number of grouting holes and group velocity of the L(0,2) mode at 30 kHz.

5. Conclusions and discussion

In this paper, the propagation ability of a low frequency longitudinal guided wave mode, the L(0,2) mode at 30 kHz, was investigated in multi-hole steel floral pipes. It was found that for the investigated, typical geometry the grouting holes have no significant effect on the amplitude of the first end-reflected echoes with a propagation distance of 8.4 m. However, with an increase of the number of grouting holes and propagation distance, the effect of the grouting holes accumulates and decreases the propagation ability of guided wave signals for long range inspection of steel floral pipes. The effect of the grouting holes on the group velocity of the L(0,2) mode at 30 kHz was analyzed. It was found using a linear least square fit that the group velocity of this mode decreases by 0.43 m/s per grouting hole.

According to the results, it has been shown that low frequency longitudinal guided wave modes have the potential for supporting quality evaluation of long steel floral pipes in civil engineering. It should be noticed that the L(0,2) mode at low frequency (30 kHz) is not the only suitable option for steel floral pipe inspection. As shown figure 2, the L(0,1) mode at a lower frequency range, below about 12 kHz, can also be used for steel floral pipe inspection as an alternative as it has long wavelength and similar, non-dispersive wave propagation characteristics. However, for the efficient excitation and reception of the L(0,1) mode at low frequency, the transducers would need to be developed further. For steel floral pipes embedded in soft strata, the attenuation of longitudinal guided wave signals due to the effect of the surrounding medium will have to be considered. Future work will include the investigation of the propagation ability and inspection potential of low frequency longitudinal guided wave modes in embedded steel floral pipes.

Acknowledgements

This work was supported by National Natural Science Foundation of China (Nos. 51075012 and 50975006), RGC-NSFC joint Project (No. 60831160523), National High Technology Research and Development Program of China (863 Program) (No. 2008AA042207) and Beijing Nova Program (No. 2008A015). The authors would also like to acknowledge Paul Fromme at University College London for his valuable suggestion and revision of this paper.

References

[1] Rose J 2002 *Mat. Eval.* **60** 53
[2] Lowe M, Alleyne D and Cawley P 1998 *Ultrasonics* **36** 147
[3] Barshinger J and Rose J 2004 *IEEE Trans. On Ultrason. Ferroelec. and Freq. Cont.* **51** 1547
[4] Liu Z, He C, Wu B, Wang X and Yang S 2006 *Ultrasonics* **44** e1135
[5] Liu Z, Wu B, He C, Wang X and Yang S 2007 *Insight* **49** 41
[6] Demma A, Cawley P, Lowe M and Roosenbrand A 2003 *J. Acoust. Soc. Am* **114** 611
[7] Pavlakovic B, Lowe M, Alleyne D and Cawley P 2000 *Rev. Prog. Quant. NDE (Ames)*, Vol 16, ed Thompson D and Chimenti D (New York: Plenum) p 185

Journal of Physics: Conference Series **353** (2012) 012014 doi:10.1088/1742-6596/353/1/012014

Modal solutions for SH guided waves radiated by an EMAT in a ferromagnetic plate

C Rouge, A Lhémery and D Ségur

CEA, LIST, F-91191 Gif-sur-Yvette, France

E.mail: alain.lhemery@cea.fr

Abstract. Electromagnetic acoustic transducers (EMAT) generate forces which are sources of elastic waves in a part without contact with it. Depending on the EMAT design, normal or tangential forces can be generated. Thanks to these capabilities, EMATs constitute an interesting alternative to piezoelectric devices in many configurations of non-destructive examination. In this paper, only shear horizontal (SH) guided waves radiated by an EMAT in a plate made of ferromagnetic material are modelled. These waves are particularly interesting to use for testing welded structures: SH waves may propagate in a weld without scattering and mode conversion phenomena. All the forces generated by an EMAT (Lorentz's, magnetosctrictive and magnetic forces) being exponentially decreasing with depth, they are re-written as series of moments, then, approximated as equivalent surface stresses. Surface stresses are then taken into account as terms of source of elastic waves. Specific features of EMAT can eventually be exploited to derive simple analytic expressions of the applied stress, leading to closed-form solutions for the modal amplitude of SH guided waves. Using these solutions, which are calculated at no computing cost, it becomes easy to study the influence of typical EMAT parameters on the modal amplitude of waves generated in the plate.

1. Introduction

Guided waves are used in NonDestructive Testing because they can propagate over long distance. They are often generated by piezoelectric transducers, but they can also be generated by Electro-Magnetic Acoustic Transducers (EMAT). Many designs of EMAT were experimentally studied and the ultrasound generation is well-known in a non-ferromagnetic metal [1]. The current in the EMAT induces currents in the part under test, known as eddy currents, which, with the superposed bias magnetization, create a force, the Lorentz force, confined in the vicinity of the part surface. In a ferromagnetic medium, EMATs create two more forces, also confined near the surface: the magnetization force and the magnetostriction force. Many EMAT designs have been studied to generate various kinds of elastic wave fields of various polarizations; for example, SH guided waves are more easily generated by means of specific EMAT design than with piezoelectric transducers. Besides, the fact that there is no contact of the EMAT with the part under test allows high speed inspection. EMATs can also work in hostile conditions such as high temperature or high pressure environments. However, those benefits are counteracted by the fact that EMAT signals are generally of much lower amplitude than that generated by piezoelectric transducers.

In weld inspection, SH waves have a specific interest because typical inner structure of welds may appear as a homogeneous material for them (figure 1) if the heterogeneous and anisotropic structure is bi-dimensional and the wave polarization is normal to it. The aim of the present paper is to develop a

Journal of Physics: Conference Series **353** (2012) 012014 doi:10.1088/1742-6596/353/1/012014

model for predicting the radiation of SH guided waves in a ferromagnetic plate-like structure by EMATs; such a model is helpful for optimizing the design of EMAT to get higher signal amplitude.

This work is based on modal description of guided waves to ease the interpretation of typical complex waveforms in guided wave inspection; one can further assume that similarly, the interpretation in terms of guided modes of the field radiated in a rather complex way by an EMAT in ferromagnetic materials shall also be made easier. Under a modal description of fields, the particle velocity and stress fields, solutions of the elastodynamic field equations governing the propagation of elastics waves, are expressed as linear combinations of the modes likely to exist. The part geometry considered herein is that of a plate made of homogeneous and isotropic ferromagnetic metal (iron, nickel). To simplify the problem, the analysis is made in 2D as shown on figure 1. As far as SH guided waves for weld inspections are concerned, this simplification makes sense since their specific interest only holds if the 2D symmetry reasonably applies.

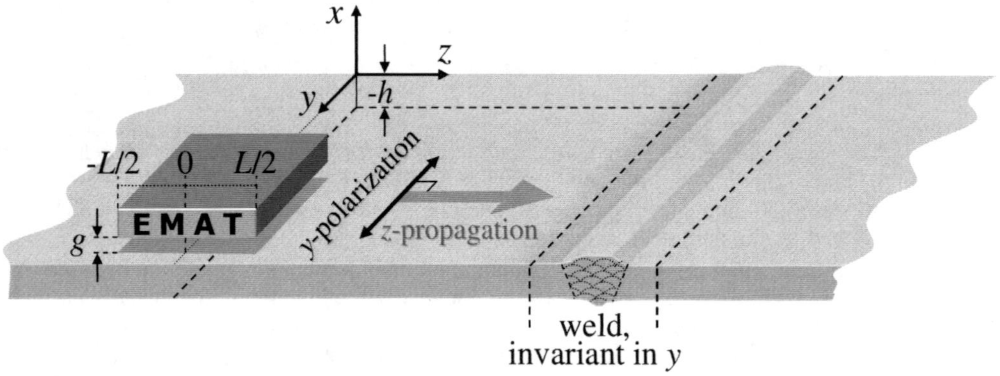

Figure 1. 3D representation of the configuration of the plate and the EMAT studied.

The present paper is organized as follows. In a first part, the theoretical model of SH wave radiation in ferromagnetic materials by EMATs is derived: closed-form solutions for the magnetic excitation field and the body forces it creates are derived; body forces are converted into surface stress sources of an elastodynamic problem solved as a modal series. In a second part, examples of application of our model illustrate some of the model capabilities.

2. Theory

The present paragraph is dedicated to the derivation of the global model for predicting SH guided waves radiated by an EMAT in a ferromagnetic plate. It is made of a succession of semi-analytical solutions of sub-problems. Semi-analytical solutions indeed can be easily translated into simulation tools allowing efficient computation works for optimizing NDT configurations.

At first, a closed-form solution for the magnetic field created by an EMAT is given by accounting for the periodicity of typical wire arrangement in classical EMAT designs. Then, body forces generated in a ferromagnetic plate are described as functions of the magnetic field. These forces are then transformed as equivalent surface stresses, for they are confined close to the surface where the magnetic field applies. A generic formula for the Green's function describing SH modes generated by a surface stress in a plate is then proposed. This last formula is eventually used in a convolution integral to predict modal amplitudes of SH guided waves generated by EMATs.

2.1. Magnetic field created by an EMAT

The magnetic field in the material can be deduced from Maxwell's equations and Ohm's law under the following assumptions:

- all quantities vary harmonically with time so that a term $\exp(i\omega t)$ can be factorized in the various equations, ω being the pulsation [rad.s^{-1}] and t is the time [s];

- time-dependency of current displacement in Maxwell-Ampere's law is neglected because, at typical EMAT working frequencies, current displacement can be considered as permanent compared to particle displacement;
- magnetostriction causes no volume change. This is true for isotropic ferromagnetic metal.

The first assumption allows the Maxwell-Faraday and Maxwell-Ampere equations to be written as:

$$\vec{\nabla} \wedge \vec{E} = -\frac{\partial \vec{B}}{\partial t}, \tag{1a}$$

$$\vec{\nabla} \wedge \vec{H} = \vec{J}. \tag{1b}$$

Besides, the relations between the magnetic field $\vec{B}$ [T], the magnetic excitation field $\vec{H}$ [A.m^{-1}] and the magnetization field $\vec{M}$ [A.m^{-1}] are:

$$\vec{B} = \mu_0(\vec{H} + \vec{M}), \tag{2a}$$

$$\vec{M} = \chi \vec{H}, \tag{2b}$$

$$\vec{B} = \mu_0 \mu \vec{H}, \tag{2c}$$

where μ_0 is the vacuum permeability (=4πx10^{-7} H.m^{-1}), μ is the magnetic permeability and χ the magnetic susceptibility. Here, for sake of simplicity, we further assume that magnetic permeability of the material μ is constant; this assumption is verified as long as $\vec{B}$ linearly varies with $\vec{H}$. It shall otherwise be accounted for as a matrix [1] whose components vary with both the frequency and the amplitude of the magnetic field [2, 3].

From now on, we assume that the current flows in the coil along the *y*-axis. With the help of Ohm's law and under the few assumptions made, the magnetic excitation field is a solution of the following system of equations:

$$\frac{\partial H_x}{\partial x} = -\frac{\partial H_z}{\partial z}, \tag{3a}$$

$$H_y = 0, \tag{3b}$$

$$\left(\frac{\partial^2}{\partial x^2} + \frac{\partial^2}{\partial z^2}\right) H_z - j\omega\sigma\mu_0\mu H_z = 0. \tag{3c}$$

The current in the electric circuit is now written as a Fourier series (as done in [4, 5]). The current distribution can be described as a sequence of 0, 1 and -1 values along the *z*-axis. This Fourier series is denoted by *S*(*z*) and is given by:

$$S(z) = \sum_{n=0}^{\infty} S_n(z) = \sum_{n=0}^{\infty} \left[S_n^c \cos(s_n z) + S_n^s \sin(s_n z) \right]. \tag{4}$$

The integral formulation of Maxwell-Ampere law is used to write boundary conditions of equation (3c) just below the electric circuit of the EMAT, at $x = g$ where g [m] is the lift-off. One gets:

$$H_z^A(g, z) = \frac{I}{2a} S(z), \tag{5}$$

where I denotes the current intensity [A] and a, the wire width [m]. Superscript "*A*" stands for quantities defined in air. With this boundary condition, equation (3c) for magnetic field is solved in the volume between the coil and the plate. One gets

$$H_z^A(x,z) = \frac{I}{2a}\sum_{n=0}^{\infty} S_n(z)\exp\left[s_n(g-x)\right] \quad \text{for } x \in \left[0, g\right]. \tag{6}$$

The solution for $\vec{H}$ in the plate is then obtained by solving equation (3c) by taking into account the continuity of tangential field at the interface between air and the plate. After simple algebra, the elementary solution of this differential equation is obtained as:

$$H_x(x,z) = \frac{I}{2a}\sum_{n=0}^{\infty}\frac{s_n}{q_n}S_n(z)\exp(-s_n g)\exp(q_n x)\,,\; H_z(x,z) = \frac{I}{2a}\sum_{n=0}^{\infty}S_n(z)\exp(-s_n g)\exp(q_n x) \tag{7}$$

with $q_n = \left(s_n^2 + 2j/\delta^2\right)^{1/2}$ and $\delta = \left(2/\omega\sigma\mu_0\mu\right)^{1/2}$.

This is a separate variable solution in x and z. Its z-dependency is a function of the coil geometry only, through the Fourier series $S(z)$. Its x-dependency is a decreasing exponential with depth that is a function of coil parameters and material characteristics through the penetration depth δ [m].

2.2. *Body forces generated by a magnetic field into a ferromagnetic medium*

As mentioned in the introduction, considering a plate made of a ferromagnetic material makes it necessary to account for two more forces than the sole Lorentz's forces an EMAT creates in non-ferromagnetic one. The three forces to consider generated by the magnetic field imposed by the EMAT are thus the Lorentz's force plus the magnetization and the magnetostriction forces. The present paragraph aims at expressing them.

2.2.1. Lorentz's force. The Lorentz's force results from the interaction of the eddy current density $\vec{J}_e = \vec{\nabla} \wedge \vec{H}$ created by the coil with the total magnetic field; the total magnetic field is the sum of the static magnetic field $\vec{B}_0$ of the magnet and the dynamic magnetic field $\mu_0\mu\vec{H}$, as calculated in the previous paragraph. The force is expressed by:

$$\vec{F}_L = (\vec{\nabla} \wedge \vec{H}) \wedge (\vec{B}_0 + \mu_0\mu\vec{H})\,. \tag{8}$$

2.2.2. Magnetization force. The interaction between the dynamic magnetic field $\vec{H}$ and the magnetization vector of the magnet $\vec{M}_0$ and that of the metal $\vec{M}$ induces the reorientation of magnetic dipoles in the metal. This results in the so-called magnetization force, expressed by:

$$\vec{F}_M = \mu_0\mu(\overline{\overline{\nabla\vec{H}}}) \times (\vec{M} + \vec{M}_0)\,. \tag{9}$$

2.2.3. Magnetostriction force. This force is generated by deformation of magnetic domains due to the interaction between the dynamic magnetic field and the static magnetic field. It is classically compared to piezoelectricity and often called "piezomagnetic" effect. The (6x3) tensor $\overline{\overline{e}}$ of piezomagnetic stress coefficients is introduced to derive the magnetostriction stress tensor as

$$\sigma_{MS}^I = -e_{Ij}H_j\,, \text{ with } I \in \{1,\dots,6\}\,, \tag{10}$$

from which the magnetostriction force is readily obtained. The force is given by:

$$\vec{F}_{MS}^{x} = \frac{\partial \sigma_{MS}^{1}}{\partial x} + \frac{\partial \sigma_{MS}^{6}}{\partial y} + \frac{\partial \sigma_{MS}^{5}}{\partial z},$$
$$\vec{F}_{MS}^{y} = \frac{\partial \sigma_{MS}^{6}}{\partial x} + \frac{\partial \sigma_{MS}^{2}}{\partial y} + \frac{\partial \sigma_{MS}^{4}}{\partial z}, \tag{11}$$
$$\vec{F}_{MS}^{z} = \frac{\partial \sigma_{MS}^{5}}{\partial x} + \frac{\partial \sigma_{MS}^{4}}{\partial y} + \frac{\partial \sigma_{MS}^{3}}{\partial z}.$$

In a more condensed form, one writes:

$$\vec{F}_{MS} = -\vec{\nabla}^{t}(\overline{\overline{e}}\vec{H}), \tag{12}$$

where the notation $\vec{\nabla}^{t}$ denotes a modified gradient as written by Auld [6].

Note that $\overline{\overline{e}}$ itself depends on the total magnetic field, sum of the static and dynamic terms as mentioned previously. The former term is of higher amplitude than the latter. As a consequence, we consider here that dynamic effects are negligible, as far as the evaluation of $\overline{\overline{e}}$ is concerned. However, variations of $\overline{\overline{e}}$ as a function of the amplitude of the static field result from quite complex microphysics phenomena. Readers interested in a detailed discussion about the possible variations of $\overline{\overline{e}}$ are referred to [7]. In what follows, we consider that the various components of $\overline{\overline{e}}$ involved in the calculation of the magnetostriction force are constant values. As a consequence, doubling frequency effect sometimes described in the literature [8, 9] cannot be accounted for.

2.3. *Transformation of body forces into equivalent surface stresses*

The three body forces created by the EMAT could be straightforwardly integrated to predict the elastic wave field they generate, in the form of a triple (volume) convolution integral of the forces with the elastodynamic Green's function for the problem in hand. However, since they were all expressed as functions of the magnetic field calculated in paragraph 2.1., they also exponentially decrease with depth, with a decreasing factor δ (penetration depth).

To avoid integrating forces onto a volume, Thompson [10] developed an approximate modelling approach to transform the volume integral over forces onto a surface integral over equivalent surface stress. The approximation used, which we re-develop in what follows, is all the more accurate since forces are confined in the vicinity of the surface where the EMAT is used.

The elastic wave equation to be solved is given by the following equation where the right-hand side is the sum of the various forces described above. We have:

$$-\mu_L \Delta \vec{u} - (\lambda_L + \mu_L)\vec{\nabla}(\nabla.\vec{u}) + \rho \frac{\partial^2 \vec{u}}{\partial t^2} = \vec{F}_L + \vec{F}_M + \vec{F}_{MS}, \tag{13}$$

where $\vec{u}$ is the particle displacement [m], ρ the material density [kg.m^{-3}] and λ_L and μ_L, the Lamé's constants [Pa] from Lamé's equation for an isotropic material.

From now on, we restrict our study to the generation of SH waves, propagating in the (x, z) plane polarized in the y-direction. Therefore, the problem in hand is bi-dimensional and we only have to predict the y-component of the particle displacement generated by forces applied in the y-direction. This restriction, which may seem severe at first glance, corresponds to actual symmetry of testing configurations of interest, in particular, for weld inspections (see figure 1); moreover, it leads to interesting simplifications in subsequent calculations. Note that the overall approach can be extended to the case of Lamb wave generation where forces to be considered apply along $\vec{x}$ or $\vec{z}$ direction and the wave particle polarization belongs to the (x, z) propagation plane.

Since the problem is 2D, the general triple integral over body force sources is restricted to a double integral in the (x, z) plane. Under the above assumptions, the SH wave displacement at an observation point $\vec{r}$ can be written in the form of the following convolution Green's integral as:

$$u_y(\vec{r}) = \int_{-L/2}^{L/2} \int_{-\infty}^{0} G_{yy}(\vec{r},\vec{r}_0) f_y(\vec{r}_0)\, dx_0\, dz_0 \,, \tag{14}$$

limited along $\vec{z}$ by the size L of the EMAT, and where G_{yy} denotes the only involved component of the Green's tensor thanks to the 2D / SH symmetry considered here. The source term reduces to

$$f_y(\vec{r}_0) = \left[(F_L + F_M + F_{MS}).\vec{e}_y\right] = -\left[e_{6x}\frac{\partial H_x}{\partial x_0} + e_{6z}\frac{\partial H_z}{\partial x_0} + e_{4x}\frac{\partial H_x}{\partial z_0} + e_{4z}\frac{\partial H_z}{\partial z_0}\right], \tag{15}$$

where only magnetostriction actually contributes, other forces being oriented in the (x, z) plane.

The detailed derivation for transforming body force convolution integral into a surface stress convolution integral is given in Appendix. It allows us to re-write equation (14) as:

$$u_y(\vec{r}) \approx \int_{-L/2}^{L/2} \left[T_y - \frac{1}{2}N_y\left(\frac{\omega^2 \rho}{C_{44}} + \frac{\partial^2}{\partial z_0^2}\right)\right] G_{yy}(\vec{r},\vec{R}_0)\, dz_0 \,, \tag{16}$$

where $\vec{R}_0 = \vec{r}_0(0, z_0)$, T_y and N_y defined in the Appendix are respectively the zeroth and the second order moments of body force f_y.

2.4. *Green's function for SH waves generated in a plate by a surface stress*

In equation (16), the Green's function, not explicitly given, stands for the elastodynamic solution of the problem of an elementary source at the surface of an elastic half-space. For solving the radiation in a plate, it is more convenient to make use of a more adapted Green's function that accounts for the finite thickness of the plate. More specifically, we want to derive such a function as a modal series of SH guided waves. This is done in the present subsection by using the integration over the plate thickness of the complex reciprocity relation [6].

Assuming a line source (along y) acting at (x_0, z_0), the reciprocity relation described in [6] leads to the following formal relation

$$4\frac{\partial}{\partial z}\sum_n A_n^{G_{yy}}(z, z_0) e^{j\beta_m^* z} P_{mn} + \left[\vec{\tilde{v}}_m^* \bar{\bar{\sigma}}_n e^{j\beta_m^* z} + \vec{\tilde{v}}_n \bar{\bar{\tilde{\sigma}}}_m^* e^{j\beta_m^* z}\right]\cdot \vec{e}_x \Big|_{x=-h}^{x=0} = 0 \,, \tag{17}$$

where amplitudes of the n modes are obtained as solutions for the line source as

$$\vec{v}(x,z) = \sum_n A_n^{G_{yy}}(z)\vec{v}_n(x), \quad \bar{\bar{\sigma}}(x,z) = \sum_n A_n^{G_{yy}}(z)\bar{\bar{\sigma}}_n(x) \,, \tag{18}$$

and where the state "m" is an arbitrary propagative mode:

$$\vec{v}_m(x,z) = \vec{\tilde{v}}_m(x) e^{-j\beta_m z} \,, \quad \bar{\bar{\sigma}}_m(x,z) = \bar{\bar{\tilde{\sigma}}}_m(x) e^{-j\beta_m z} \,. \tag{19}$$

The factor P_{mn} arises from the complex orthogonality relation and is given by [6, 11]:

$$P_{mn} = -\frac{1}{4}\int_{-h}^{0} (\vec{\tilde{v}}_n^* \bar{\bar{\tilde{\sigma}}}_m + \vec{\tilde{v}}_m \bar{\bar{\tilde{\sigma}}}_n^*).\vec{z}\, dx \,. \tag{20}$$

The symbol * stands for conjugate complex values. Besides, the boundary conditions and the initial conditions are:

$$\bar{\bar{\sigma}}_m(x=0,z)\cdot\vec{e}_x = \bar{\bar{\sigma}}_m(x=-h,z)\cdot\vec{e}_x = 0 \,, \quad \bar{\bar{\sigma}}_n(x=-h,z) = \bar{\bar{0}} \,, \quad \bar{\bar{\sigma}}_n(x=0,z) = -\begin{pmatrix} 0 & 1 & 0 \\ 1 & 0 & 0 \\ 0 & 0 & 0 \end{pmatrix}\delta(z - z_0) \,, \tag{21}$$

where h is the plate thickness and δ the Dirac function. Equation (17) becomes:

$$4\sum_n A_n^{G_{yy}}(z,z_0)\,j\beta_m^* e^{j\beta_m^* z}P_{mn} + 4\sum_n \frac{\partial A_n^{G_{yy}}(z,z_0)}{\partial z} e^{j\beta_m^* z}P_{mn} = \tilde{v}_{y_m}^*(x=0)e^{j\beta_m^* z}\delta(z-z_0)\,, \tag{22}$$

where $\tilde{v}_{y_m}(x=0)$ is the amplitude of the particle velocity along the y-direction for the "m mode" evaluated at the top surface of the plate and z is the distance between the line-source and the observation point.

Here, only real propagative modes are studied, inhomogeneous and evanescent modes vanishing far from the source. Therefore, conjugate complex values are equal to the real ones. Orthogonality relation implies "$m = n$" [6]. Thus, modal amplitudes verify the following first order differential equation:

$$4P_{mm}\left(\frac{\partial}{\partial z} + j\beta_m\right)A_m^{G_{yy}}(z,z_0) = \tilde{v}_{y_m}(x=0)\delta(z-z_0)\,, \tag{23}$$

which is easily solved as:

$$A_m^{G_{yy}}(z,z_0) = \frac{e^{-j\beta_m z}}{4P_{mm}}\tilde{v}_{y_m}(x=0)\int_{-\infty}^{+\infty} e^{j\beta_m \eta}\delta(\eta - z_0)d\eta = \frac{e^{-j\beta_m z}}{4P_{mm}}\tilde{v}_{y_m}(x=0)e^{j\beta_m z_0}\,. \tag{24}$$

This modal amplitude is that of the m-th propagative SH mode in a waveguide excited by a line source of tangential stress. Their wavenumber β_n is given analytically by:

$$\beta_n = \left(\frac{\rho\omega^2}{\mu_L} - \left(\frac{n}{h}\right)^2\right)^{1/2}. \tag{25}$$

2.5. Modal amplitude of SH guided modes generated by an EMAT

Thanks to the modal decomposition, the displacement $\vec{u}_G$ generated by a tangential stress line source in the plate is now written as:

$$\vec{u}_G(x,z,[z_0]) = \sum_m A_m^{G_{yy}}(z,z_0)\vec{u}_m(x) = \sum_m \frac{e^{-j\beta_m z}}{4P_{mm}}\tilde{v}_{y_m}(x=0)e^{j\beta_m z_0}\vec{u}_m(x)\,. \tag{26}$$

This solution is now used for predicting the field radiated by forces created by an EMAT. Body forces have been transformed into equivalent surface stress that can be extracted from equation (16) in the form of an operator as

$$\sigma_{eq}(z_0) = \left[T_y - \frac{1}{2}N_y\left(\frac{\omega^2\rho}{C_{44}} + \frac{\partial^2}{\partial z_0^2}\right)\right]. \tag{27}$$

Equation (16) was derived in the Green's function convolution integral formalism assuming the Green's function shall verify free boundary condition at $x = 0$ (half-space solution). The Green's function derived in the previous subsection as a modal series also verifies the same boundary condition plus a similar condition at the boundary $x = -h$. At this surface, sources of body force may be considered as not contributing; this is true while $h \gg \delta$, a relation which is almost always verified for typical frequencies and materials of interest in NDT applications. Therefore, Green's function for the elastic plate can be substituted for the original half-space solution in equation (16). By doing so, equation (16) straightforwardly becomes the solution for SH guided waves since the new kernel was built to account for the guided propagation. The particle displacement radiated at (x, z) by an EMAT is then found after rather simple algebraic calculations that follow:

$$
\begin{aligned}
u_y(x,z) &= \int_{-L/2}^{L/2} \sigma_{eq}(z_0) G_{yy}(x,z,[z_0])\,dz_0 = \int_{-L/2}^{L/2} \sigma_{eq}(z_0)\vec{u}_G(x,z,[z_0])\,dz_0 \\
&= \sum_m \vec{\tilde{u}}_m(x)\frac{e^{-j\beta_m z}}{4P_{mm}}\tilde{v}_{y_m}(x=0)\int_{-L/2}^{L/2}\left[T_y - \frac{1}{2}N_y\left(\frac{\omega^2\rho}{C_{44}}+\frac{\partial^2}{\partial z_0^2}\right)\right]e^{j\beta_m z_0}dz_0 \\
&= \sum_m \vec{\tilde{u}}_m(x)\frac{e^{-j\beta_m z}}{4P_{mm}}\tilde{v}_{y_m}(x=0)\int_{-L/2}^{L/2}\left[T_y - \frac{1}{2}N_y\left(\frac{\omega^2\rho}{C_{44}}-\beta_m^2\right)\right]e^{j\beta_m z_0}dz_0 \\
&= \sum_m \vec{\tilde{u}}_m(x)\frac{e^{-j\beta_m z}}{4P_{mm}}\tilde{v}_{y_m}(x=0)\int_{-L/2}^{L/2}\left[T_y - \frac{1}{2}\left(\frac{m}{h}\right)^2 N_y\right]e^{j\beta_m z_0}dz_0 .
\end{aligned}
\tag{28}
$$

The modal amplitude for the problem of the tangential surface stress generated by an EMAT is finally given by:

$$
A_m(z) = \frac{e^{-j\beta_m z}}{4P_{mm}}\tilde{v}_{y_m}(x=0)\int_{-L/2}^{L/2}\left[T_y - \frac{1}{2}\left(\frac{m}{h}\right)^2 N_y\right]e^{j\beta_m z_0}dz_0 . \tag{29}
$$

It is analytically related to EMAT characteristics through the moments of body forces, and also related to material characteristics, again through the moments but also through the propagation terms.

2.6. Brief summary of the theoretical model derived

The technical content of this section led to a semi-analytical solution for the modal amplitude of SH modes radiated by an EMAT into a ferromagnetic plate. To summarize, the first step consists in obtaining the magnetic field created by the specific wire configuration of the EMAT which writes as a Fourier series given by equation (7). The second step consists in calculating the body force generated by the EMAT which depends on the magnetic field; in the present configuration considered, the force to take into account is only the magnetostriction expressed by equation (12). In the third step, the zeroth and second order moments of this force are calculated by means of equation (A3). The fourth and final step consists in computing the convolution integral given by equation (29) to get the amplitude of the SH guided modes that exist at the working frequency considered. By Fourier synthesis, it is then possible to predict the transient field radiated by the EMAT.

In what follows, modal amplitudes for various EMATs are computed using the present model.

3. Two examples of application

3.1. Radiation by a meander-coil EMAT

A most studied EMAT is the so-called meander-coil. In this first example, the various calculation steps with intermediate results are shown. Current in the coil can be described as a function of z as shown by figure 2. This function is decomposed as a Fourier series given by:

$$
S(z) = \sum_{n=0}^{\infty}\frac{4}{(2n+1)\pi}\cos\left(\frac{2(2n+1)\pi z}{D}\right)\sin\left(\frac{a(2n+1)\pi}{D}\right). \tag{30}
$$

This EMAT is graphically described by figure 2.

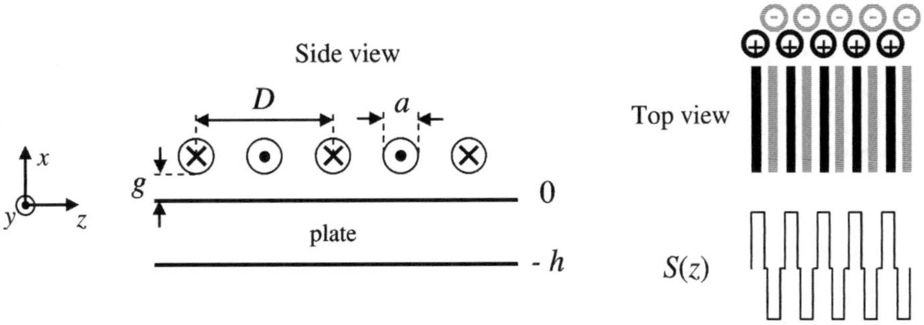

Figure 2. Left: Side view of a meander-coil EMAT. Right-top: Top view of the same EMAT. Right-bottom: a representation of the current passing through the coil.

In this example, the characteristics of the EMAT, of the material (iron) and the sampling in space and frequency are recorded in table 1: *n* is the number of wires, *nS* is the number of terms in the Fourier series, σ is the electrical conductivity and μ is the magnetic permeability of the material, n_z is the number of calculation points along the *z*-axis, n_x is the number of calculation points along the *x*-axis (the sampling is of 101 points for a depth of ten times the penetration depth δ, then of 10 points in the remaining thickness), *fmin* is the first calculation frequency, *fmax* the last, and finally, *nf* is the number of calculation frequencies.

Table 1. Characteristics of the EMAT, the material and the sampling in space and frequency.

n	17	-	ρ	7700	kg.m^{-3}	n_z	251	-
a	0.0005	m	***h***	0.003	m	n_x	101+10	-
D	0.002	m	μ_L	7.617 10^{10}	Pa	***fmin***	0.1	MHz
nS	50	-	σ	9.93 10^6	S.m^{-1}	***fmax***	5	MHz
g	0.0001	m	μ	10000	-	***nf***	501	-

The magnetostrictive tensor $\bar{\bar{e}}$ [N.A^{-1}] is considered as not varying with the dynamic magnetic field. In our computations, we take:

$$\bar{\bar{e}}^t = \begin{bmatrix} - & - & - & 1.10^4 & - & 2.10^4 \\ - & - & - & - & - & - \\ - & - & - & 1.10^4 & - & 2.10^4 \end{bmatrix}. \quad (31)$$

Most of the values are not given because they are not involved in equation (15). A more detailed study of this tensor [7, 12] shows that most of these coefficients actually equal zero and that the others are expressed as functions of other material parameters. This has not been used here; it may describe better magnetostriction in particular circumstances. Figure 3 displays the current Fourier series *S*(*z*).

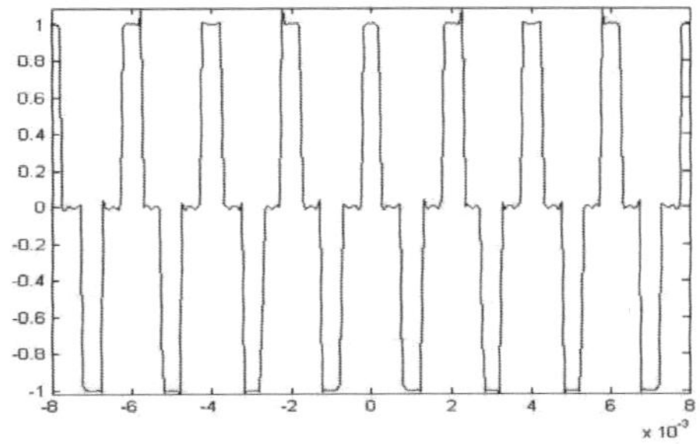

Figure 3. Current Fourier series *S*(*z*) of a meander-coil.

It is a good approximation of the real current spatial distribution, with limited Gibbs phenomenon. The magnetic field created in the plate for an excitation frequency of 1 MHz is then calculated and shown by figure 4. The penetration depth effect is clearly visible: the field amplitude drastically decreases at very short distance of the surface. The body force is calculated and plotted on figure 5 (left). The corresponding surface stress is shown on figure 5 (right). Note that these plots display the absolute values of complex quantities. In practice, at a given frequency, the tangential surface stress acts on the plate surface with alternating positive and negative sign along the transducer length. Finally, the modal amplitude for different modes as a function of frequency are calculated and plotted on figure 6 (left). On figure 6 right, only the modal amplitude of the SH0 mode is plotted.

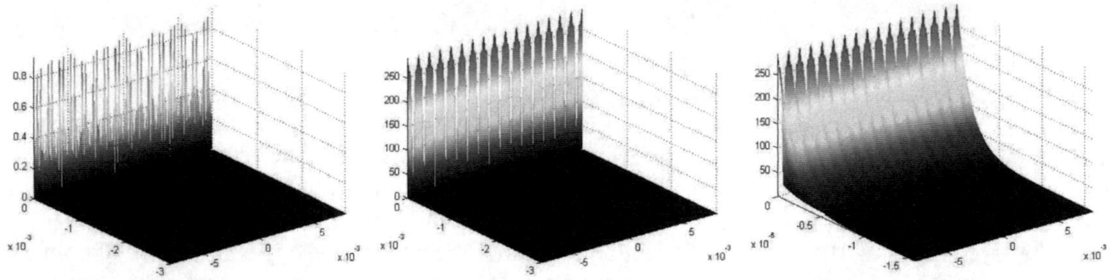

Figure 4. Magnetic field as a function of *x* (depth) and *z* (transducer length) created in an iron plate by a meander-coil EMAT at a frequency of 1 MHz. Left: *x*-component. Middle: *z*- component. Right: zoom on the *z*- component in the region near the top surface of the plate.

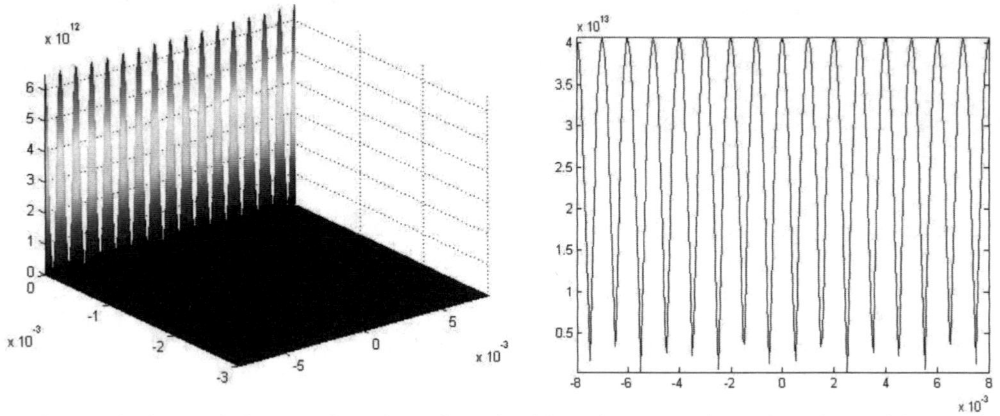

Figure 5. Body force (left) as a function of *x* (depth) and *z* (transducer length) and its equivalent surface stress (right) created by a meander-coil EMAT in an iron plate at 1 MHz. The equivalent surface stress is given for the SH0 mode only.

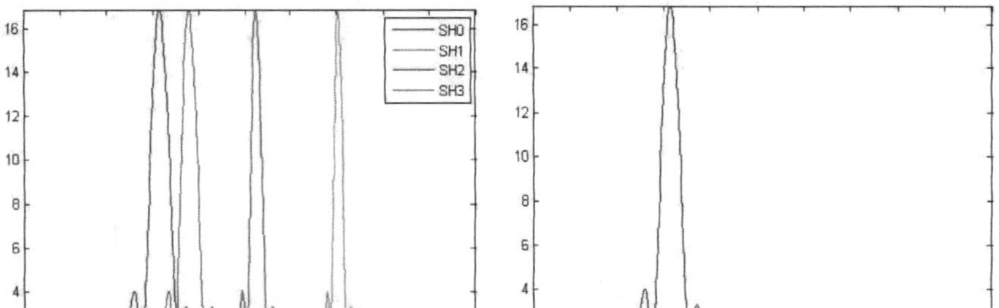

Figure 6. Left: modal amplitude over the frequency (in MHz) of the first four SH modes generated by the meander-coil. Right: same as left but only the SH0 mode is shown.

The modal amplitude variations with frequency for the different modes have a rather similar overall shape. The peak amplitudes are roughly equal and a resonance effect is clearly seen for all the modes considered. This last effect is due to the presence of the function *sinc* in the analytical expression of the modal amplitude typically of the form: $\mathrm{sinc}\ (\beta_m - s_n)L/2$. When the spatial periodicity of the coil, represented by the term s_n, matches the wavelength β_m, the function *sinc* takes its maximal value. This arises at a frequency which differs from one mode to another.

3.2. *Radiation by a uniform EMAT*
A uniform EMAT is now considered as shown by figure 7.

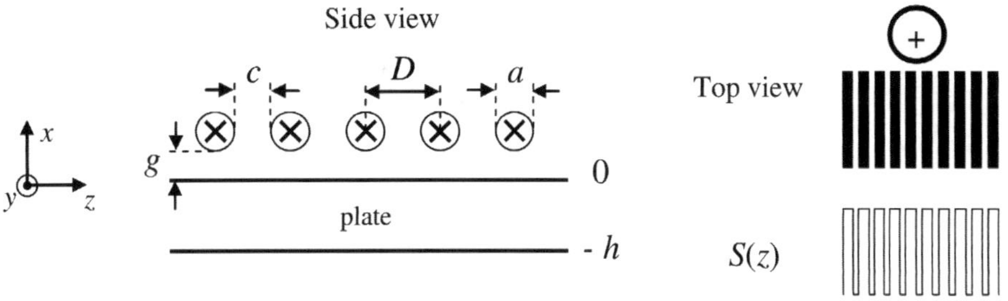

Figure 7. Left: Side view of a uniform EMAT. Right-top: Top view of the same EMAT. Right-bottom: a representation of the current passing through the wires.

In the following calculations, the two parameters *a* and *c* are equal. Other parameters (material etc.) are the same as those used for the meander-coil. On figure 8 are plotted modal amplitudes of the first four SH modes radiated for the uniform EMAT (left) from which is extracted that of the SH0 mode (right). A similar plot was presented for the meander-coil EMAT in figure 6.

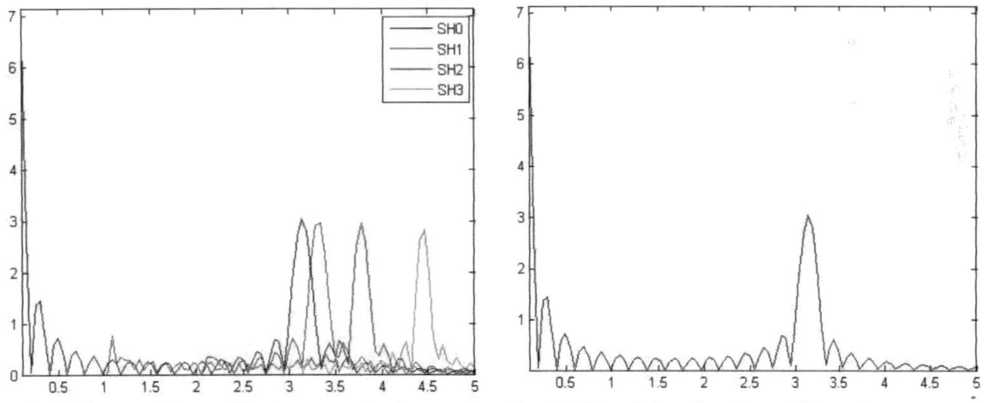

Figure 8. Left: modal amplitude over the frequency (in MHz) of the first four SH modes generated by the uniform EMAT, with c=a. Right: same as left but only the SH0 mode is shown.

This time, modal amplitudes vary with frequency from one mode to another. The first mode has its maximal value at null frequency; this continuous component was likely to appear with a uniform loading whereas it does not exist with spatially alternating loading of a meander-coil. Again, a modulation of amplitude is visible as for the meander-coil case; similarly, peak values correspond to a resonance when wire spacing coincides with the wavelength.

4. Conclusions
An analytical solution of the modal amplitude of SH guided modes generated by EMAT in a ferromagnetic plate has been developed. In the chain of sub-models developed for deriving the final formula, several offer more general capabilities to address other configurations than that of SH guided waves. In the present case, the formula developed allows fast and easy parametric studies if required

(such as those necessary for optimizing the design of a transducer). Experimental validations, extension to the 3D case and radiation of both surface and bulk waves will be further investigated.

5. Appendix. Transformation of body force into surface stress through a Taylor's series

The detailed derivation of the transformation of body force convolution integral into a surface stress convolution integral is complex and requires several steps of calculation. This calculation starts with the expression of the particle displacement as given by equation (14). Because the spatial variations of the Green's function are very low compared to those of body forces, the Green's function can be approximated by a Taylor series at the surface of the half-space. Here, the first three orders are taken into account:

$$G_{yy}(\vec{r},\vec{r}_0) = G_{yy}(\vec{r},\vec{r}_0\big|_{x_0=0}) + x_0 \frac{\partial G_{yy}(\vec{r},\vec{r}_0\big|_{x_0=0})}{\partial x_0} + \frac{1}{2} x_0^2 \frac{\partial^2 G_{yy}(\vec{r},\vec{r}_0\big|_{x_0=0})}{\partial x_0^2} + o\ x_0^2 \ . \tag{A1}$$

To simplify notations, the point $\vec{r}_0\big|_{x_0=0}$ is denoted $\vec{R}_0$ in what follows. Substituting equation (A1) into equation (14) yields

$$u_y(\vec{r}) \approx \int_{-L/2}^{L/2} \left[G_{yy}(\vec{r},\vec{R}_0) T_y(\vec{R}_0) + \frac{\partial G_{yy}(\vec{r},\vec{R}_0)}{\partial x_0} M_y(\vec{R}_0) + \frac{1}{2} \frac{\partial^2 G_{yy}(\vec{r},\vec{R}_0)}{\partial x_0^2} N_y(\vec{R}_0) \right] dz_0 \ , \tag{A2}$$

where T_y, M_y and N_y denote moments of the body force f_y, defined by:

$$T_y(\vec{R}_0) = \int_{-\infty}^{0} f_y(\vec{r}_0) dx_0 \ , \quad M_y(\vec{R}_0) = \int_{-\infty}^{0} x_0 f_y(\vec{r}_0) dx_0 \quad \text{and} \quad N_y(\vec{R}_0) = \int_{-\infty}^{0} x_0^2 f_y(\vec{r}_0) dx_0 . \tag{A3}$$

The following development aims at simplifying the derivative of the Green's function G_{yy}. This requires returning momentarily in 3D space. Using the definition of the strain tensor and the isotropy of the material, the local equilibrium Lamé's law is written as

$$\sum_j K_{ij}(\vec{r}_0) u_j(\vec{r}_0) = f_i(\vec{r}_0) \ , \tag{A4}$$

where $K_{ij}(\vec{r}_0) =$

$$\begin{pmatrix} -\omega^2\rho - C_{11}\frac{\partial^2}{\partial x_0^2} - C_{44}\left(\frac{\partial^2}{\partial y_0^2} + \frac{\partial^2}{\partial z_0^2}\right) & -(C_{12}+C_{44})\frac{\partial^2}{\partial x_0 \partial y_0} & -(C_{12}+C_{44})\frac{\partial^2}{\partial x_0 \partial z_0} \\ -(C_{12}+C_{44})\frac{\partial^2}{\partial x_0 \partial y_0} & -\omega^2\rho - C_{11}\frac{\partial^2}{\partial y_0^2} - C_{44}\left(\frac{\partial^2}{\partial x_0^2} + \frac{\partial^2}{\partial z_0^2}\right) & -(C_{12}+C_{44})\frac{\partial^2}{\partial y_0 \partial z_0} \\ -(C_{12}+C_{44})\frac{\partial^2}{\partial x_0 \partial z_0} & -(C_{12}+C_{44})\frac{\partial^2}{\partial y_0 \partial z_0} & -\omega^2\rho - C_{11}\frac{\partial^2}{\partial z_0^2} - C_{44}\left(\frac{\partial^2}{\partial x_0^2} + \frac{\partial^2}{\partial y_0^2}\right) \end{pmatrix} \tag{A5}$$

and where the components of the compliance matrix C are:

$$C_{12} = \lambda_L \ ; \quad C_{44} = \mu_L \ ; \quad C_{11} = C_{12} + 2C_{44} = \lambda_L + 2\mu_L . \tag{A6}$$

The next steps will not solve the equations for the Green's function but develop some conditions that will help us to simplify equation (A2). By multiplying equation (A4) by G_{ki}, then, summing over i and finally integrating over the space, one gets:

$$\sum_{i,j} \int_{-\infty}^{+\infty}\int_{-\infty}^{+\infty}\int_{-\infty}^{0} G_{ki}(\vec{r},\vec{r}_0) K_{ij}(\vec{r}_0) u_j(\vec{r}_0) dy_0 dz_0 dx_0 = \sum_i \int_{-\infty}^{+\infty}\int_{-\infty}^{+\infty}\int_{-\infty}^{0} G_{ki}(\vec{r},\vec{r}_0) f_i(\vec{r}_0) dy_0 dz_0 dx_0 \ . \tag{A7}$$

Using the definition of the displacement as a spatial convolution between the Green's functions and the body forces, the displacement is now re-written as:

$$u_k(\vec{r}) = \sum_{i,j} \int_{-\infty}^{+\infty} \int_{-\infty}^{+\infty} \int_{-\infty}^{0} G_{ki}(\vec{r},\vec{r}_0) K_{ij}(\vec{r}_0) u_j(\vec{r}_0) dy_0 dz_0 dx_0 . \tag{A8}$$

Moreover, free boundary conditions are assumed on the surface of the half-space. Integrating by parts each term of the right-hand side of this last equation, using the boundary conditions and the fact that $\bar{\bar{G}}$ and $\vec{u}$ tends to zero while the integration point tends to infinity, equation (A8) becomes

$$u_k(\vec{r}) = \sum_{i,j} \int_{-\infty}^{+\infty} \int_{-\infty}^{+\infty} \left[\begin{array}{l} \displaystyle\int_{-\infty}^{0} u_j(\vec{r}_0) K_{ij}(\vec{r}_0) G_{ki}(\vec{r},\vec{r}_0)\, dx_0 \\ \displaystyle + u_x(\vec{R}_0) \left[C_{11} \frac{\partial G_{kx}(\vec{r},\vec{R}_0)}{\partial x_0} + C_{12} \left(\frac{\partial G_{ky}(\vec{r},\vec{R}_0)}{\partial y_0} + \frac{\partial G_{kz}(\vec{r},\vec{R}_0)}{\partial z_0} \right) \right] \\ \displaystyle + u_y(\vec{R}_0) C_{44} \left[\frac{\partial G_{ky}(\vec{r},\vec{R}_0)}{\partial x_0} + \frac{\partial G_{kx}(\vec{r},\vec{R}_0)}{\partial y_0} \right] \\ \displaystyle + u_z(\vec{R}_0) C_{44} \left[\frac{\partial G_{kx}(\vec{r},\vec{R}_0)}{\partial z_0} + \frac{\partial G_{kz}(\vec{r},\vec{R}_0)}{\partial x_0} \right] \end{array} \right] dy_0 dz_0 . \tag{A9}$$

The conditions for the existence of $\bar{\bar{G}}$ are then:

$$\begin{gathered} \sum_i K_{ij}(\vec{r}_0) G_{ki}(\vec{r},\vec{r}_0) = \delta^3(\vec{r}-\vec{r}_0)\delta_{kj} , \\ C_{11} \frac{\partial G_{kx}(\vec{r},\vec{R}_0)}{\partial x_0} + C_{12} \left(\frac{\partial G_{ky}(\vec{r},\vec{R}_0)}{\partial y_0} + \frac{\partial G_{kz}(\vec{r},\vec{R}_0)}{\partial z_0} \right) = 0 , \\ \frac{\partial G_{ky}(\vec{r},\vec{R}_0)}{\partial x_0} + \frac{\partial G_{kx}(\vec{r},\vec{R}_0)}{\partial y_0} = 0 \text{ and } \frac{\partial G_{kx}(\vec{r},\vec{R}_0)}{\partial z_0} + \frac{\partial G_{kz}(\vec{r},\vec{R}_0)}{\partial x_0} = 0 , \end{gathered} \tag{A10}$$

where δ_{kj} is the Krönecker symbol.

Using equations (A10) and (A4), the conditions for $k = y$ are:

$$\begin{aligned} K_{xy}(\vec{r}_0) G_{yx}(\vec{r},\vec{r}_0) + K_{yy}(\vec{r}_0) G_{yy}(\vec{r},\vec{r}_0) + K_{zy}(\vec{r}_0) G_{yz}(\vec{r},\vec{r}_0) = -(C_{12}+C_{44}) \left(\frac{\partial^2 G_{yx}(\vec{r},\vec{r}_0)}{\partial x_0 \partial y_0} + \frac{\partial^2 G_{yz}(\vec{r},\vec{r}_0)}{\partial y_0 \partial z_0} \right) \\ - \left[\omega^2 \rho + C_{11} \frac{\partial^2}{\partial y_0^2} + C_{44} \left(\frac{\partial^2}{\partial x_0^2} + \frac{\partial^2}{\partial z_0^2} \right) \right] G_{yy}(\vec{r},\vec{r}_0) = 0 , \end{aligned} \tag{A11a}$$

$$C_{11} \frac{\partial G_{yx}(\vec{r},\vec{R}_0)}{\partial x_0} + C_{12} \left(\frac{\partial G_{yy}(\vec{r},\vec{R}_0)}{\partial y_0} + \frac{\partial G_{yz}(\vec{r},\vec{R}_0)}{\partial z_0} \right) = 0 , \tag{A11b}$$

$$\frac{\partial G_{yy}(\vec{r},\vec{R}_0)}{\partial x_0} + \frac{\partial G_{yx}(\vec{r},\vec{R}_0)}{\partial y_0} = 0 \text{ and } \frac{\partial G_{yx}(\vec{r},\vec{R}_0)}{\partial z_0} + \frac{\partial G_{yz}(\vec{r},\vec{R}_0)}{\partial x_0} = 0 . \tag{A11c and d}$$

Then, differentiating the second condition of equation (A11) with respect to y_0 leads to:

$$\frac{\partial^2 G_{yx}(\vec{r},\vec{R}_0)}{\partial y_0 \partial x_0} = -\frac{C_{12}}{C_{11}}\left(\frac{\partial^2 G_{yy}(\vec{r},\vec{R}_0)}{\partial^2 y_0} + \frac{\partial^2 G_{yz}(\vec{r},\vec{R}_0)}{\partial y_0 \partial z_0}\right). \tag{A12}$$

With the condition (A11a), the second derivative of G_{yy} with respect to x_0 is:

$$\begin{aligned}\frac{\partial^2 G_{yy}(\vec{r},\vec{R}_0)}{\partial x_0^2} = &-\left[\frac{\omega^2\rho}{C_{44}} + \frac{C_{11}^2 - C_{12}^2 - C_{12}C_{44}}{C_{11}C_{44}}\frac{\partial^2}{\partial y_0^2} + \frac{\partial^2}{\partial z_0^2}\right] G_{yy}(\vec{r},\vec{R}_0) \\ &- \frac{(C_{12}+C_{44})(C_{11}-C_{12})}{C_{11}C_{44}}\frac{\partial^2 G_{yz}(\vec{r},\vec{R}_0)}{\partial y_0 \partial z_0}.\end{aligned} \tag{A13}$$

Finally, using the third condition (A11c), the first order derivative of G_{yy} with respect to x_0 is:

$$\frac{\partial G_{yy}(\vec{r},\vec{R}_0)}{\partial x_0} = -\frac{\partial G_{yx}(\vec{r},\vec{R}_0)}{\partial y_0}. \tag{A14}$$

The first and the second derivatives of G_{yy} with respect to x_0, given by equations (A14) and (A13), are substituted into equation (A2). Returning in 2D, derivatives with respect to y_0 vanish so that equation (A2) eventually becomes:

$$u_y(\vec{r}) \approx \int_{-L/2}^{L/2}\left[T_y - \frac{1}{2}N_y\left(\frac{\omega^2\rho}{C_{44}} + \frac{\partial^2}{\partial z_0^2}\right)\right] G_{yy}(\vec{r},\vec{R}_0)\,dz_0\,, \tag{A15}$$

which corresponds to equation (16) in the body of the paper.

References

[1] Hirao M and Ogi H 2003 *EMATS for Science and Industry: Noncontacting Ultrasonic Measurements* (Boston: Kluwer Academic Publishers)
[2] Landau L and Lifshits E 2008 On the theory of the dispersion of magnetic permeability in ferromagnetic bodies *Ukr. J. Phys.* **53** 14-22 reprinted from 1935 *Phys. Zeitsch. der Sow.* **8** 153
[3] Ahmadi B, Chazal H, Waeckerlé T and Roudet J 2010 *IEEE Trans. Magn.* **46** 4001
[4] Ogi H 1997 *J. Appl. Phys.* **82** 3940
[5] Thompson R B 1990 Physical principles of measurements with EMAT transducers *Physical Acoustics: Ultrasonic Measurement Methods* vol 19 chapter 4 pp 157–200
[6] Auld B A 1973 *Acoustics fields and waves in solids* vol 2 (Malabar, Florida: Krieger publishing company)
[7] Ogi H, Goda E and Hirao M 2003 *Jpn. J. Appl. Phys.* **42** 3020
[8] Laguerre L, Aime J-C and Brissaud M 2002 *Ultrasonics* **39** 503
[9] Hueter T F and Bolt R H *Sonics: Techniques for the Use of Sound and Ultrasound in Engineering and Science* (New York: John Wiley and Sons inc.)
[10] Thompson R B 1980 *J. Nond. Eval.* **1** 79
[11] Thompson R B 1973 *IEEE Trans. Son. Ultra.* **SU-20** 340
[12] Ribichini R, Cegla F, Nagy P B and Cawley P 2010 *IEEE Trans. Ultra. Ferro. Freq. Cont.* **57** 2808

Application of the physical theory of diffraction in ultrasonic non-destructive evaluation

V Zernov[1], L Fradkin[1] and M Darmon[2]

[1] Sound Mathematics Ltd., Cambridge, CB4 2AS, UK
[2] CEA, LIST, NDE Department, Gif-sur-Yvette, F-91191, France.

E-mail: v.zernov@soundmathematics.com

Abstract. We study two canonical problems of interest in non-destructive evaluation, diffraction of a plane elastic wave by a thin crack and diffraction of a plane elastic wave by a wedge, both in the high-frequency regime. In applications this regime is usually treated using the so-called Kirchhoff approximation. It is very easy to implement but there are situations when it is known to give distorted results. We discuss an easy correction procedure, which is applicable not only in geometrical regions but inside penumbrae as well. It involves a version of the Physical Theory of Diffraction that relies on the Geometrical Theory of Diffraction rather than the full solution of the corresponding canonical problem.

1. Introduction

The fields scattered by straight edged obstacles have a complicated structure, particularly in the so-called radiating near field, the first Fresnel zone outside the classical near field zone. In optics, acoustics, electrodynamics and elastodynamics such fields can be modelled well using the high-frequency asymptotics known as the Geometrical Optics/Acoustics/Electrodynamics/ Elastodynamics (GO/GA/GE) [1] and Geometrical Theory of Diffraction (GTD) [2], all employing the concept of a ray. The former theories describe waves carried by incident and reflected rays and the latter, waves diffracted by edges, corners and vertices as well as waves carried by rays which are grazing obstacles. The regions that support different types of waves are classified as either geometrical or transition zones. In a geometrical region, each point is reached by an isolated ray or else by a finite number of isolated rays carrying the same type of wave (the rays are considered coalescing—as opposed to isolated—if the difference between the phases they carry is less than say π). A transition zone is a boundary layer that lies between an irradiated geometrical region and the related geometrical shadow. When elastic waves are scattered by straight edged cracks the only types of geometrical regions are those irradiated by longitudinal or transverse incident or reflected waves or else regions irradiated by head waves. Therefore, the only transition zones are the associated penumbrae which surround incident and reflected shadow boundaries; neighbourhoods of rays of grazing incidence and reflection; and finally, neighbourhoods of critical rays that mark the boundaries of the head wave regions. GO/GA/GE and GTD combine to give a perfectly adequate description of geometrical regions but are inapplicable inside transition zones. A more sophisticated, Uniform, GTD is required to complete the description. The radiating near field of obstacles with curved edges require additional asymptotics to describe caustic and focal zones (e.g. [3], [4], [5], [6]).

The near radiating field of the scatterer is of practical importance in applications, such as Non-Destructive Evaluation (NDE) of industrial components and structures. When dealing with obstacles of a complicated shape, instead of combining GO/GA/GE and GTD, NDE modellers often simulate the high-frequency scattered fields using the so-called Kirchhoff approximation. This is also referred to as the physical optics approximation and relies on a numerical evaluation of the Kirchhoff-Rayleigh integral (e.g. [7], [8], [9]). Unlike the classical GTD, the Kirchhoff approximation works in the far field and in the near radiating field, it is regular inside penumbrae. The resulting leading terms describe the reflected waves, and the terms which describe diffracted waves have the same phases as the GTD asymptotics. However, away from the directions normal to the obstacle edge the Kirchhoff amplitudes of the diffracted waves are quite different to those predicted by GTD [10],[11], particularly in elastodynamics. The elastodynamic Kirchhoff approximation does not describe the contribution of the Rayleigh waves or head waves either.

The Physical Theory of Diffraction (PTD) has been developed to overcome some of these limitations [11]. In PTD, the total field on a scatterer is described similarly to the Kirchhoff approximation, as an GO/GA/GE field, while near the scatterer edge the total field is corrected by adding to it the difference between the solution of the relevant canonical problem and GO/GA/GE. A procedure similar to PTD was developed by McMaken [12] to describe diffraction of elastic waves by embedded planar cracks. However, the canonical displacements on the planar cracks and other elastic wedges are particularly difficult to compute [12], [13], [14]. To overcome this difficulty we offer a modification of PTD and correct the Kirchhoff approximation by employing GTD rather than the full solution of the corresponding canonical problem. The GTD diffraction coefficients can be computed in an efficient manner using algorithms described in [15] and [16]. In this paper we show how this approach can be employed to simulate diffraction by a planar crack and elastic wedge. The proposed method is applicable inside geometrical regions and penumbrae but not other transition zones.

2. Scatter by a semi-infinite planar crack

Let us assume that a uniform elastic solid contains a semi-infinite planar crack S. Let us introduce the associated Cartesian coordinate system (x_1, x_2, x_3) (see Fig. 1) and cylindrical polar coordinates (r, θ, x_3), such that we have $x_1 = r\cos\theta$, $x_2 = r\sin\theta$. Then the crack can be described as $S = \{\mathbf{x} : x_1 \geq 0, x_2 = 0\}$, with $\mathbf{x} = (x_1, x_2, x_3)$—a point in space and the upper and lower faces of the crack designated, respectively, by S^+ and S^-.

Let the crack be irradiated by an incident time-harmonic plane wave. This means that the incident field and all scattered fields contain the common factor $\exp(-\mathrm{i}\omega t)$, where ω denotes the circular frequency, t is time and i is the imaginary unity. For simplicity of presentation we omit this factor throughout.

It is well known that the amplitude $\mathbf{U}(\mathbf{x})$ of the time-harmonic displacement vector field satisfies the reduced equation of motion

$$\nabla \times (\nabla \times \mathbf{U}(\mathbf{x})) + \kappa^2 \nabla \cdot (\nabla \cdot \mathbf{U}(\mathbf{x})) + k_T^2 \mathbf{U}(\mathbf{x}) = \mathbf{0}, \qquad (1)$$

where $\kappa = c_L/c_T$, c_L and c_T are the respective longitudinal and transverse wave speeds; and $k_L = \omega/c_L$ and $k_T = \omega/c_T$ are the respective longitudinal and transverse wave numbers.

Let us assume that the faces of the crack satisfy the standard traction free boundary conditions

$$\sigma_{12}(\mathbf{x})|_S = \sigma_{22}(\mathbf{x})|_S = \sigma_{23}(\mathbf{x})|_S = 0. \qquad (2)$$

Above, $\sigma(\mathbf{x}) = \{\sigma_{ij}(\mathbf{x})\}_{i,j=1}^3$ is the stress tensor, which according to Hooke's law is related to the displacement $\mathbf{U}(\mathbf{x})$ via

$$\sigma_{ij}(\mathbf{x}) = \lambda \delta_{ij} \nabla \cdot \mathbf{U}(\mathbf{x}) + \mu \left(U_{i,j}(\mathbf{x}) + U_{j,i}(\mathbf{x}) \right), \quad i, j = 1, 2, 3, \qquad (3)$$

with $\lambda = \rho(c_L^2 - 2c_T^2)$ and $\mu = \rho c_T^2$—the Lamé constants and ρ—solid density.

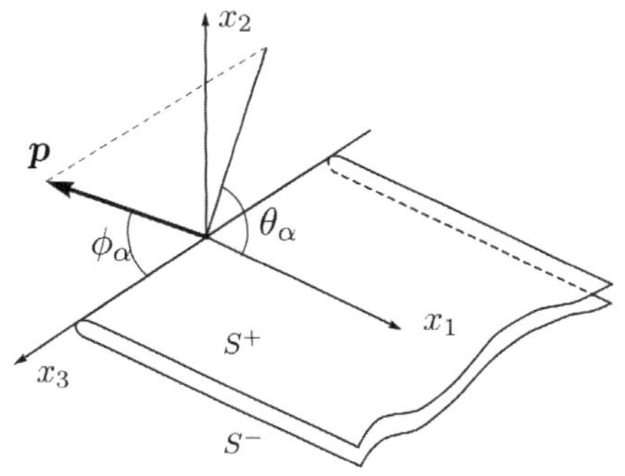

Figure 1. The geometry of the plane crack.

The incident time-harmonic plane wave can be described by

$$\mathbf{U}^{\mathrm{Inc}}(\mathbf{x}) = \boldsymbol{d}^{\alpha} \mathrm{e}^{\mathrm{i}k_\alpha \boldsymbol{p}\cdot\mathbf{x}}. \qquad (4)$$

Here we follow the tradition and omit the dependence of $\mathbf{U}(\mathbf{x})$ on $\boldsymbol{p}$ and k_α. The index $\alpha = L, TV$ or TH refers to the longitudinal, transverse-vertical or transverse-horizontal polarisation of incident wave, respectively. Note that it is a subscript when describing scalars, and superscript otherwise. The vector

$$\boldsymbol{p} = (p_1, p_2, p_3) = (\sin\phi_\alpha \cos\theta_\alpha, \sin\phi_\alpha \sin\theta_\alpha, \cos\phi_\alpha). \tag{5}$$

denotes the direction of wave propagation and is specified by two angles, ϕ_α and θ_α (Fig. 1). It defines the polarisation vectors

$$\boldsymbol{d}^L = \boldsymbol{p}, \quad \boldsymbol{d}^{TH} = \frac{\boldsymbol{e_2} \times \boldsymbol{p}}{|\boldsymbol{e_2} \times \boldsymbol{p}|}, \quad \boldsymbol{d}^{TV} = \boldsymbol{d}^{TH} \times \boldsymbol{p}. \tag{6}$$

2.1. The high-frequency approximations

Using the well known Green's theorem the above problem can be reformulated as an integral equation which relates the unknown scattered field $\mathbf{U}^{\mathrm{Scat}}(\mathbf{x})$ at an arbitrary point outside the crack to the total field $\mathbf{U}(\mathbf{x}) = \mathbf{U}^{\mathrm{Scat}}(\mathbf{x}) + \mathbf{U}^{\mathrm{Inc}}(\mathbf{x})$ on its surface which, of course, is also unknown. The component representation of this relationship is given by

$$U_k^{\mathrm{Scat}}(\mathbf{x}) = -\int_S \tau_{i2;k}^G(\mathbf{x} - \mathbf{x}')U_i(\mathbf{x}')ds(\mathbf{x}'), \; i, k = 1, 2, 3. \tag{7}$$

Above, $\mathbf{x}'$ is an arbitrary point on the surface of the crack, $ds(\mathbf{x}')$ is an elementary area surrounding $\mathbf{x}'$ and τ^G is the Green's stress tensor [17]. The integral on the right of (7) is known as the Rayleigh integral.

The problem under consideration can be solved analytically [17]. The high-frequency asymptotics of the resulting scattered field $\mathbf{U}^{\mathrm{Scat}}(\mathbf{x})$, which are valid outside penumbrae, critical and grazing regions have been shown to be [17]

$$\mathbf{U}^{\mathrm{Scat}}(\mathbf{x}) = \mathbf{U}^{\mathrm{GE}}(\mathbf{x}) + \mathbf{U}^{\mathrm{Rayleigh}}(\mathbf{x}) + \mathbf{U}^{\mathrm{GTD}}(\mathbf{x}) + O([k_T r]^{-1}), \tag{8}$$

where the geometrical elastodynamics field $\mathbf{U}^{\mathrm{GE}}(\mathbf{x})$ and Rayleigh field $\mathbf{U}^{\mathrm{Rayleigh}}(\mathbf{x})$ comprise reflected waves and surface waves, respectively, and the GTD field is

$$\mathbf{U}^{\mathrm{GTD}}(\mathbf{x}) = \sum_{\beta = L,TV,TH} D_{\alpha\beta}^{\mathrm{GTD}}(\theta, \phi_\alpha, \theta_\alpha) \frac{\mathrm{e}^{\mathrm{i}k_\beta(x_3 \cos\phi_\beta + r \sin\phi_\beta)}}{\sqrt{k_\beta r}} \boldsymbol{d}^\beta(\boldsymbol{p}^\beta). \tag{9}$$

Above, $D_{\alpha\beta}^{\mathrm{GTD}}$ are the so-called GTD diffraction coefficients and $\boldsymbol{p}^\beta$ are the directions of diffracted waves. The recipes for calculating $\mathbf{U}^{\mathrm{GE}}(\mathbf{x})$, $\mathbf{U}^{\mathrm{Rayleigh}}(\mathbf{x})$ and $D_{\alpha\beta}^{\mathrm{GTD}}$ can be found in [17] but need not be reproduced here.

As mentioned in the Introduction, many numerical high-frequency NDE models employ another high-frequency approximation: assuming the boundary S^+ to be irradiated and boundary S^- to lie in the geometric shadow ($-\pi < \theta_\alpha < 0$), the total field $\mathbf{U}(\mathbf{x})$ in (7) is assumed to be

$$\mathbf{U}(\mathbf{x}')\big|_{S+} \approx \mathbf{A}^\alpha \mathrm{e}^{\mathrm{i}k_\alpha(\sin\phi_\alpha \cos\theta_\alpha x_1' + \cos\phi_\alpha x_3')}, \quad \mathbf{U}(\mathbf{x}')\big|_{S^-} \approx \mathbf{0}. \tag{10}$$

where the vector amplitudes $\mathbf{A}^\alpha$ are approximations to the crack opening displacements given by (A.3) and (A.5). Substituting then the known fields (10) into (7) the scattered fields $\mathbf{U}^{\mathrm{Scat}}(\mathbf{x})$ are evaluated in the Kirchhoff approximation,

$$U_k^{\mathrm{Kir}}(\mathbf{x}) = -\int_{S^+} \tau_{i2;k}^G(\mathbf{x}-\mathbf{x}')A_i^\alpha \mathrm{e}^{\mathrm{i}k_\alpha(x_1'\sin\phi_\alpha\cos\theta_\alpha + x_3'\cos\phi_\alpha)}ds(\mathbf{x}'). \tag{11}$$

The Kirchhoff-Rayleigh integrals on the right of (11) are uaually evaluated numerically.

As also mentioned in the Introduction, there are geometrical regions where the Kirchhoff approximation deteriorates and GTD performs better. We will now show that the Kirchhoff approximation can be improved by evaluating the Kirchhoff-Rayleigh integral asymptotically and comparing the result to (8). Let us introduce auxiliary matrices

$$B_{ij}^L(\widehat{\mathbf{x}}) = 2\kappa^{-2}\widehat{x}_i\widehat{x}_j + (1-2\kappa^{-2})\delta_{ij},$$
$$B_{ij}^{TH}(\widehat{\mathbf{x}}) = \widehat{x}_j d_i^{TH} + \widehat{x}_i d_j^{TH}, \quad B_{ij}^{TV}(\widehat{\mathbf{x}}) = \widehat{x}_j d_i^{TV} + \widehat{x}_i d_j^{TV}, \tag{12}$$

where $\mathrm{x} = |\mathbf{x}|$ and $\widehat{\mathbf{x}}$ is a unit vector in the direction of $\mathbf{x}$, $\widehat{\mathbf{x}} = \mathbf{x}/\mathrm{x}$. It is convenient to represent the phase of the incident α wave on S in terms of the scattered β wave,

$$k_\alpha(x_1'\sin\phi_\alpha\cos\theta_\alpha + x_3'\cos\phi_\alpha) = k_\beta(x_1'\cos\psi_\beta^{\mathrm{GE}} + x_3'\cos\phi_\beta), \tag{13}$$

where $\alpha, \beta = L, TV, TH$ and the projections of the reflected β ray onto the edge and inner normal to this edge are, respectively,

$$\cos\phi_\beta = \frac{k_\alpha}{k_\beta}\cos\phi_\alpha, \quad \cos\psi_\beta^{\mathrm{GE}} = \frac{k_\alpha}{k_\beta}\sin\phi_\alpha\cos\theta_\alpha. \tag{14}$$

When the observation point lies outside all penumbrae, the Kirchhoff-Rayleigh integral (11) can be evaluated asymptotically using the standard stationary phase method [4] to give

$$\mathbf{U}^{\mathrm{Kir}}(\mathbf{x}) = \mathbf{U}^{\mathrm{GE}}(\mathbf{x}) + \sum_{\beta=L,TV,TH} D_{\alpha\beta}^{\mathrm{Kir}}(\theta,\phi_\alpha,\theta_\alpha)\boldsymbol{d}^\beta \frac{\mathrm{e}^{\mathrm{i}k_\beta(x_3\cos\phi_\beta + r\sin\phi_\beta)}}{\sqrt{k_\beta r}} + O([k_T r]^{-1}), \tag{15}$$

where a Kirchhoff diffraction coefficient describing a diffracted β wave generated by an incident α waves is

$$D_{\alpha\beta}^{\mathrm{Kir}} = -\frac{\mathrm{e}^{\mathrm{i}\frac{\pi}{4}}}{2\sqrt{2\pi\sin\phi_\beta}} \cdot \frac{B_{i2}^\beta(\boldsymbol{p}^\beta)A_i^\alpha}{\cos\psi_\beta^{\mathrm{GE}} - \cos\psi_\beta^{\mathrm{Diff}}}, \tag{16}$$

with $\psi_\beta^{\mathrm{Diff}}$—the angle between the diffracted β ray and inner normal to the edge. Comparing (8) with (15), outside the transition zones we have

$$\mathbf{U}^{\mathrm{Scat}}(\mathbf{x}) = \mathbf{U}^{\mathrm{Kir}}(\mathbf{x}) + \mathbf{U}^{\mathrm{Rayleigh}}(\mathbf{x}) + \sum_{\beta=L,TV,TH} D_{\alpha\beta}^{\mathrm{Corr}}(\theta,\phi_\alpha,\theta_\alpha)\boldsymbol{d}^\beta \frac{\mathrm{e}^{\mathrm{i}k_\beta(x_3\cos\phi_\beta + r\sin\phi_\beta)}}{\sqrt{k_\beta r}} + O([k_T r]^{-1}), \tag{17}$$

where the correction to the Kirchhoff diffraction coefficient is

$$D_{\alpha\beta}^{\mathrm{Corr}}(\theta,\phi_\alpha,\theta_\alpha) = D_{\alpha\beta}^{\mathrm{GTD}}(\theta,\phi_\alpha,\theta_\alpha) - D_{\alpha\beta}^{\mathrm{Kir}}(\theta,\phi_\alpha,\theta_\alpha). \tag{18}$$

It is possible to show that this correction applies inside all penumbrae as well as inside geometrical regions.

3. The scatter by a two-dimensional elastic wedge

Let us now consider scatter of a plane wave by a two-dimensional infinite wedge of angle 2Ω (Fig. 2). Let us introduce the associated Cartesian coordinate system (x_1, x_2) and polar coordinates (r, θ), such that we have $x_1 = r\cos\theta$ and $x_2 = r\sin\theta$. Then we can say that the elastic wedge occupies the region $\{(r,\theta) : -\Omega \leq \theta \leq \Omega\}$ and designate its boundary by $S = S^+ \bigcup S^-$, with S^+ and S^+—its upper face and lower face, respectively. As above, we assume that the wedge faces satisfy the traction free boundary conditions and let the time-harmonic incident wave be of the form (4). This problem can be reduced to an integral equation, which can be solved numerically [13], [14], [16] and then the integral representing the scattered field $\mathbf{U}_\Omega^{\mathrm{Scat}}(\mathbf{x})$ can be evaluated asymptotically. When the observation point lies outside all transition zones, the evaluation can be carried out using the standard high-frequency asymptotics [4] to give

$$\mathbf{U}_\Omega^{\mathrm{Scat}}(\mathbf{x}) = \mathbf{U}_\Omega^{\mathrm{GE}}(\mathbf{x}) + \mathbf{U}_\Omega^{\mathrm{Rayleigh}}(\mathbf{x}) + \mathbf{U}_\Omega^{\mathrm{GTD}}(\mathbf{x}) + O([k_T r]^{-1}), \tag{19}$$

where the geometrical elastodynamics term $\mathbf{U}_\Omega^{\mathrm{GE}}(\mathbf{x})$ is the sum of all multiply reflected plane waves arriving at the observation point, $\mathbf{U}_\Omega^{\mathrm{Rayleigh}}(\mathbf{x})$ is the sum of two scattered Rayleigh waves and the edge diffraction term $\mathbf{U}_\Omega^{\mathrm{GTD}}(\mathbf{x})$ is

$$\mathbf{U}_\Omega^{\mathrm{GTD}}(\mathbf{x}) = \sum_{\beta=L,TV} D_{\alpha\beta,\Omega}^{\mathrm{GTD}}(\theta,\theta_\alpha)\frac{\mathrm{e}^{ik_\beta r}}{\sqrt{k_\beta r}}\boldsymbol{d}^\beta, \tag{20}$$

with the polarisation vectors are $\boldsymbol{d}^L = (\cos\theta, \sin\theta)$ and $\boldsymbol{d}^{TV} = (-\sin\theta, \cos\theta)$.

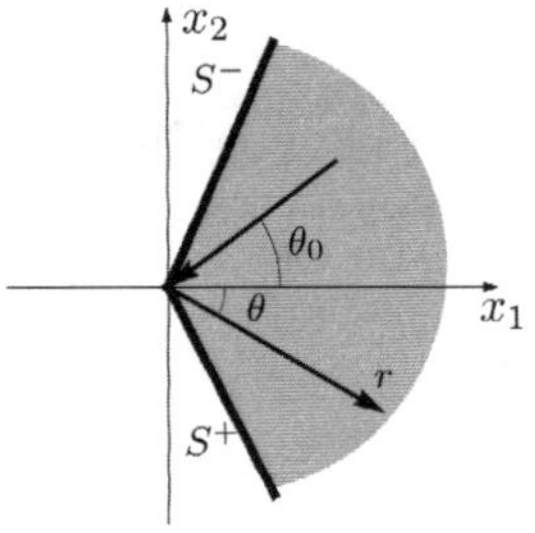

Figure 2. Geometry of the wedge problem.

Similarly to the previous section, in NDE, the most widely used approach to calculating the scattered field in the high-frequency regime is the Kirchhoff

Approximation. As above, this means that the problem is reformulated using the Green's Theorem and the wedge surface displacement in the Rayleigh integral is approximated as the geometrical elastodynamics displacement $\mathbf{U}_{\Omega}^{\mathrm{GE}}(\mathbf{x})$, that is, as the sum of a finite number of plane waves multiply reflected by the wedge faces $S^{\pm}$. These waves can be evaluated by using ray tracing and amplitude calculations to give

$$C_{ij}^{\beta}\boldsymbol{d}^{\beta}\mathrm{e}^{\mathrm{i}k_{\beta}r\cos(\theta-\theta_{ij}^{\beta})}, \quad i=1,2, \quad \beta=L,TV, \quad j=1,2,...,N_i^{\beta}, \tag{21}$$

where the index β describes the wave polarisation, the first subscript i indicates whether the plane wave is impinging on the irradiated wedge face S^+ $(i=1)$ or the other face S^- $(i=2)$ and the second subscript j specifies the number of underlying reflections, with N_1^{β} (N_2^{β}) being the maximum number of multiply reflected β-plane waves which impinge on the face S^+ (S^-). Each of these waves is characterised by its polar angle θ_{ij}^{β} and constant amplitude C_{ij}^{β}. This implies that the wedge Kirchhoff approximation $\mathbf{U}_{\Omega}^{\mathrm{Kir}}(\mathbf{x})$ can be represented as a sum of the Kirchhoff-Rayleigh integrals of type (11) evaluated over the wedge faces. Using the non-uniform expansion (15), outside all transition zones, we can write

$$\mathbf{U}_{\Omega}^{\mathrm{Kir}}(\mathbf{x}) = \mathbf{U}_{\Omega}^{\mathrm{GE}}(\mathbf{x}) + \sum_{\beta=L,TV} D_{\alpha\beta,\Omega}^{\mathrm{Kir}}(\theta,\theta_{\alpha})\boldsymbol{d}^{\beta}\frac{\mathrm{e}^{ik_{\beta}r}}{\sqrt{k_{\beta}r}} + O([kr]^{-1}), \tag{22}$$

where the Kirchhoff diffraction coefficients are

$$D_{\alpha\beta,\Omega}^{\mathrm{Kir}}(\theta,\theta_{\alpha}) = \sum_{i=1}^{2}\sum_{j=1}^{N_i^{\beta}} C_{ij}^{\beta} D_{\alpha\beta}^{\mathrm{Kir}}(\theta,\theta_{\beta,i}^{+}). \tag{23}$$

Again, there are geometrical regions where the Kirchhoff approximation deteriorates and GTD performs better. The Kirchhoff approximation can be improved by comparing (22) to (20). This comparison shows that outside all transition regions we have

$$\mathbf{U}_{\Omega}^{\mathrm{Scat}}(\mathbf{x}) = \mathbf{U}_{\Omega}^{\mathrm{Kir}}(\mathbf{x}) + \mathbf{U}_{\Omega}^{\mathrm{Rayleigh}}(\mathbf{x}) + \\ + \sum_{\beta=L,TV} D_{\alpha\beta,\Omega}^{\mathrm{Corr}}(\theta,\theta_{\alpha})\frac{\mathrm{e}^{ik_{\beta}r}}{\sqrt{k_{\beta}r}} + O([k_T r]^{-1}), \tag{24}$$

where the corrected Kirchhoff diffraction coefficients are

$$D_{\alpha\beta,\Omega}^{\mathrm{Corr}}(\theta,\theta_{\alpha}) = D_{\alpha\beta,\Omega}^{\mathrm{GTD}}(\theta,\theta_{\alpha}) - D_{\alpha\beta,\Omega}^{\mathrm{Kir}}(\theta,\theta_{\alpha}).$$

It is possible to show that this correction applies inside all penumbrae as well as inside geometrical regions.

Journal of Physics: Conference Series **353** (2012) 012015 doi:10.1088/1742-6596/353/1/012015

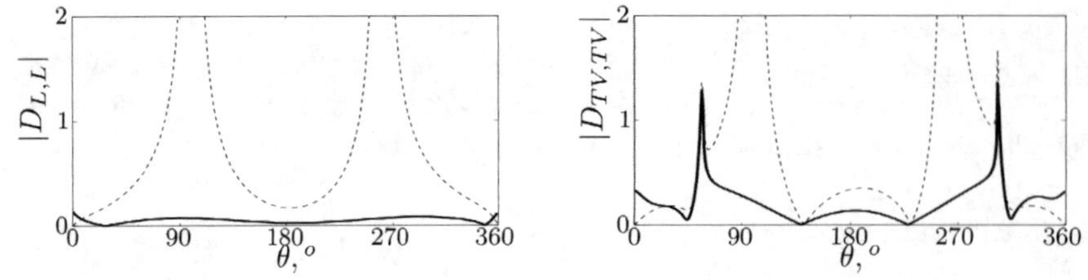

Figure 3. Diffraction coefficients for the planar crack. Dashed line - D^{GTD}, solid line - D^{Corr}, $\phi_\alpha = 90^o$, $\theta_\alpha = 100^o$,

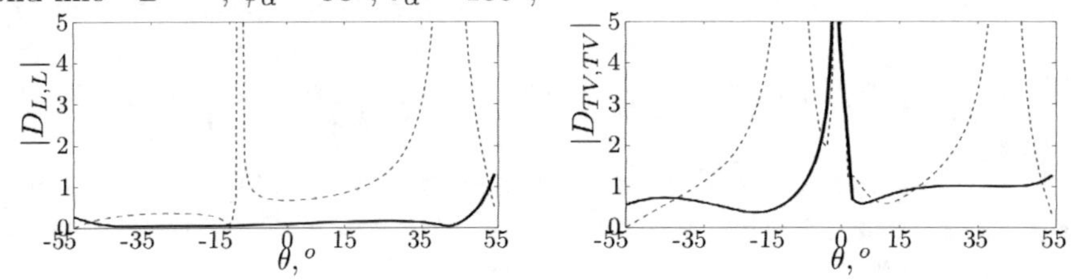

Figure 4. Diffraction coefficients for the wedge. Wedge angle $2\Omega = 110^o$. Dashed line - D^{GTD}, solid line - D^{Corr}, $\theta_\alpha = 20^o$,

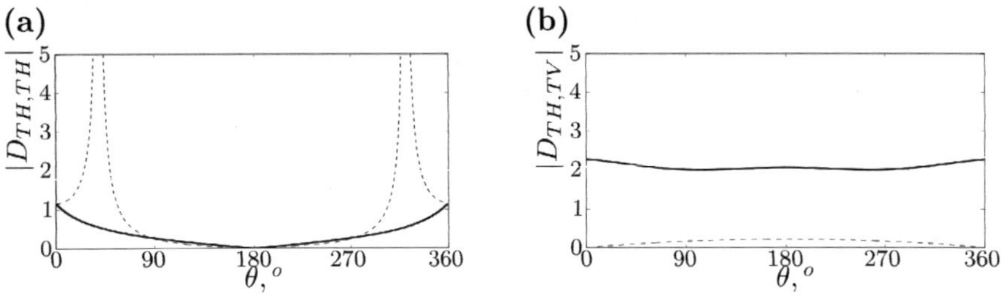

Figure 5. Diffraction coefficients for the planar crack. Dashed line - D^{GTD}, solid line - D^{Corr}, (a) $\phi_\alpha = 90^o$, $\theta_\alpha = 40^o$, (b) $\phi_\alpha = 10^o$, $\theta_\alpha = 40^o$,

4. Numerical results

The GTD diffraction coefficients and corrected Kirchhoff diffraction coefficients for the planar crack and 110° wedge are presented in Figs. 3 and 4, respectively. The wave speeds are assumed to be those of steel, $c_L = 5890\,m/s$ and $c_T = 3210\,m/s$.

As can be seen form the formulae in main text, inside penumbrae both Kirchhoff and GTD coefficients are irregular. Figs. 3–5 show that the corrected Kirchhoff diffraction coefficient which is the difference between the two is regular there, that is, the singularities cancel. However, these figures also show that the corrected coefficients for the diffracted transverse waves exhibit unrealistic spikes in regions surrounding critical rays. These could be smoothed

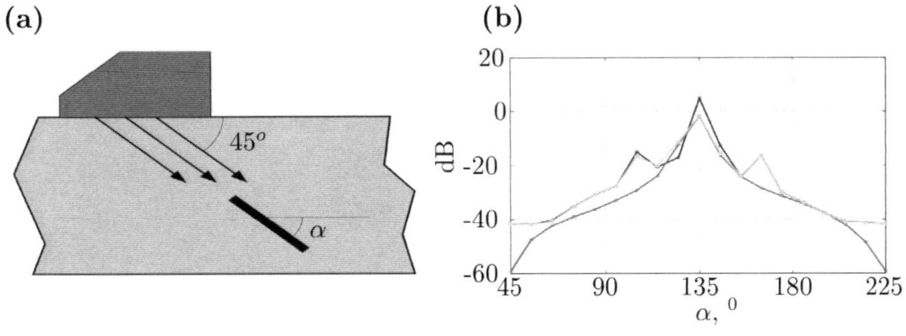

Figure 6. SV45 pulse echo configuration. a) Orientation of the crack (thick black line) is specified by the angle α. b) Maximum displacements calculated using GTD (blue line), Kirchhoff approximation (red line) and corrected Kirchhoff approximation (green line).

by developing a fully uniform version of GTD. The problem is more challenging than the one discussed here.

Numerical results presented in Figs. 3–5 confirm that the Kirchhoff-Rayleigh integral gives a good approximation when simulating longitudinal diffracted waves but, as expected, the quality of approximation deteriorates in the regions surrounding critical rays and when the incidence is near grazing (Fig. 5b).

In Fig. 6 we study an SV45 pulse echo configuration. The underlying geometry is shown in Fig. 6a). A transducer beam (indicated by arrows) irradiates a finite sized crack (presented as a thick black line) whose orientation is specified by angle α. In Fig. 6b) we plot maximum amplitudes of the received signal (in dB) against the crack orientation angle α. We compare three different approximations, GTD (blue line), Kirchhoff (red line) and corrected Kirchhoff (green line). As expected, the corrected Kirchhoff approximation coincides with the Kirchhoff approximation at the angles of specular reflection and with GTD at the angles of oblique incidence.

This study has been motivated by the need to refine the Kirchhoff approximation as used in ultrasonic NDE [7]. The refinement is now implemented in CIVA, a commercial NDE simulation package produced by CEA-LIST. It allows for benchmarking and modelling a wider class of realistic NDE configurations, thus increasing the scope for CIVA's experimental validation.

5. Acknowledgement
This work has been carried out within the framework of the CIVAMONT 2012 collaborative project.

References
[1] Deschamps G A 1972 *Proc. IEEE* **60** 1022
[2] Keller J B 1962 *J. Opt. Soc.* **52** 116
[3] Knott E F and Senior T B A 1974 *Proc. IEEE* **62** 1468

[4] Borovikov V A 1994 *Uniform Stationary Phase Method* (London: IEE)
[5] Fradkin L and Stacey R 2010 *Ultrasonics* **50**(4-5) 529
[6] Borovikov V A and Fradkin L Ju 2009 Russian J. Math. Phys. **16**(2) 139
[7] Mahaut S, Darmon M, Chatillon S, Jenson F and Calmon P, 2009 *Insight* **51**(2) 78
[8] Chapman R K 1990 *J. NDE* **9** 197
[9] Schmerr Jr L W and Song J S 2007 *Ultrasonic Nondestructive Evaluation Systems: Models and Measurements* (New York: Plenum Press)
[10] Borovikov V A and Kinber B Ye 1994 *Geometrical Theory of Diffraction* (London: IEE)
[11] Ufimtsev P Ya 2007 *Fundamentals of the Physical Theory of Diffraction* (Chicester: Wiley-Blackwell)
[12] McMaken H 1984 *J. Acoust. Soc. Am.* **75** 1352
[13] Kamotski V, Fradkin L, Babich V M, Borovikov V A, and Samokish B A 2006 *SIAM J. Appl. Math.* **67**(1) 235
[14] Gautesen A 2002 *Wave Motion* **36**(4) 417
[15] Gautesen A 2010 *Wave Motion* **47** 327
[16] Gautesen A and Fradkin L 2011 *SIAM J. Appl. Math.* **70**(8) 3065
[17] Achenbach A, Gautesen A, and McMaken H 1979 *Ray Methods for Waves in Elastic Solids; With Applications to Scattering by Cracks* (New York: Pitman)

Appendix A. The Geometrical Elastodynamics

On the illuminated surface S^+ the geometrical elastodynamics term can be represented as a product of the constant vector $\mathbf{A}^\alpha$ and rapidly oscillating exponent (10). Let us introduce a new Cartesian coordinate system (y_1, y_2, y_3), where the y_2-axis coincides with the x_2-axis, the (y_1y_2)-plane is the incidence plane and the y_3-axis lies in the (x_1x_3)-plane. Let ϑ_α be the angle which the incident wave vector makes with the y_2-axis and let φ_α denote the angle between the y_1-axis and x_1-axis. The angles $(\vartheta_\alpha, \varphi_\alpha)$ are related to the angles $(\phi_\alpha, \theta_\alpha)$ via

$$\cos\vartheta_\alpha = |\sin\phi_\alpha \sin\theta_\alpha|, \quad \cos\varphi_\alpha = \frac{\sin\phi_\alpha\cos\theta_\alpha}{\sqrt{1+\sin^2\phi_\alpha\sin^2\theta_\alpha}}. \tag{A.1}$$

Let us denote by $\vartheta_\beta, \beta = L, T$ the angles which the reflected β wave vector makes with the y_2-axis, so that we have

$$\sin\vartheta_L = \frac{k_\alpha}{k_L}\sin\vartheta_\alpha, \quad \sin\vartheta_T = \frac{k_\alpha}{k_T}\sin\vartheta_\alpha. \tag{A.2}$$

When the incident wave type is longitudinal the constant vector amplitude $\mathbf{A}^L$ can be written as

$$\mathbf{A}^L = \frac{2\kappa^2\cos\vartheta_L}{R(\vartheta_L,\vartheta_T)}\begin{pmatrix}\sin 2\vartheta_T\cos\varphi_\alpha\\ -\cos 2\vartheta_T\\ \sin 2\vartheta_T\sin\varphi_\alpha\end{pmatrix}, \tag{A.3}$$

where the Rayleigh function $R(\vartheta_L, \vartheta_T)$ is given by

$$R(\vartheta_L,\vartheta_T) = \sin 2\vartheta_T \sin 2\vartheta_L + \kappa^2\cos^2 2\vartheta_T. \tag{A.4}$$

Similarly, when the incident wave is transverse the constant vector amplitudes $\mathbf{A}^{TV}$ and $\mathbf{A}^{TH}$ can be represented, respectively, by

$$\mathbf{A}^{SV} = \frac{-2\cos\vartheta_T}{R(\vartheta_L,\vartheta_T)}\begin{pmatrix}\kappa^2\cos 2\vartheta_T\cos\varphi_\alpha\\ \sin 2\vartheta_L\\ \kappa^2\cos 2\vartheta_T\sin\varphi_\alpha\end{pmatrix}, \quad \mathbf{A}^{SH} = \begin{pmatrix}-2\sin\varphi_\alpha\\ 0\\ 2\cos\varphi_\alpha\end{pmatrix}. \tag{A.5}$$

Analytical modelling of acoustic emission from buried or surface-breaking cracks under stress

W Ben Khalifa[1], K Jezzine[1], G Hello[2] and S Grondel[3]

[1] CEA, LIST, 91191 Gif-sur-Yvette CEDEX, France
[2] Laboratoire Roberval, Université de Technologie de Compiègne, Centre de Recherches de Royallieu, BP 20529, 60205 Compiègne
[3] Département OAE, IEMN, UMR CNRS 8520, Université de Valenciennes et du Hainaut Cambrésis, Le Mont Houy, 59313 Valenciennes Cedex 9

E-mail: warida.ben-khalifa@cea.fr

Abstract. Acoustic emission (AE) is a non-destructive testing method used in various industries (aerospace, petrochemical and pressure-vessel industries in general, power generation, civil engineering, mechanical engineering, etc...) for the examination of large structures subjected to various stresses (e.g. mechanical loading).The energy released by a defect under stress (the AE phenomenon) can propagate as guided waves in thin structures or as surface Rayleigh waves in thick ones. Sensors (possibly permanently) are positioned at various locations on the structure under examination and are assumed to be sensitive to these waves. Then, post-processing tools typically based on signal processing and triangulation algorithms can be used to inverse these data, allowing one to estimate the position of the defect from which emanates the waves measured. The French Atomic Energy Commission is engaged in the development of tools for simulating AE examinations. These tools are based on specific models for the AE sources, for the propagation of guided or Rayleigh waves and for the behaviour of AE sensors. Here, the coupling of a fracture mechanics based model for AE source and surface/guided wave propagation models is achieved through an integral formulation relying on the elastodynamic reciprocity principle. As a first approximation, a simple piston-like model is used to predict the sensitivity of AE sensors. Predictions computed by our simulation tool are compared to results from the literature for validation purpose.

1. Introduction

Acoustic emission from a crack under stress is composed of body waves as well as surface waves in thick structures. In the case of thin structures, acoustic emission is composed of guided waves that can propagate over long distance without loss of energy.

In this paper, we consider only thick structures and we assume that only the radiated surface waves need to be considered. In the far field, surface waves dominate body waves. In a two-dimensional geometry surface waves do not decay, while body waves decay as $r^{-1/2}$.In a three-dimensional geometry, surface wave decay as $r^{-1/2}$ and body waves decay as r^{-1}, where r is the distance between the source and an observation point.

The objective is to model the acoustic emission monitoring from the propagation to the detection.

The geometry is an isotropic half space which is subjected to an arbitrary stress. We consider a two-dimensional configuration and we assume that a crack with an arbitrary orientation under stress will propagate and the released energy will produce transient surface waves.

We have, as input data, an acoustic emission model provided in the frame of the ANR project MACSIM by the University of Compiegne UTC; this fracture mechanics based model gives the evolution of the spatial-temporal elastodynamic fields (displacements and stresses) on the crack.

On the other hand we consider an analytical surface waves propagation model given by [1].

Following a method proposed by [2], the surface wave generated by acoustic emission from a crack under stress will be obtained by an application of the elastodynamic reciprocity principle. This principle can be written for a body of volume V and surface S:

$$\int_V \left[f_i^A u_i^B - f_i^B u_i^A \right] dV = \int_S \left[\tau_{ij}^B u_j^A - \tau_{ij}^A u_j^B \right] n_j dS \tag{1}$$

where n_i are the components of the outward normal to S, f_i , u_i and τ_{ij} are the components of body forces, displacements and stresses. This theorem connects two different elastodynamic states, state A and state B. We defined state A as the desired solution for the surface wave generated by a crack under stress while state B is an auxiliary solution involving a virtual surface wave.

2. Coupling of an AE model and an analytical Rayleigh wave propagation model

2.1 Rayleigh wave in 2D geometry

We consider an isotropic homogenous half-space relative to the x, z coordinate system.
In this case, the displacements and stresses of the Rayleigh wave can be written as:

$$u_x^A(x,z) = iV^R(z)\exp(-ik_R x) \tag{2}$$

$$u_z^A(x,z) = -W^R(z)\exp(-ik_R x) \tag{3}$$

$$\tau_{xx}^A(x,z) = -T_{xx}{}^R(z)\exp(-ik_R x) \tag{4}$$

$$\tau_{xz}^A(x,z) = iT_{xz}{}^R(z)\exp(-ik_R x) \tag{5}$$

$$\tau_{zz}^A(x,z) = -T_{zz}{}^R(z)\exp(-ik_R x), \tag{6}$$

where

$$V^R(z) = d_1 \exp(-pz) + d_2 \exp(-qz) \tag{7}$$

$$W^R(z) = d_3 \exp(-pz) - \exp(-qz) \tag{8}$$

$$T_{xx}{}^R(z) = \mu\left[d_4 \exp(-pz) + d_5 \exp(-qz)\right] \tag{9}$$

$$T_{xz}{}^R(z) = \mu\left[d_6 \exp(-pz) + d_7 \exp(-qz)\right] \tag{10}$$

$$T_{zz}{}^R(z) = \mu\left[d_8 \exp(-pz) + d_9 \exp(-qz)\right] \tag{11}$$

d_1, d_2 d_3, d_4, d_5, d_6, d_7, d_8 and d_9 are defined by:

$$d_1 = -\frac{1}{2}\frac{k_R^2 + q^2}{k_R p} \tag{12}$$

$$d_2 = \frac{q}{k_R} \tag{13}$$

$$d_3 = \frac{1}{2}\frac{k_R^2 + q^2}{k_R} \tag{14}$$

$$d_4 = \frac{1}{2}(k_R^2 + q^2)\frac{2p^2 + k_R^2 - q^2}{pk_R^2} \tag{15}$$

$$d_5 = -2q \tag{16}$$

$$d_6 = \frac{k_R^2 + q^2}{k_R p} \tag{17}$$

$$d_7 = -\frac{k_R^2 + q^2}{k_R p} \tag{18}$$

$$d_8 = -\frac{1}{2}\frac{k_R^2 + q^2}{k_R^2 p}(k_R^2 + q^2) \tag{19}$$

$$d_9 = 2q \tag{20}$$

The quantities p and q are defined by:

$$p^2 = k_R^2 - k_l^2 \tag{21}$$

$$q^2 = k_R^2 - k_t^2 \tag{22}$$

The Rayleigh wave number is defined by $k_R = k_t/\eta_R$ where $k_t = \omega\left(\frac{\rho}{\mu}\right)^{\frac{1}{2}}$ and $k_l = \omega\left(\frac{\rho}{(\lambda + 2\mu)}\right)^{\frac{1}{2}}$ are the wave numbers of transverse and longitudinal bulk waves and

$$\eta_R = \frac{0.87 + 1.12\nu}{1 + \nu} \tag{23}$$

ρ, ν, λ and μ are respectively the density of the medium, Poisson ratio and the elastic Lamé constants defined by: $\lambda = \frac{E\nu}{(1+\nu)(1-2\nu)}$ and $\mu = \frac{E}{2(1+\nu)}$, where E is the Young's modulus.

2.2. State A

We consider a crack with arbitrary orientation under an arbitrary stress. The acoustic emission from this crack will generate surface waves whose displacements and stresses can be written as:

$$u_x^A(x,z) = iA(\omega)V^R(z)\exp(\pm ik_R x) \tag{24}$$

$$u_z^A(x,z) = \pm A(\omega)W^R(z)\exp(\pm ik_R x) \tag{25}$$

$$\tau_{xx}^A(x,z) = \pm A(\omega)T_{xx}{}^R(z)\exp(\pm ik_R x) \tag{26}$$

$$\tau_{xz}^A(x,z) = iA(\omega)T_{xz}{}^R(z)\exp(\pm ik_R x) \tag{27}$$

$$\tau_{zz}^A(x,z) = \pm A(\omega)T_{zz}{}^R(z)\exp(\pm ik_R x) \tag{28}$$

The plus and minus signs, in equations (24)-(28), are for the propagation in the negative x direction and positive x direction and $A(\omega)$ is the amplitude of the Rayleigh wave (unknown of the radiation problem).

2.3. State B

The second state is an auxiliary solution; we define this state as a virtual surface wave that propagates in the positive x direction, in the absence of the crack, the displacements and stresses of the Rayleigh wave can be written as:

$$u_x^B(x,z) = iV^R(z)\exp(-ik_R x) \tag{29}$$

$$u_z^B(x,z) = W^R(z)\exp(-ik_R x) \tag{30}$$

$$\tau_{xx}^B(x,z) = T_{xx}{}^R(z)\exp(-ik_R x) \tag{31}$$

$$\tau_{xz}^B(x,z) = iT_{xz}{}^R(z)\exp(-ik_R x) \tag{32}$$

$$\tau_{zz}^{B}(x,z) = T_{zz}^{\ R}(z)\exp(-ik_R x) \tag{33}$$

2.4. Application of the elastodynamic reciprocity theorem

In equation (1), the first integral on the volume vanishes, since there are no body forces in states A and B. The reciprocity principle can be written as:

$$\int_S \left[\tau_{ij}^{B} u_j^{A} - \tau_{ij}^{A} u_j^{B}\right] n_j dS = 0 \tag{34}$$

We consider the region of the half-plane limited by the line $x=a$ and the line $x=b$. The acoustic emission source, in the half plane, is an arbitrary oriented crack. l is the final length of the crack located at v_0 with respect to the v axis of the local basis (u, v) defined in figure 2. The crack is considered as a surface of displacement and stress discontinuity evolving with time. The plane of the crack makes an angle β with the z axis. The surface S of the reciprocity problem is composed by the surface limited by the blue contour in the half-plane and the crack surface, n is the local normal to S.

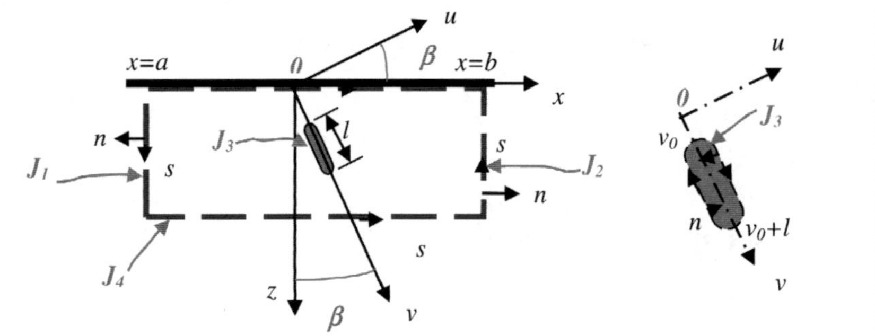

Figure 1. Geometry of the arbitrary oriented crack and contours of integration for the application of the reciprocity theorem

J_3 is the integral on the surface of the crack, J_1 is the integral along the line $x = a, 0 \le z < \infty$, J_2 is the integral along the line $x = b, 0 \le z < \infty$ and J_4 is the integral along the line z =constant$\to \infty$, where the amplitude of the Rayleigh wave tends to zero.

There is no contribution from the integral on the free surface and on z =constant$\to \infty$, and the integral along J_1 and J_2 only yields contributions from counter-propagating wave [3].

For $(x = b, 0 \le z < \infty)$ we have:

$$J_2 = \int_0^{\infty} \left[\tau_{ij}^{B} u_i^{A} - \tau_{ij}^{A} u_i^{B}\right]\Big|_{x=b} n_i dS = 0 \tag{35}$$

Along $(x = a, 0 \le z < \infty)$, we have:

$$J_1 = -\int_0^{\infty} F_{AB}(x,z)\Big|_{x=a} dz \tag{36}$$

where

$$F_{AB}(x,z) = \tau_{xx}^{B} u_x^{A} + \tau_{xz}^{B} u_z^{A} - \tau_{xx}^{A} u_x^{B} - \tau_{xz}^{A} u_z^{B} \tag{37}$$

We replace the displacements and stresses by their expressions:

$$J_1 = 2iAI \tag{38}$$

with

$$I = \int_0^\infty \left[T_{xx}^R(z) V^R(z) - T_{xz}^R(z) W^R(z) \right] dz \tag{39}$$

We divide the integral along the surface of the crack into two parts:

The first integral J_3^+ is along the line ($u=0^+$, $v_0 \leq v < v_0 + l$) and the second integral J_3^- is along the line ($u=0^-$, $v_0 \leq v < v_0 + l$).

Along ($u=0^+$, $v_0 \leq v < v_0 + l$), J_3^+ can be written as:

$$J_3^+ = -\int_{v0}^{l+v0} E_{AB}(u,v)\Big|_{u=0^+} dv \tag{40}$$

where

$$E_{AB}(u,v) = \tau_{uu}^B u_u^A + \tau_{uv}^B u_v^A - \tau_{uu}^A u_u^B - \tau_{uv}^A u_v^B \tag{41}$$

Along ($u=0^-$, $v_0 \leq v < v_0 + l$), J_3^- can be written as:

$$J_3^- = \int_{v0}^{l+v0} E_{AB}(u,v)\Big|_{u=0^-} dv \tag{42}$$

The integral along the surface of the crack can be written as

$$J_3 = J_3^+ + J_3^- = \int_{v0}^{l+v0} \left\{ E_{AB}(u,v)\Big|_{u=0^-} - E_{AB}(u,v)\Big|_{u=0^+} \right\} dv \tag{43}$$

$$J_3 = \int_{v0}^{l+v0} \left\{ -\tau_{uu}^B(u,v)\Big|_{u=0} \Delta u_u - \tau_{uv}^B(u,v)\Big|_{u=0} \Delta u_v + \Delta\tau_{uu}^A u_u^B(u,v)\Big|_{u=0} + \Delta\tau_{uv}^A u_v^B(u,v)\Big|_{u=0} \right\} dv \tag{44}$$

where

$$\Delta u_u = u_u(0^+,v) - u_u(0^-,v) \tag{45}$$

$$\Delta u_v = u_v(0^+,v) - u_v(0^-,v) \tag{46}$$

$$\Delta\tau_{uv}^A = \left[\tau_{uv}^A(0^+,v) - \tau_{uv}^A(0^-,v) \right] \tag{47}$$

$$\Delta\tau_{uv}^A = \left[\tau_{uv}^A(0^+,v) - \tau_{uv}^A(0^-,v) \right] \tag{48}$$

The amplitude of the Rayleigh wave can be calculated from equation (49):

$$J_1 + J_3 = 0 \tag{49}$$

We replace J_1 and J_3 by their expressions from equations (44) and (38):

$$2iAI = \int_{v0}^{l+v0} \left\{ \tau_{uu}^B(0,v)\,\Delta u_u + \tau_v^B(0,v)\,\Delta u_v - \Delta\tau_{uu}^A u_u^B - \Delta\tau_{uv}^A u_v^B \right\} dv \tag{50}$$

To express the different quantities of the local basis in the global basis, we note

$$\begin{cases} C = \cos\beta \\ S = \sin\beta \end{cases} \tag{51}$$

The rotation matrix can be written as:

$$M = \begin{pmatrix} C & -S \\ S & C \end{pmatrix} \tag{52}$$

The displacements and stresses in the local crack basis are defined by:

$$\begin{pmatrix} u_u \\ u_v \end{pmatrix} = M \begin{pmatrix} u_x \\ u_z \end{pmatrix} \tag{53}$$

$$\begin{pmatrix} \tau^B_{uu} & \tau^B_{uv} \\ \tau^B_{vu} & \tau^B_{vv} \end{pmatrix} = M^{-1} \begin{pmatrix} \tau^B_{xx} & \tau^B_{xz} \\ \tau^B_{zx} & \tau^B_{zz} \end{pmatrix} M \tag{54}$$

After replacing the displacements and stresses in equation (50) we obtain:

$$2i\,A\,I = \int_{v0}^{l+v0} \left\{ P(0,v)\,\Delta u_u + Q(0,v)\,\Delta u_v - \Delta\tau^A_{uu}\tilde{V}(0,v) - \Delta\tau^A_{uv}\tilde{W}(0,v) \right\} dv \tag{55}$$

where

$$P = [C^2\,T^R_{xx} + 2C\,S\,i\,T^R_{xz} + S^2\,T^R_{zz}]\,e^{ik_R S v} \tag{56}$$

$$Q = [-C\,S\,T^R_{xx} + (C^2 - S^2)\,i\,T^R_{xz} + S\,C\,T^R_{zz}]\,e^{ik_R S v} \tag{57}$$

(58)

$$\tilde{W} = [-Si\,V + C\,W]\,e^{ik_R S v} \tag{59}$$

The amplitude of the emitted Rayleigh wave is:

$$A = \frac{-i}{2I} \int_{v0}^{l+v0} \left\{ P(0,v)\,\Delta u_u + Q(0,v)\,\Delta u_v - \Delta\tau^A_{uu}\tilde{V}(0,v) - \Delta\tau^A_{uv}\tilde{W}(0,v) \right\} dv \tag{60}$$

In the case of a tensile stress crack the amplitude of the Rayleigh wave can be written as:

$$A = \frac{-i}{2I} \int_{z_0}^{l+z_0} \{ T_{xx}(0,z)\,\Delta u_x \}\, dz \tag{61}$$

By substituting A in equations (24) and (25), the displacement components at z=0 are :

$$u^A_z(x,z_1) = \pm\frac{-i}{2I} \int_{z_0}^{l+z_0} \{ T_{xx}(0,z)\,\Delta u_x \}\, dz\, W^R(z_1) \exp(\pm i k_R x) \tag{62}$$

$$u^A_x(x,z_1) = \frac{1}{2I} \int_{z_0}^{l+z_0} \{ T_{xx}(0,z)\,\Delta u_x \}\, dz\, V^R(z_1) \exp(\pm i k_R x) \tag{63}$$

As it is shown by Achenbach, this result is equivalent to the displacement due to the presence of a distribution of dipoles along $z_0 \le z \le z_0 + l$ [4]

2.5. Acoustic emission source

The COD (crack opening displacement) for a crack of length *2a* along the *z* axis subjected to uniaxial tension σ^{∞} is:

$$\Delta u_x(z) = \sigma^{\infty} \frac{\kappa+1}{4\mu} \sqrt{a^2 - z^2} \tag{64}$$

where $\kappa = (3 - 4\upsilon)$ in the case of plane strain.

We consider a crack of length *l* subjected to uniaxial tension, figure 2 shows the crack shape. We discretise the crack as shown in figure 3.

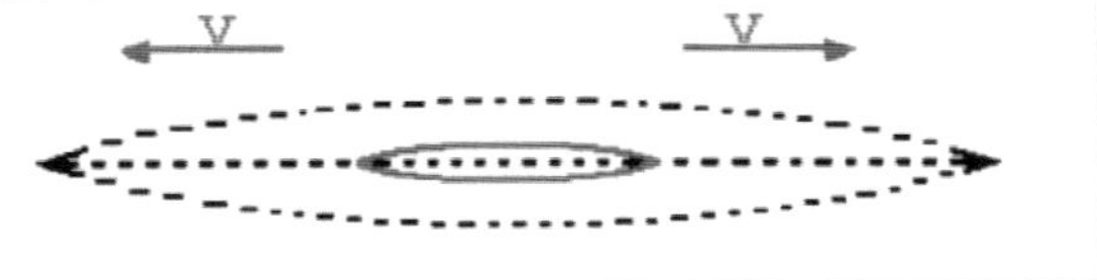

Figure 2. Crack shape

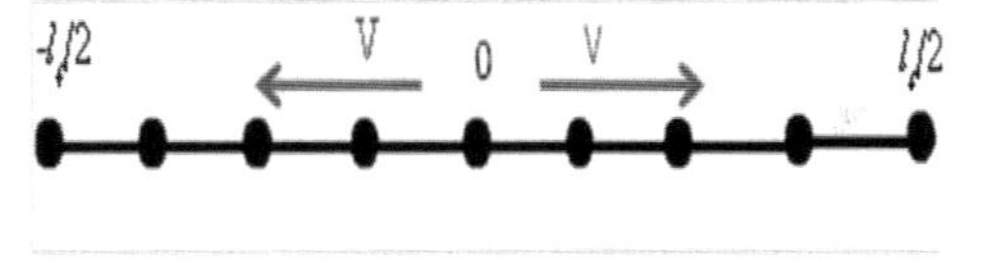

Figure 3. Discretisation of the crack

We assume that the crack length evolves from $l_0 = 1\text{mm}$ to $l = 5\text{mm}$ at a velocity *V* during *T=0.8µs*, we consider a sampling frequency $F_e = 50\,MHz$.

We define z_j and t_n as:

$$z_j = jV\Delta t \tag{65}$$

$$t_n = t_0 + n\frac{1}{Fe} \tag{66}$$

j varies from *–H/2* to *H/2*, and *n* varies from *0* to *N* where:

$$\Delta t = 1/F_e \tag{67}$$

$$N = \frac{T}{\Delta t} \tag{68}$$

$$H = \frac{l}{V\Delta t} \tag{69}$$

Equation (64) can be discretised as follows:

$$\Delta u_x\left(z_j, t_n\right) = \sigma^{\infty} \frac{\kappa+1}{4\mu} \sqrt{((l_0 + nV\Delta t)/2)^2 - z_j^2} \tag{70}$$

for $\left|z_j\right| \le (l_i + nV\Delta t)/2$

$$\Delta u_x(z_j) = 0 \tag{71}$$

for $(l_0 + nV\Delta t)/2 \le \left|z_j\right| \le l/2$

Journal of Physics: Conference Series **353** (2012) 012016 doi:10.1088/1742-6596/353/1/012016

We take the Fourier transform of the COD to obtain the displacement and stress fields of the emitted Rayleigh wave in the frequency domain.

Displacement and stress fields in the time domain are obtained by the inverse Fourier Transform.

3. Simulation and comparison with literature

We have simulated the acoustic emission from a buried crack under a uniform stress with different crack propagation velocities *V* (500m/s, 1000m/s, 2000m/s and 2500m/s). We assume that the crack, of final length *l* located at a distance *d* from the free surface, propagates at a velocity *V* and the crack length evolves from 1mm to 5mm under 200 MPa.

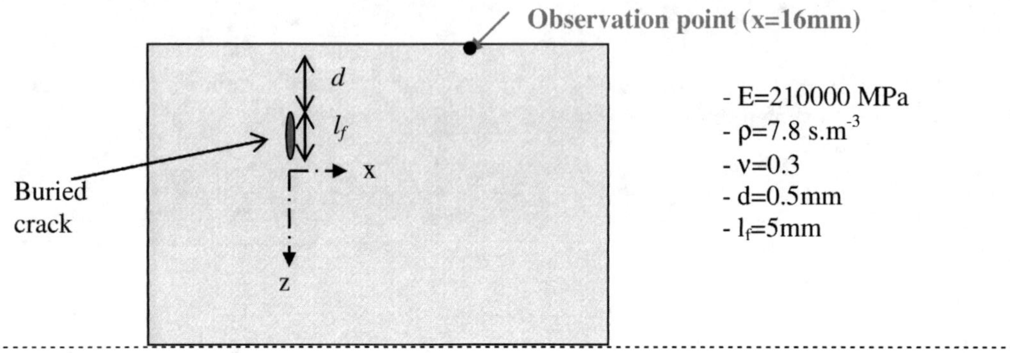

Figure 4. Geometry of the crack used for the simulation and position of the observation point

Figure 5 to 8 show the evolution of the crack opening displacement, obtained from the fracture mechanics based crack model, at different propagation times (*t*=0μs, *t*=0.2μs, *t*=0.4μs and *t*=0.8μs) with a constant crack-tip velocity *V*=2500m/s.

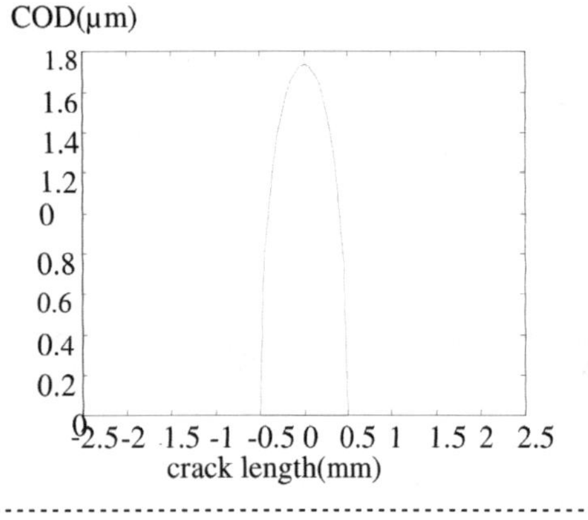

Figure 5. The initial crack opening displacement, *V*=2500m/s

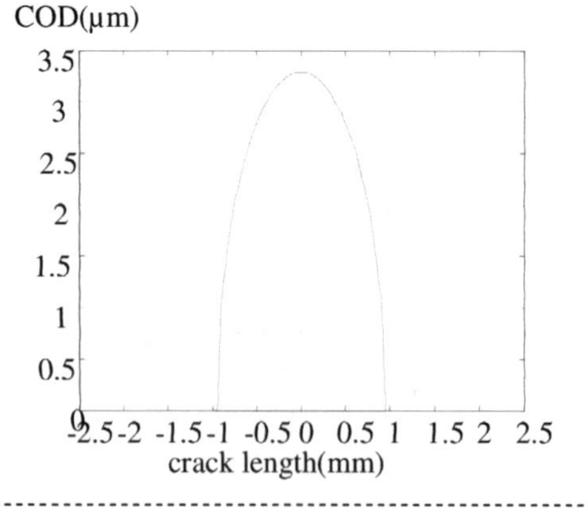

Figure 6. Crack opening displacement at *t*=0.2μs, *V*=2500m/s

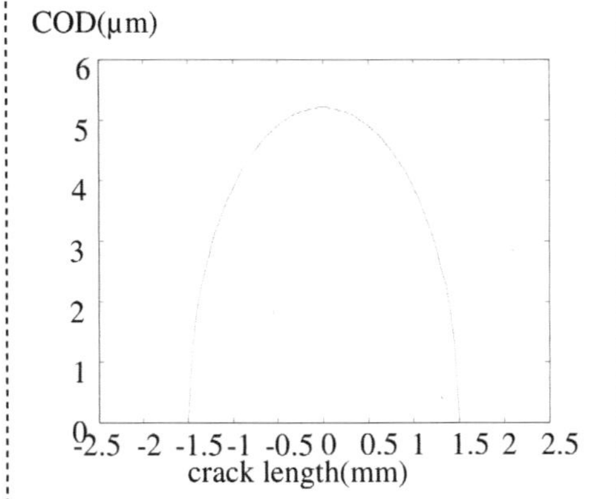

Figure 7. Crack opening displacement at t=0.4μs, V=2500m/s

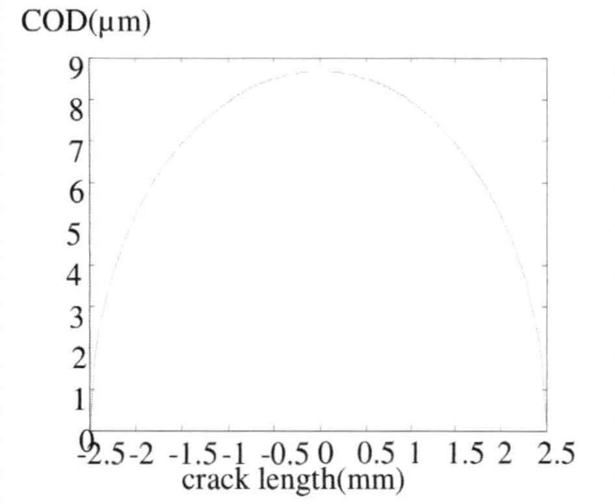

Figure 8. Crack opening displacement at t=0.8μs, V=2500m/s

The normal displacement of the emitted wave was simulated with different crack-tip speeds.
Figure 9 to 12 shows the influence of the crack propagation speed on the wave front and the Rayleigh wave arrival time.

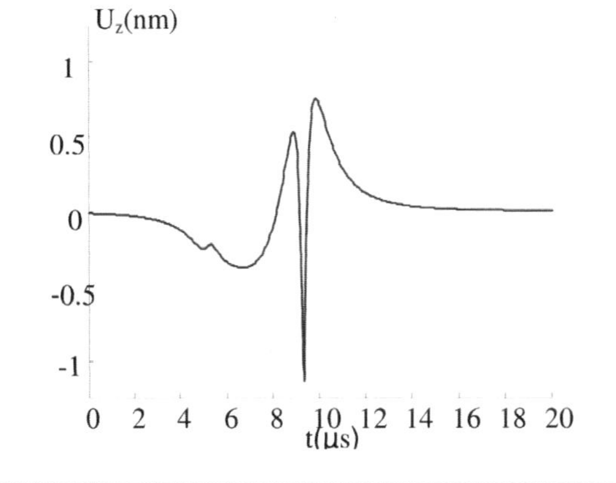

Figure 9. Surface normal displacement in nm at x=16mm,V=500m/s

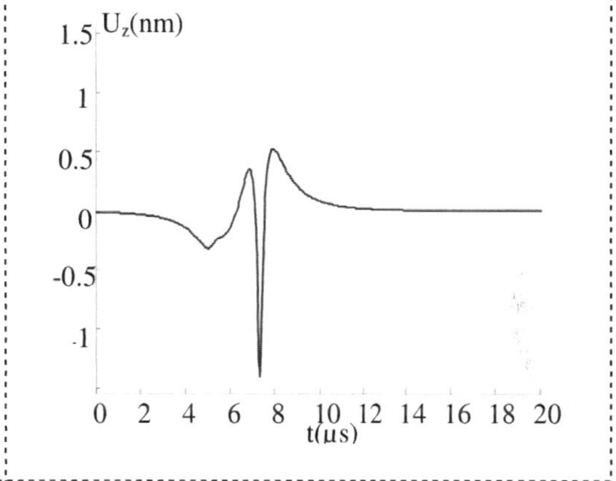

Figure 10. Surface normal displacement in nm at x=16mm,V=1000m/s

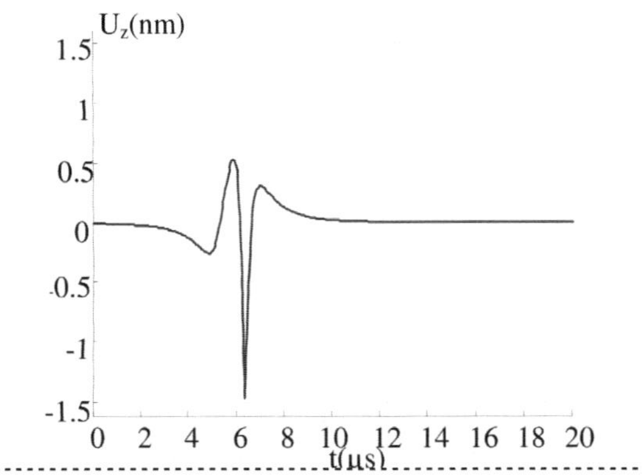

Figure 11. Surface normal displacement in nm at x=16mm,V=2000m/s

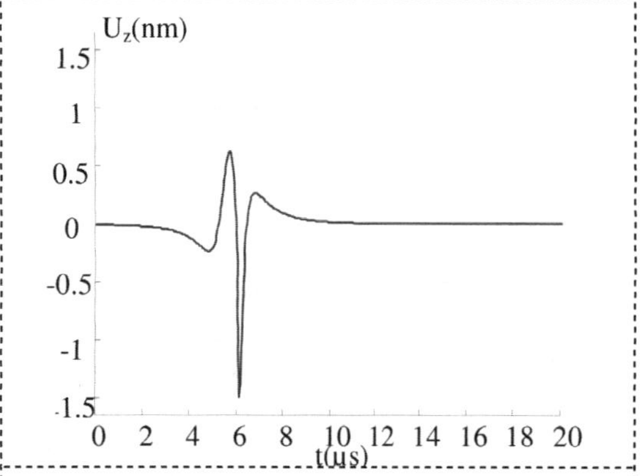

Figure 12. Surface normal displacement in nm at x=16mm,V=2500m/s

Harris *et al.* [5] developed an analytical model for acoustic emission from a buried crack in which he takes into account the displacement and the velocity of the crack-tip. He used an approximation of the Fourier transform near the arrival time of the Rayleigh wave. Figure 13 shows the approximation of the normalized Rayleigh wave velocity near the arrival time of the Rayleigh wave obtained by [5].

In figure 14 we present the simulation of the Rayleigh wave velocity emitted by a crack with the same length and location. Our 2D source model (the crack is considered infinite in the y direction) takes into account the crack opening displacement at all the points of the crack along its length, whereas the 3D model presented in [5] is a circular crack and it is based on the displacement at small distances ahead of the original crack tip.

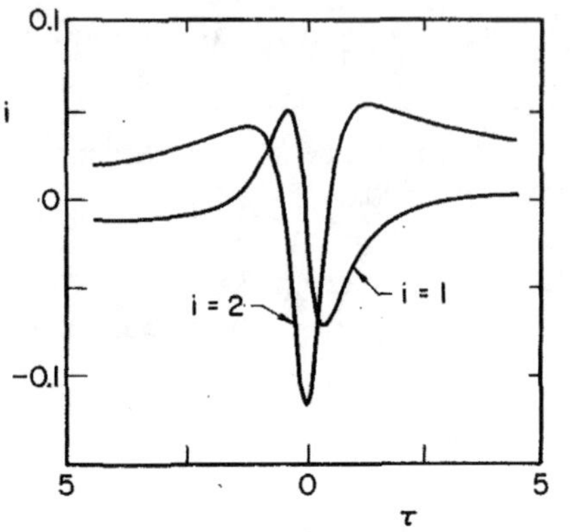

Figure 13. Normalised Rayleigh wave particle velocity v_i from [5] as a function of the normalized time τ, i=2 corresponds to the vertical velocity and i=1 corresponds to the horizontal velocity, $\tau = t/(d/v_R) - x/d$

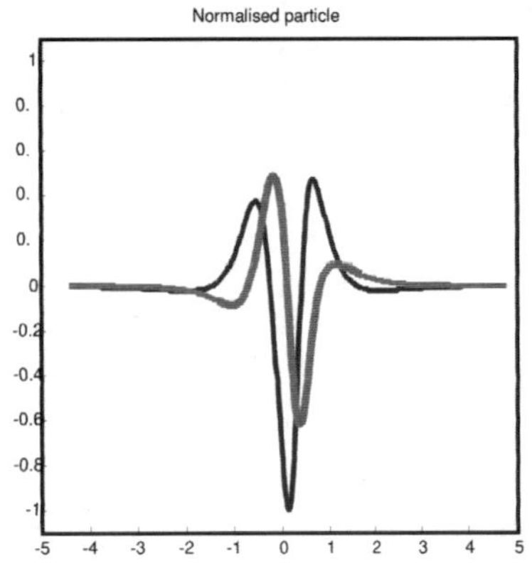

Figure 14. Normalised Rayleigh wave particle velocity, V=1666m/s

For these reasons, we can compare our simulation, only qualitatively, with the results of Harris et al. We notice that the curves are similar near τ=0. The difference is due to the dimension of the AE source model and the propagation model (3D or 2D), to the fracture mechanic solution of the crack problem asymptotic solution used by [5] or analytical solution used in our formulation. In addition, curves in figure 13 are computed with the approximation of the Fourier transform near the arrival time of the Rayleigh wave.

A 3D Rayleigh wave model is under development, this model will take into account a 3D AE source model.

4. Example of application

We have simulated the reception by an AE sensor; we use a piston-like model as a first approximation, in which we take into account only the normal velocity of the Rayleigh wave.

In the case of a 2D piston-like model, corresponding to a rectangular transducer whose width is considered infinite, the delivered voltage can be written as [6]:

$$E_{0R} = \frac{1}{L}\int_0^L v_z(x)m(x)dx \tag{62}$$

where L is the length of the transducer, v_z is the vertical velocity of the Rayleigh wave and m the spatial sensibility.

We consider transducer with 10mm length and a constant spatial sensibility $m(x)=1$, placed at a distance of 100mm from the plane of the crack as shown in figure 13.

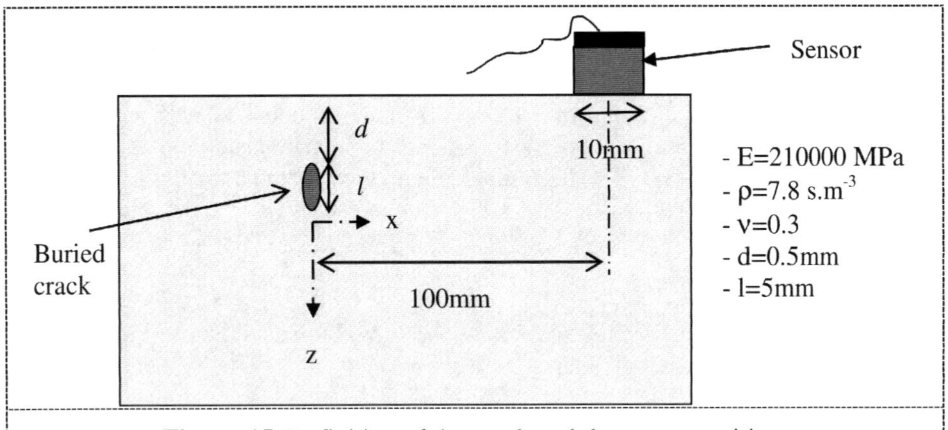

Figure 15. Definition of the crack and the sensor positions

The delivered voltage by the transducer is presented in figure 16.

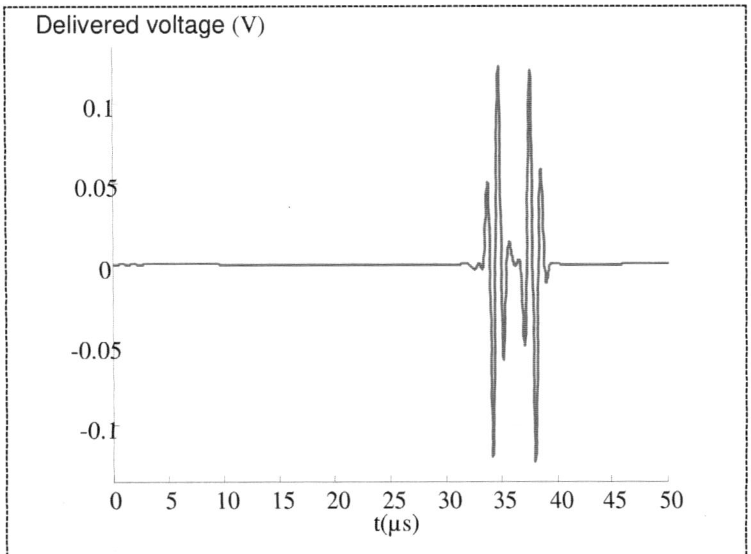

Figure 16. Delivered voltage by the AE sensor after the acoustic emission of the Rayleigh wave (V=1666m/s).

The delivered voltage signal comprises two waveforms which correspond to the difference between the phase terms at the ends of the transducer. In equation (62) the integral is reduced to an integral along the transducer length and to a simple difference of the exponential terms at the transducer ends.

5. Conclusion

The Rayleigh wave emitted by a crack under stress was simulated in the far field. In fact, the surface wave is the dominant disturbance because of its slower geometrical decay rate.

We have shown that the Rayleigh wave displacement depends on the crack-tip speed.

Simulation of the AE of the Rayleigh wave shows the influence of the AE source model on the prediction of the emitted wave. We have compared qualitatively our simulation with the result of Harris *et al.* The differences between the results can be explained by the dimension of the acoustic source and propagation models, the fracture mechanic solutions and the Fourier transform approximation near the Rayleigh wave arrival time. The delivered voltage at a transducer output was simulated with a piston-like model, as a first approximation, considering only the surface normal displacement of the Rayleigh wave.

Because of their thicknesses, AE sensors present also radial modes and this will be simulated by finite elements in order to take into account the radial displacements of the Rayleigh wave in addition to the vertical displacements. Results will be compared with those obtained with the piston-like model.

Acoustic emission from crack under stress in 3D geometry will be modeled using a virtual cylindrical surface wave.

6. Acknowledgement

G. Hello's work concerning the acoustic emission source model was supported by the ANR in the frame of the project MACSIM ANR-08-COSI-005-01.

References

[1] Viktorov I A 1967 *Rayleigh and Lamb waves* (New York: Plenium Press)
[2] Achenbach J D 2008 Acoustic emission from a surface-breaking crack under cyclic loading *Acta Mech* **195** 61
[3] Achenbach J D 2003 *Reciprocity in elastodynamics* (Cambridge: Cambridge University Press)
[4] Achenbach J D 2005 Combination of a virtual wave and the reciprocity theorem to analyse surface wave generation on a transversely isotropic solid *Philosophical Magazine* **85** 4143
[5] Harris J G and Pott J 1984 Surface motion excited by acoustic emission from a buried crack *J. Appl. Mech.* **51** 77
[6] Goujon L and Baboux J C 2003 Behaviour of acoustic emission sensors using broadband calibration techniques *Meas. Sci. Technol.* **14** 903

10th Anglo-French Physical Acoustics Conference (AFPAC 2011) IOP Publishing
Journal of Physics: Conference Series **353** (2012) 012017 doi:10.1088/1742-6596/353/1/012017

Aspects of diffraction of a creeping wave by a back-wall crack

V Zernov[1], L Fradkin[1], A Gautesen[2] and M Darmon[3]

[1] Sound Mathematics Ltd., Cambridge, CB4 2AS, UK
[2] Dept. of Mathematics, Iowa State University and Ames Laboratory, Ames, IA 50011, U. S. A.
[3] CEA, LIST, 91191 Gif-sur-Yvette CEDEX, France.

E-mail: v.zernov@soundmathematics.com

Abstract. In this paper we address the issue of a creeping wave by investigating a two-dimensional configuration which involves a surface-braking crack in a traction free planar back-wall. We assume that this back-wall is irradiated by a shear Gaussian beam which is incident at a critical angle and hits the wall away from the crack edge. We show that the generated wave - known in the engineering literature as "creeping", because it propagates near the surface - consists of non-geometrical waves, the longitudinal and transverse components of the head wave and a longitudinal Goodier-Bishop type wave whose amplitude grows linearly with the distance to its propagation path. The "creeping" wave interacts with the crack edge but the resulting field cannot be simulated using the Kirchhoff Approximation, because this does not embrace the higher order effects. We show that the diffracted field can be simulated using a modification of the Uniform Geometrical Theory of Diffraction which involves a relatively simple limiting procedure.

1. Introduction

Interaction of an ultrasonic beam with a surface-breaking planar faces crack is a well known problem in NDE (Non-Destructive Evaluation). Two-dimensional configurations in which a beam irradiates the whole crack can be modelled using the UGDT as developed for diffraction by half-planes and wedges [1], [2], [3]. In this paper we study an even more challenging problem of scatter of a high frequency Gaussian beam by a crack whose edge lies outside the beam - see Fig. 1. In the engineering literature this situation is usually referred to as interaction between a crack and "creeping wave" that the beam generates. There are several high-frequency approximations which offer effective computational algorithms for many scattering problems. The simplest and most versatile among those is the Kirchhoff Approximation which is based on the Geometrical Elastodynamics. However the Geometrical Elastodynamics is correct only to the leading order and cannot be used to describe the next order effects. For this reason the Kirchhoff approximation fails when the incident transversal beam hits the inspection surface at an angle close to critical. In this paper we show how to conduct efficient simulations of the above configuration using a simple modification of another well-known high-frequency approximation, the Geometrical Theory of Diffraction (GTD). First we develop approximations to the field generated by the incident beam near the wall - the "creeping" wave - because it is only this field that interacts with the crack edge. The paper is organised as follows: In Section 2 we introduce the problem, first assuming the incident field to be a cylindrical wave and then generalising that to a Gaussian beam, which is a good approximation to a realistic transducer beam. In Section 3, we use a near boundary approximation to show that the scattered "creeping" wave can be described as a combination of the T (transverse) portion of the head wave, its L (longitudinal) portion propagating along the surface and a somewhat unusual L wave whose amplitude grows proportionally to the distance to the path of propagation The latter is known in the literature as a Goodier-Bishop wave [4]. In Section 4 we justify a limiting procedure that allows us to use a canonical GTD problem of plane wave diffraction by a wedge to simulate wedge diffraction of the "creeping" wave. In Section 5 we present numerical results and in Section 6, discussion of some controversial issues. We finish by describing what has been achieved and issues that require further consideration.

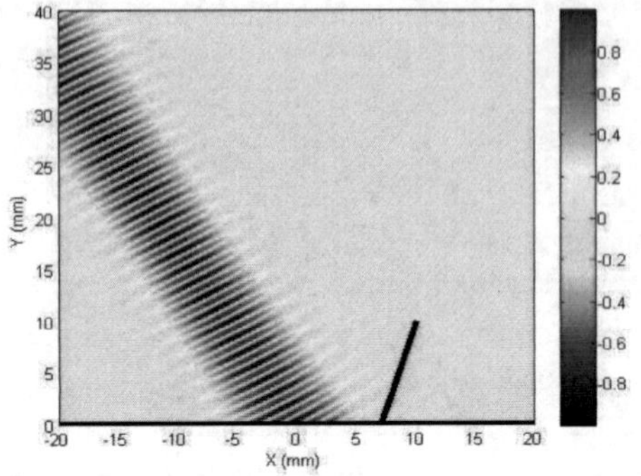

Figure 1. A back-wall crack (solid black line) irradiated by a Gaussian beam.

2. The problem statement

We consider an isotropic and homogeneous elastic medium occupying a half-space and assume a two-dimensional configuration, with the associated Cartesian basis $\{\mathbf{e}_1, \mathbf{e}_2\}$ such that $\mathbf{e}_1$ runs along the boundary, $\mathbf{e}_2$ is its inner normal and the third axis is implied but omitted everywhere below. Therefore all points $\boldsymbol{x}$ referred to below lie in the same plane and are characterised only by two coordinates x_1 and x_2. If the circular frequency is ω, then whatever time harmonic displacement field $\boldsymbol{u}\exp(i\omega t)$ is excited inside this medium, its amplitude $\boldsymbol{u} = \boldsymbol{u}(\boldsymbol{x})$ can be written in terms of the longitudinal potential $\phi = \phi(\boldsymbol{x})$ and transverse potential $\psi = \psi(\boldsymbol{x})$,

$$\boldsymbol{u} = \nabla\phi + \nabla^{\perp}\psi, \tag{1}$$

where the nabla operators are $\nabla = (\partial_1, \partial_2)$, $\nabla^{\perp} = (\partial_2, -\partial_1)$ and ∂_i is a symbol of partial differentiation with respect to the spatial coordinate x_i, $i = 1, 2$. Everywhere below we use the following nomenclature: a field is designated by superscript ^{creep}, ^{GB}, ^{H}, ^{inc}, ^{ref}, ^{scat} or ^{tot} to indicate whether it is creeping, Goodier-Bishop type, head, incident, reflected, scattered or total wave, respectively. Note that the words "scattered field" are used to describe a combination of diffracted and reflected fields. One section below introduces a concept of a remainder field and utilises the subscript ^{rem} explained there. The accompanying subscript describes the nature of the incidence $_{\text{cyl}}$, $_{\text{plane}}$ and $_{\text{Gauss}}$ for cylindrical wave, plane wave and Gaussian beam, respectively. In addition, when describing displacement fields this subscript also includes letters $_L$, $_T$ and $_C$ for longitudinal, transverse and creeping waves, respectively. In the absence of body forces, the total potentials satisfy the following Helmholtz equations

$$\Delta\phi^{\text{tot}} + k_L^2\phi^{\text{tot}} = 0, \quad \Delta\psi^{\text{tot}} + k_T^2\psi^{\text{tot}} = 0, \tag{2}$$

where the Laplacian $\Delta = \partial_1^2 + \partial_2^2$, $k_\alpha = \omega/c_\alpha$, $\alpha = L, T$, with k_L,k_T,c_L and c_T - the wave numbers and speeds of the longitudinal (L) and transverse (T) waves, respectively. Let $\sigma^{\text{tot}} = \{\sigma_{ij}^{\text{tot}}\}$ be the amplitude of the related stress tensor,

$$\sigma_{ij}^{\text{tot}}(\boldsymbol{u}) = \lambda\delta_{ij}\nabla\cdot\boldsymbol{u}^{\text{tot}} + \mu\left(u_{i,j}^{\text{tot}} + u_{j,i}^{\text{tot}}\right), \quad i, j = 1, 2, \tag{3}$$

where λ and μ are the Lamé constants, and the comma before an index denotes differentiation with respect to the corresponding spatial coordinate, so that e.g. $u_{i,j} = \partial_j u_i$. Then the traction free boundary condition can be written as

$$\sigma_{12}^{\text{tot}}|_{x_2=0} = \sigma_{22}^{\text{tot}}|_{x_2=0} = 0. \tag{4}$$

We impose the usual radiation condition at infinity and assume that a line source $-Ck_T^{-2}\delta(x_1 - a_1)\delta(x_2 - a_2)$ running through a point $\boldsymbol{a} = (a_1, a_2)$ excites transverse potential $\psi_{\text{cyl}}^{\text{inc}}(\boldsymbol{x})$. It is well known that the above incident potential is proportional to the outward propagating cylindrical function, Hankel function of the first kind,

$$\psi_{\text{cyl}}^{\text{inc}}(\boldsymbol{x}) = \frac{\mathrm{i}C}{4k_T^2} H_0^{(1)}(k_T r), \tag{5}$$

where $r = \sqrt{(x_1 - a_1)^2 + (x_2 - a_2)^2}$ is the distance from observation point to excitation source. Let us now present the coordinates a_1 and a_2 in the form

$$a_1 = a\cos\varphi^{\text{inc}}, \quad a_2 = a\sin\varphi^{\text{inc}}, \tag{6}$$

where φ is the polar angle associated with the Cartesian coordinates (x_1, x_2). Let us allow the distance a from the source to the coordinate origin to be complex, with $\operatorname{Im} a \geq 0$. Let us further introduce an auxiliary Cartesian coordinate system (x_1', x_2') with the origin at $(\operatorname{Re} a\cos\varphi^{\text{inc}}, \operatorname{Re} a\sin\varphi^{\text{inc}})$ and the axis x_1' running in the direction of vector $(\cos\varphi^{\text{inc}}, \sin\varphi^{\text{inc}})$. Then the incident cylindrical wave turns into a Gaussian beam,

$$\psi_{\text{Gauss}}^{\text{inc}}(\boldsymbol{x}) = \frac{\mathrm{i}C}{4k_T^2} H_0^{(1)}(k_T r) \approx \frac{\mathrm{i}C}{4k_T^2}\sqrt{\frac{2}{\mathrm{i}\pi k_T r}}\,\mathrm{e}^{k_T\left[\operatorname{Im} a - \frac{(x_2')^2}{2\omega^2}\right]}\mathrm{e}^{\mathrm{i}k_T\left[x_1' + \frac{(x_2')^2}{2\rho}\right]}, \quad |k_T r| \gg 1, \tag{7}$$

where the beam radius of curvature ϱ and beam spot size w are given, respectively, by

$$\varrho = x_1'\left[1 + \left(\frac{\operatorname{Im} a}{x_1'}\right)^2\right], \quad w^2 = \operatorname{Im} a\left[1 + \left(\frac{x_1'}{\operatorname{Im} a}\right)^2\right]. \tag{8}$$

Below we impose the condition

$$C = \mathrm{e}^{-\operatorname{Im} a k_T} \tag{9}$$

and assume that the incident field is a Gaussian beam. We choose this type of incidence, because it is a good approximation to a realistic transducer beam and because it allows for an analytical treatment. To simplify our calculations, we introduce the double Fourier transform

$$\widetilde{f}(\boldsymbol{\xi}) = \int_{-\infty}^{+\infty}\int_{-\infty}^{+\infty} f\mathrm{e}^{-\mathrm{i}(\xi_1 x_1 + \xi_2 x_2)}dx_1 dx_2, \tag{10}$$

where $\boldsymbol{\xi} = (\xi_1, \xi_2)$ is an arbitrary wave vector, and also the single Fourier transform in x_1

$$\widehat{f}(\xi_1, x_2) = \int_{-\infty}^{+\infty} f\mathrm{e}^{-\mathrm{i}\xi_1 x_1}dx_1. \tag{11}$$

Introducing the line source into the right hand side of the second equation in (2) and applying to it the double Fourier transform, $\widetilde{\psi}_{\text{cyl}}^{\text{inc}}(\boldsymbol{\xi})$ satisfies the following algebraic equation:

$$(k_T^2 - \xi_1^2 - \xi_2^2)\widetilde{\psi}(\boldsymbol{\xi}) = -Ck_T^{-2}\mathrm{e}^{-\mathrm{i}\xi_1 a_1}\mathrm{e}^{-\mathrm{i}\xi_2 a_2} \tag{12}$$

and therefore its inverse Fourier transform in ξ_2 is

$$\widehat{\psi}^{\rm inc}_{\rm cyl}(\xi_1, x_2) = \frac{C\mathrm{e}^{-\mathrm{i}\xi_1 a_1}}{2\pi k_T^2}\int_{-\infty}^{+\infty}\frac{\mathrm{e}^{\mathrm{i}\xi_2(x_2-a_2)}}{\xi_2^2-\eta_T^2}d\xi_2. \tag{13}$$

Above and below we use the notation

$$\eta_\alpha = \sqrt{k_\alpha^2 - \xi_1^2},\ \alpha = L,\, T. \tag{14}$$

As explained in the Introduction, we are interested only in the total field near the traction-free boundary. Imposing the corresponding condition, the integral in (13) can be evaluated on the basis of the Cauchy Residue Theorem, provided the contour of integration is distorted to pass the pole $\xi_2 = \eta_T$ from below and $\xi_2 = -\eta_T$ from above. It can be shown that provided $\mathrm{Im}\, a_2 = 0$ or the Gaussian beam is not grazing the back-wall, under the source $(\mathrm{Re}\, a_1, \mathrm{Re}\, a_2)$

$$\widehat{\psi}^{\rm inc}_{\rm Gauss}(\xi_1, x_2) = \frac{\mathrm{i}C}{2\eta_T k_T^2}\mathrm{e}^{-\mathrm{i}[\xi_1 a_1 + \eta_T(x_2-a_2)]},\quad x_2 < \mathrm{Re}\, a_2. \tag{15}$$

3. Interaction of the shear Gaussian beam with the traction-free planar surface

The Helmholtz equations (1) are satisfied, respectively, by both the total potentials $\phi^{\rm tot}_{T(\rm Gauss)} = \phi^{\rm scat}_{T(\rm Gauss)}$, $\psi^{\rm tot}_{T(\rm Gauss)} = \psi^{\rm inc}_{\rm Gauss} + \psi^{\rm scat}_{T(\rm Gauss)}$ and incident potential $\psi^{\rm inc}_{\rm Gauss}$. Therefore they are satisfied by the potential $\phi^{\rm scat}_{T(\rm Gauss)}$ which is scattered by the medium boundary. Applying the Fourier transform in x_1 to (1) we obtain ordinary differential equations of the second order. This means that the Fourier transforms in x_1 of the scattered (outgoing) potentials that satisfy the radiation condition at infinity have the form

$$\widehat{\phi}^{\rm scat}_{T(\rm Gauss)}(\xi_1, x_2) = R_{TL}\widehat{\psi}^{\rm inc}_{\rm Gauss}(\xi_1, 0)\mathrm{e}^{\mathrm{i}\eta_L x_2},\quad \widehat{\psi}^{\rm scat}_{T(\rm Gauss)}(\xi_1, x_2) = R_{TT}\widehat{\psi}^{\rm inc}_{\rm Gauss}(\xi_1, 0)\mathrm{e}^{\mathrm{i}\eta_T x_2}. \tag{16}$$

Substituting (16) into the boundary condition (4) which is first re-formulated for the scattered potentials and is then Fourier transformed in x_1 gives us the reflection coefficients R_{TL} and R_{TT},

$$R_{TL} = -\frac{2\bar{\eta}_T^2\eta_T\xi_1}{\mathcal{R}(\xi_1)},\ R_{TT} = -\frac{\bar{\mathcal{R}}(\xi_1)}{\mathcal{R}(\xi_1)}, \tag{17}$$

where $\bar{\eta}_T = \sqrt{k_T^2/2 - \xi_1^2}$ and the Rayleigh function $\mathcal{R}$ and its conjugate $\bar{\mathcal{R}}$ are given by

$$\mathcal{R}(\xi_1) = \bar{\eta}_T^4 + \xi_1^2\eta_L\eta_T,\ \bar{\mathcal{R}}(\xi_1) = \bar{\eta}_T^4 - \xi_1^2\eta_L\eta_T. \tag{18}$$

The expressions for scattered elastodynamic potentials $\widehat{\phi}^{\rm scat}_T$ and $\widehat{\psi}^{\rm scat}_T$ which are valid near the traction-free planar surface can be found by applying the inverse Fourier transform in ξ_1 to (15) and (16), respectively. We perform this operation in the next subsections.

3.1. The scattered T component of the head wave - a near surface approximation

It is well known that a ray incident on the plane at a critical angle gives rise to a head wave (see [5] and references therein). The T portion of the head wave can be described by a contribution to the T component of the inverse Fourier transform in ξ_1 of a branch point in η_L. Let us start this calculation by rewriting R_{TT} as

$$R_{TT} = -\frac{\mathrm{i}C}{2k_T^2\eta_T}\left(\frac{\bar{\eta}_T^8 + \xi_1^4\eta_L^2\eta_T^2}{\bar{\eta}_T^8 - \xi_1^4\eta_L^2\eta_T^2} - \eta_L\frac{2\xi_1^2\bar{\eta}_T^4\eta_T}{\bar{\eta}_T^8 - \xi_1^4\eta_L^2\eta_T^2}\right)\mathrm{e}^{\mathrm{i}(-\xi_1 a_1 + \eta_T a_2)}. \tag{19}$$

Only the second term contains a branch point at $\eta_L = 0$. The corresponding part of the inverse Fourier transform is

$$\psi^H_{T(\mathrm{Gauss})} = \frac{\mathrm{i}C}{k_T^2}\int_{-\infty}^{\infty} \eta_L \frac{\xi_1^2\bar{\eta}_T^4}{\bar{\eta}_T^8 - \xi_1^4\eta_L^2\eta_T^2} \mathrm{e}^{\mathrm{i}[\xi_1(x_1-a_1)+\eta_T(a_2+x_2)]} d\xi_1. \tag{20}$$

Provided the distance to the boundary is small, $k_T x_2 \approx 0$, this potential can be shown to be

$$\psi^H_{T(\mathrm{Gauss})} = \frac{E_c}{(k_L d_1)^{3/2}}\psi^0, \quad \psi^0 = \frac{4\mathrm{i}\gamma \mathrm{e}^{\mathrm{i}(k_L x_1 + k_T \sin\varphi^{head} x_2)}}{k_T(2\gamma^2-1)}, \tag{21}$$

where the critical angle $\varphi^{\mathrm{head}} = \cos^{-1}\gamma$, the distance d_1 is defined graphically in Fig. 2 and the constant E_c is given by

$$E_c = \frac{\sqrt{2\pi\mathrm{i}}\, C\gamma^3}{k_T(2\gamma^2-1)} \mathrm{e}^{\mathrm{i}(-k_L a_1 + k_T\sqrt{1-\gamma^2}a_2)}. \tag{22}$$

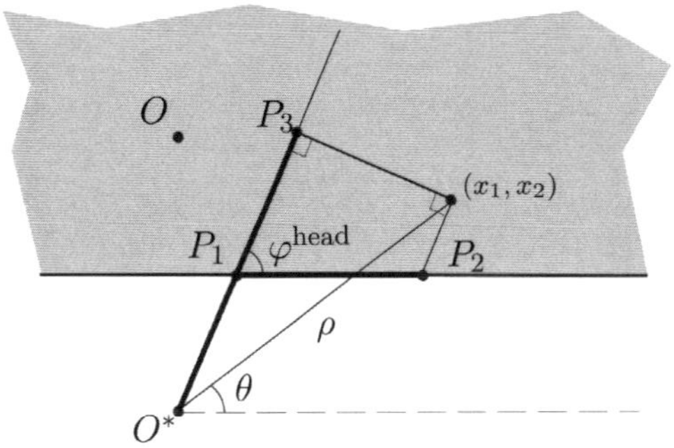

Figure 2. Geometry associated with the scattered T wave. $O = (\mathrm{Re}\, a_1, \mathrm{Re}\, a_2)$ - the point source, $O^* = (\mathrm{Re}\, a_1, -\mathrm{Re}\, a_2)$ the imaginary source, (x_1, x_2) - the observation point. $d_0 = O^*P_3$, $d_1 = P_1P_2$.

3.2. The scattered L waves - a near surface approximation

Combining (16) and (17) and applying the inverse Fourier transform in ξ_1, the scattered longitudinal potential is

$$\phi^{\mathrm{scat}}_{T(\mathrm{Gauss})} = -\frac{\mathrm{i}C}{k_T^2}\int_{-\infty}^{\infty} \frac{\xi_1\bar{\eta}_T^2}{\mathcal{R}(\xi_1)} \mathrm{e}^{\mathrm{i}[\xi_1(x_1-a_1)+\eta_T a_2 + \eta_L x_2]} d\xi_1. \tag{23}$$

Let us assume that the observation point lies far from the point where the axis of the incident Gaussian beam hits the boundary $k_L\widetilde{d} \gg 1$, where the distance $\widetilde{d}$ is defined graphically in Fig. 3). It can be shown that in this approximation, we have

$$\phi^{\mathrm{creep}}_{T(\mathrm{Gauss})} = \frac{\sqrt{2\pi\mathrm{i}}C\gamma^2\mathrm{e}^{\mathrm{i}\Phi(\widetilde{\vartheta})}}{k_T^2}\left[\frac{A(\widetilde{\vartheta})}{(k_L\widetilde{d})^{1/2}} - \frac{\mathrm{i}A''(\widetilde{\vartheta})}{2(k_L\widetilde{d})^{3/2}}\right] + O\left((k_L\widetilde{d})^{-5/2}\right), \tag{24}$$

where

$$A(\widetilde{\vartheta}) = \frac{2\widetilde{\vartheta}}{1-2\gamma^2} + O(\widetilde{\vartheta}^2), \quad A''(\widetilde{\vartheta}) = -\frac{16\gamma^3\sqrt{1-\gamma^2}}{(1-2\gamma^2)^3} + O(\widetilde{\vartheta}) \tag{25}$$

and the angle $\widetilde{\vartheta}$ is defined graphically in Fig. 3. Then near the boundary $\widetilde{d\vartheta} \approx x_2$ and we obtain

$$\phi_{T(\text{Gauss})}^{\text{creep}} = \frac{E_c}{(k_L\widetilde{d})^{3/2}}\phi^0, \quad \phi^0 = -\left[\frac{8\mathrm{i}\gamma^3\sqrt{1-\gamma^2}}{k_L(1-2\gamma^2)^2} + 2x_2\right]\mathrm{e}^{\mathrm{i}k_L x_1}, \tag{26}$$

where the coefficient E_c is defined by (22). It follows that the longitudinal component is a sum of two waves, one ordinary and another - the Goodier-Bishop type wave whose amplitude grows linearly with the distance to its propagation path [4].

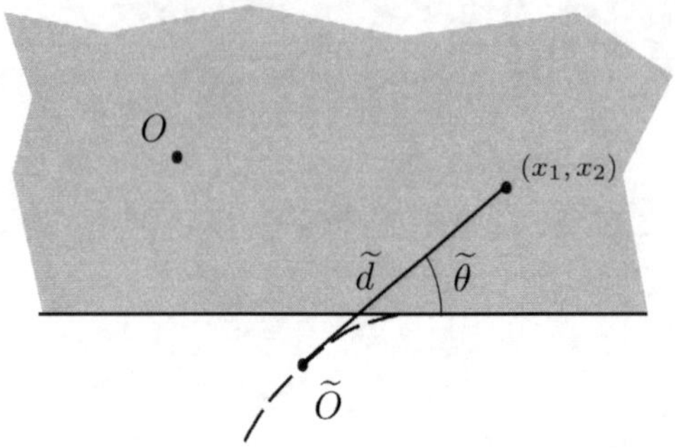

Figure 3. Geometry associated with the scattered L wave. $O = (\mathrm{Re}\, a_1, \mathrm{Re}\, a_2)$ - the point source, $\widetilde{O} = (\mathrm{Re}\,\widetilde{a}_1, \mathrm{Re}\,\widetilde{a}_2)$ - an imaginary L source. Dashed line - locus of imaginary sources for the observation angles $\widetilde{\vartheta}$ that vary between 0 to $\pi/2$.

4. Interaction of the "creeping" wave with a planar faces crack

Let us now assume that the back-wall contains an infinite planar surface-breaking crack. Formulae (21) and (26) show that near the boundary the field generated by an incident shear Gaussian beam and known as the "creeping" wave consists of one T wave and two L waves. If this combination reaches a crack before its amplitude becomes negligible the interference between resulting reflected and diffracted waves creates patches of significant amplitude. In this paper we concentrate on diffraction of the "creeping" wave and in order to calculate its amplitude we first investigate an auxiliary problem of diffraction of a plane L wave.

Let the incident L wave

$$\boldsymbol{u}_{L(\text{plane})}^{\text{inc}} = \frac{1}{\mathrm{i}k_L}\nabla \mathrm{e}^{\mathrm{i}k_L(x_1\cos\vartheta_L - x_2\sin\vartheta_L)}, \tag{27}$$

irradiate the crack and let the angle it forms with the face ($x_2 = 0+$, $x_1 < 0$) be near grazing, $\vartheta_L \approx 0$. In the second quadrant ($x_1 \leq 0$, $x_2 \geq 0$) the resulting total field can be written as

$$\boldsymbol{u}_{L(\text{plane})}^{\text{tot}} = \boldsymbol{u}_{L(\text{plane})}^{\text{inc}} + \boldsymbol{u}_{L(\text{plane})}^{\text{ref}} + \boldsymbol{u}_{L(\text{plane})}^{\text{rem}}. \tag{28}$$

where $\boldsymbol{u}_{L(\text{plane})}^{\text{ref}}$ is the sum of the first reflected L and T waves and $\boldsymbol{u}^{\text{rem(aux)}}$ is the remainder. The reflected term is

$$\boldsymbol{u}_{L(\text{plane})}^{\text{ref}} = \frac{R_{LL}}{\mathrm{i}k_L}\nabla \mathrm{e}^{\mathrm{i}k_L(x_1\cos\vartheta_L + x_2\sin\vartheta_L)} + \frac{R_{LT}}{\mathrm{i}k_T}\nabla^{\perp}\mathrm{e}^{\mathrm{i}k_T(x_1\cos\vartheta_T + x_2\sin\vartheta_T)}, \tag{29}$$

where R_{LL} and R_{LT} are the respective reflection coefficients [8] and the angle ϑ_T is determined by the Snell's law, $k_L\cos\vartheta_L = k_T\cos\vartheta_T$. For $\vartheta_L \ll 1$ these reflection coefficients, can be expanded in $\vartheta_L \ll 1$ to give

$$R_{LL} = -1 + \frac{8\gamma^3\sqrt{1-\gamma^2}}{(1-2\gamma^2)^2}\vartheta_L + O(\vartheta_L^2), \quad R_{LT} = \frac{4\gamma}{(1-2\gamma^2)}\vartheta_L + O(\vartheta_L^2). \tag{30}$$

As $\vartheta_L \to 0$, $R_{LL} \to -1$ and $R_{LT} \to 0$ and the sum of the first two terms in (28) - which in the grazing limit becomes the field incident on the crack - vanishes,

$$\lim_{\vartheta_L \to 0} \left(\boldsymbol{u}^{\rm inc}_{L({\rm plane})} + \boldsymbol{u}^{\rm ref}_{L({\rm plane})} \right) = 0. \tag{31}$$

Since the wedge diffraction problem has an unique solution this implies

$$\lim_{\vartheta_L \to 0} \boldsymbol{u}^{\rm rem}_{L({\rm plane})} = 0. \tag{32}$$

It follows that as $\vartheta_L \to 0$ diffraction coefficients $D_{L\beta}(\vartheta_L; \vartheta), \beta = L, T$ vanish too, that is, we have

$$\lim_{\vartheta_L \to 0} D_{L\beta}(\vartheta_L; \vartheta) = 0, \beta = L, T. \tag{33}$$

for all observation angles ϑ.

To proceed, we note that there is a simple relation between the normalised creeping waves described in the previous section

$$\boldsymbol{u}^{\rm inc}_0 \equiv \nabla \phi^0 + \nabla^{\perp} \psi^0 \tag{34}$$

and the first two terms in (28):

$$\boldsymbol{u}^{\rm inc}_0 = \left. \frac{\partial \left(\boldsymbol{u}^{\rm inc}_{L({\rm plane})} + \boldsymbol{u}^{\rm ref}_{L({\rm plane})} \right)}{\partial \vartheta_L} \right|_{\vartheta_L = 0}. \tag{35}$$

Differentiating (28) with respect to ϑ_L and using (35) we obtain

$$\left. \frac{\partial \boldsymbol{u}^{\rm tot}_{L({\rm plane})}}{\partial \vartheta_L} \right|_{\vartheta_L = 0} = \boldsymbol{u}^{\rm inc}_0 + \left. \frac{\partial \boldsymbol{u}^{\rm rem}_{L({\rm plane})}}{\partial \vartheta_L} \right|_{\vartheta_L = 0}. \tag{36}$$

Since the differentiation symbols commute the left-hand side of (36) satisfies the equations of motion and the boundary conditions. Also, if an outgoing wave is differentiated with respect to ϑ_L the result is another outgoing wave. This differentiation does not change the nature of the corresponding tip condition either. It follows that $\partial \boldsymbol{u}^{\rm tot}_{L({\rm plane})}/\partial \vartheta_L$ is the solution of the same wedge diffraction problem as $\boldsymbol{u}^{\rm tot}_{L({\rm plane})}$ but under another incidence. Since the waves in $\partial \boldsymbol{u}^{\rm rem}_{L({\rm plane})}/\partial \vartheta_L$ are outgoing, this incident field is $\boldsymbol{u}^{\rm inc}_0$ and the left-hand side of (36) is $\boldsymbol{u}^{\rm tot}_0$, the total field generated by the "creeping" wave incident on the wedge. We can now write

$$\boldsymbol{u}^{\rm tot}_0 = \left. \frac{\partial \boldsymbol{u}^{\rm tot}_{L({\rm plane})}}{\partial \vartheta_L} \right|_{\vartheta_L = 0} = \lim_{\vartheta_L \to 0} \frac{\boldsymbol{u}^{\rm tot}_{L({\rm plane})}}{\vartheta_L}. \tag{37}$$

Eq. (37) suggests the following formula for calculation of the diffraction coefficients for the normalised "creeping" wave,

$$D_{0,\beta}(\vartheta) = \frac{\partial D_{L\beta}(0; \vartheta)}{\partial \vartheta_L} = \lim_{\vartheta_L \to 0} \frac{D_{L\beta}(\vartheta_L; \vartheta)}{\vartheta_L}. \tag{38}$$

Formulae (21) and (26) show that near the wedge tip the "creeping wave" has the amplitude $E_c(k_L d_0)^{-3/2}$, with d_0 defined graphically in Fig. 2. Therefore in order to calculate the diffraction coefficients $D^{\beta}_{T({\rm Gauss})}(\vartheta), \beta = L, T$ which are due to the transverse Gaussian beam incident at the critical angle we need to multiply (38) by this additional coefficient to get

$$D^{\beta}_{T({\rm Gauss})}(\vartheta) = \frac{E_c}{(k_L d_0)^{3/2}} D_{0,\beta}(\vartheta), \quad \beta = L, T \tag{39}$$

where we used the fact that in the far field $d_0 \approx \tilde{d}$.

The last formula suggests a limiting procedure that allows us to utilise the standard GTD to evaluate non-uniform asymptotics for the "creeping" wave diffracted off a planar faces crack. All we need to do given a wedge angle is to evaluate the standard diffraction coefficients using the existing numerical codes for a sequence of small incidence angles ϑ_L, each time dividing the result by the corresponding angle and multiplying by a constant and terminate the procedure on reaching a specified accuracy.

The resulting diffraction coefficients have singularities at the shadow boundaries associated with reflected waves as well as critical rays, where the non-uniform asymptotics are inapplicable. Similarly to the case of plane incidence [15] it is possible to introduce the uniform asymptotics that are valid inside penumbra associated with the Goodier-Bishop type waves as well as standard body waves. Moreover, the above considerations can be used to evaluate scattered head waves. We will discuss both topics in future publications.

5. Numerical results

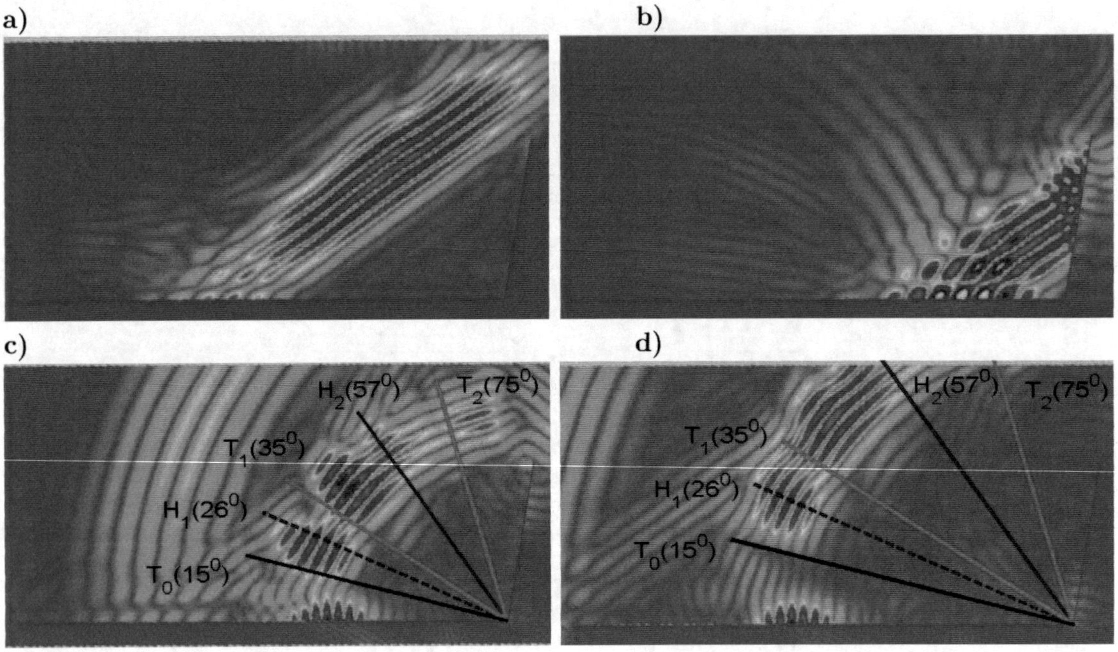

Figure 4. ATHENA simulations with a shear beam incident on the wall at the supercritical angle of 55^0. The crack angle is 100^0. The incident beam irradiates whole crack. a) and b) are consecutive snapshots of incident field. c) and d) are consecutive snapshots of scattered field.

Some features of the above models have been simulated numerically using ATHENA, a finite element commercial code. The simulation parameters are $c_L = 5890\,km/s$, $c_T = 3210\,km/s$ and the frequency is $2\,MHz$. We restrict all discussions to the surface-breaking crack which forms a 100° angle with the back-wall and to the beam that is incident at 55^0 to this wall. In Figs. 4 - 6 we consider three different positions of the transducer as it moves away from the crack and study the changes in the respective scattered fields. Note that in ATHENA simulations it is assumed that the medium below the wall is air or water. For this reason non-zero displacements are indicated in Figs. 5 - 7 underneath the wall. The two consecutive screen shots presented in Fig. 4 a) and b) have been obtained for a transducer positioned so that when the beam hits the back-wall it irradiates the whole crack. The two consecutive screen shots presented in Fig. 6 a) and b) have been taken for a transducer moved so far from the crack that the beam hits the back-wall in front

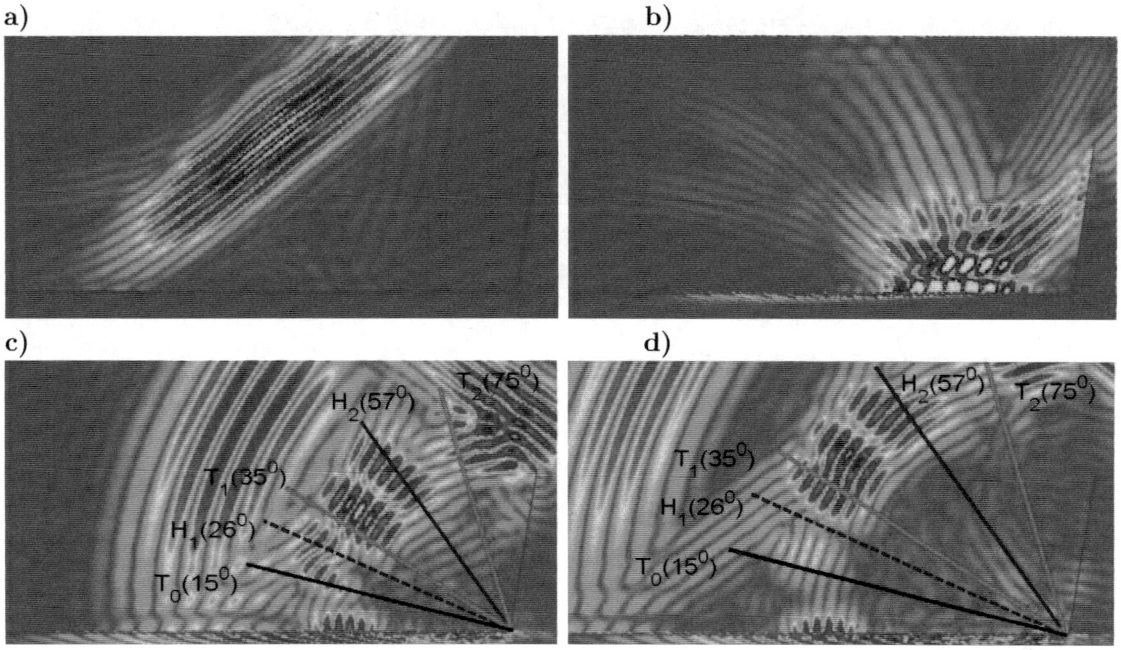

Figure 5. ATHENA simulations with a shear beam incident on the wall at the supercritical angle of 55^0. The crack angle is 100^0. A portion of the incident beam irradiates the crack. a) and b) are consecutive snapshots of incident field. c) and d) are consecutive snapshots of scattered field.

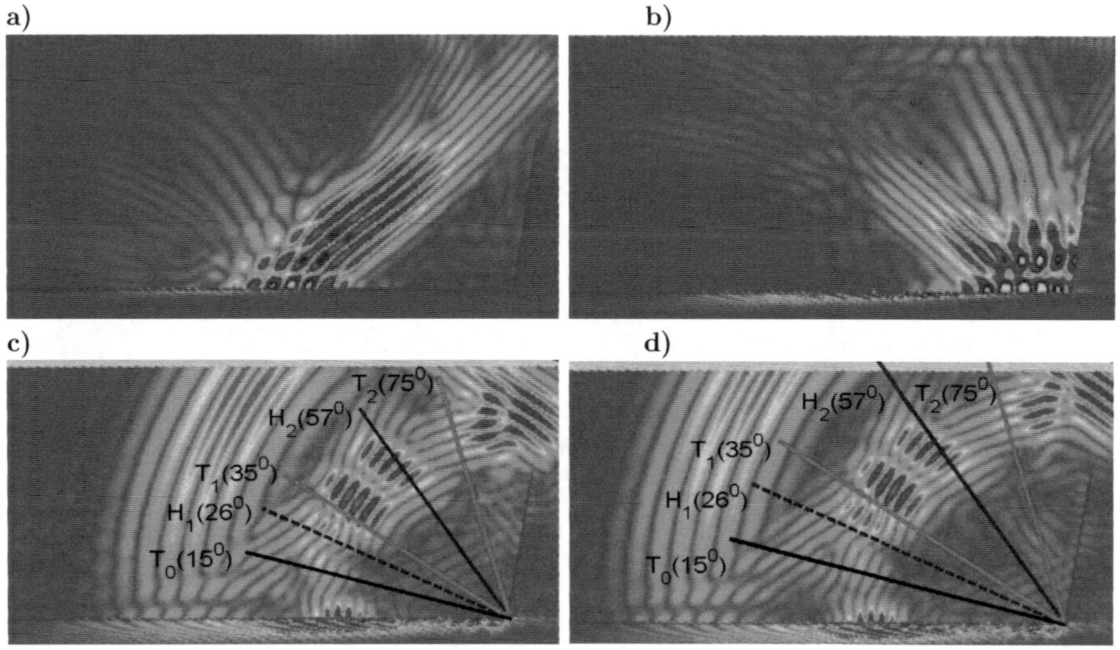

Figure 6. ATHENA simulations with a shear beam incident on the wall at the supercritical angle of 55^0. The crack angle is 100^0. The incident beam hits the wall in front of crack. a) and b) are consecutive snapshots of incident field. c) and d) are consecutive snapshots of scattered field.

of the crack and generates a "creeping" wave, which is the only wave that reaches the crack tip. Fig. 5 is an intermediate case and the two consecutive screen shots presented in a) and b) show

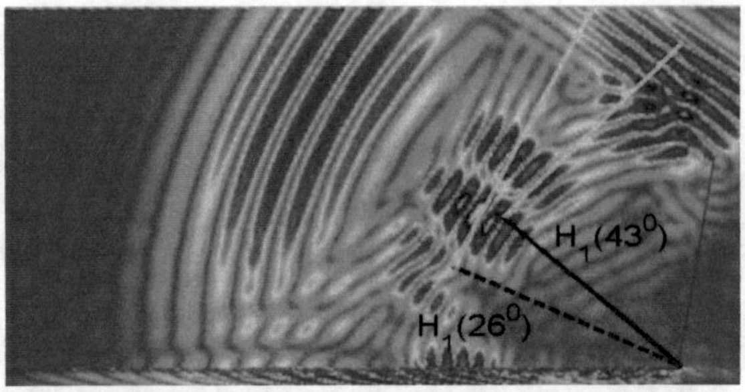

Figure 7. ATHENA simulations of the incidence of a shear beam at the supercritical angle of 55^0. The crack angle is 100^0. The incident beam irradiates the whole crack. Solid blue line - theoretical critical ray, solid green line - corresponding head wave front, Dashed green line - simulated head wave front, dahed blue line - corresponding critical ray.

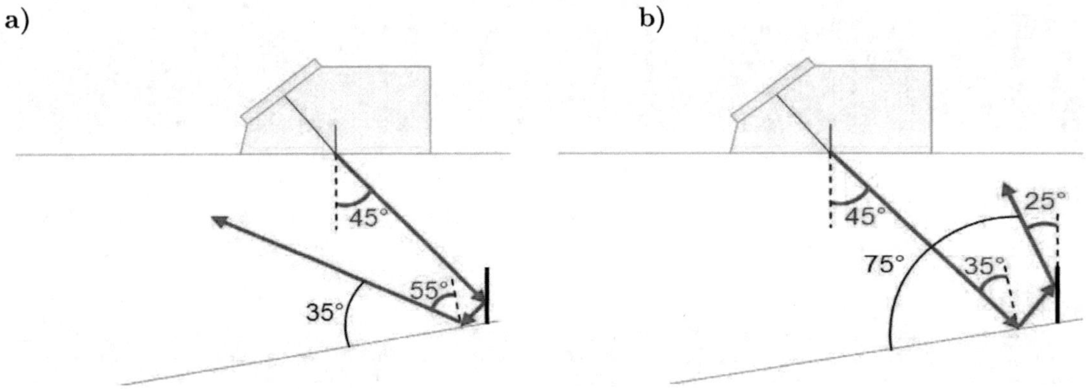

Figure 8. Geometry of two double reflected T waves.

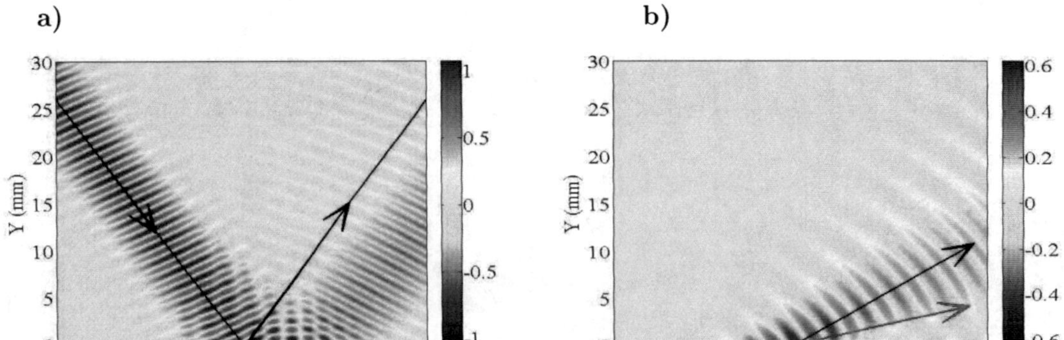

Figure 9. A Gaussian beam incident at the critical angle on a traction free surface. a) The transverse components of the incident and scattered fields. The black arrows indicate the direction of propagation. b) The longitudinal component of the scattered field. Black line - the direction of the scattered longitudinal beam $\approx 30^0$ to the back-wall. Red line - the direction of the scattered longitudinal lobe in [13]. Red line makes 16^0 with the back-wall.

that in this transducer position the crack is irradiated by both the "creeping" wave and a portion of the incident beam.

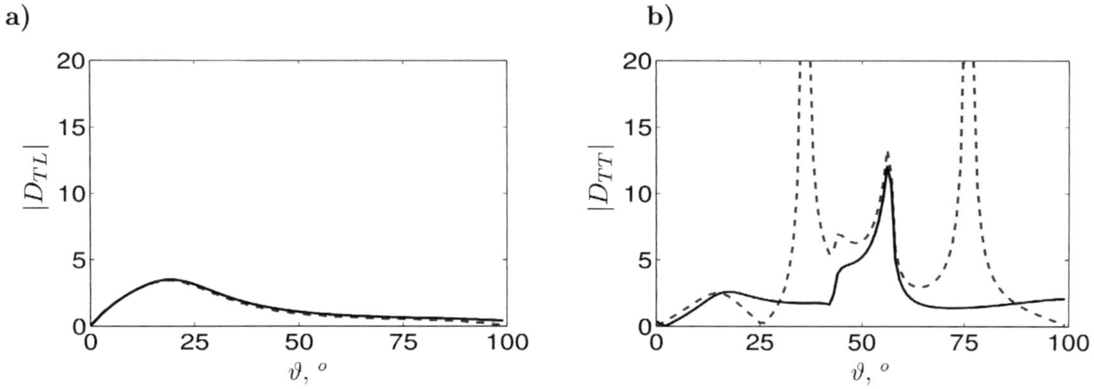

Figure 10. Diffraction coefficients vs the observation angle ϑ. The wedge angle - 100^0. The incident wave - the plane T wave incident at 56^0 to the wall (supercritical configuration). a) $D_{0,L}$ b) $D_{0,T}$. Dashed line - GTD coefficients, solid line - corrected coefficients as defined in [15].

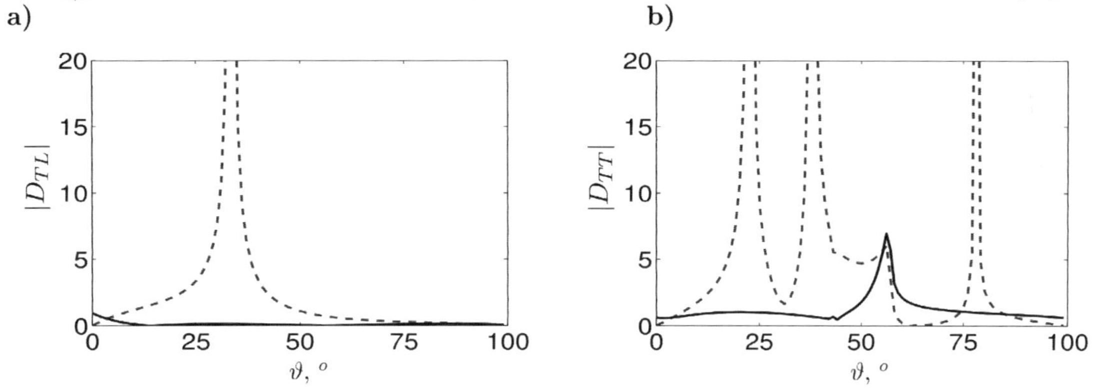

Figure 11. Diffraction coefficients vs the observation angle ϑ. The wedge angle - 100^0. The incident wave - the plane T wave incident at 58^0 to the wall (subcritical configuration). a) $D_{0,L}$ b) $D_{0,T}$. Dashed line - GTD coefficients, solid line - corrected coefficients as defined in [15].

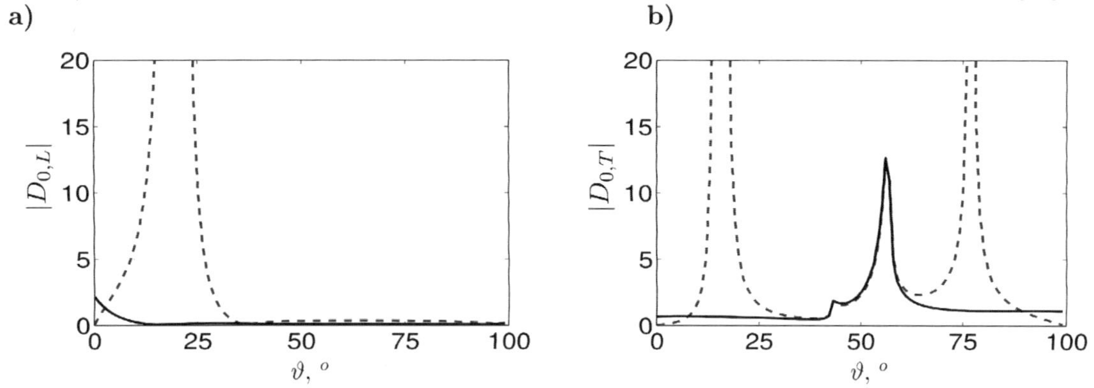

Figure 12. Diffraction coefficients vs the observation angle ϑ. The wedge angle - 100^0. The incident wave - the normalised creeping wave propagating along the wall. a) $D_{0,L}$ b) $D_{0,T}$. Dashed line - GTD coefficients, solid line - corrected coefficients as defined in [15].

Let us first discuss geometry of the resulting scattered fields presented in c) and d) portions of Figs. 4 - 6. They all feature two circular wave fronts, T and L, centered at the crack corner as well as the corresponding head wave fronts as well as the following T rays,

- T_0 (at 15^0 to the wall) - the T reflection of the L component of the wall creeping wave,
- T_1 (at 35^0 to the wall) - the shadow boundary associated with the twice reflected T wave,
- H_1 (at 26^0 to the wall) - the critical ray associated with the diffracted head wave,
- H_2 (at 57^0 to the wall) - the critical ray associated with the diffracted head wave,
- T_2 (at 75^0 to the wall) - the shadow boundary associated with the twice reflected T wave.

The angles of propagation of the waves T_1 and T_2 are justified by reference to Figs. 8 a) and b), respectively. The angle of propagation of T_0, the T reflection of the L portion of the creeping wave is found using the Snell's law. The situation with the head waves is more complicated: had the vertical crack face been of infinite extent the critical rays would each form $\cos^{-1}(\gamma) \approx 57^0$ with the corresponding crack face. In the ATHENA simulations presented in Figs. 4 - 6, the critical ray H_2 corresponding to the lower head wave front indeed appears to run at $\approx 57^0$ to the back-wall. However, the upper head wave front in these simulations is inclined at such an angle that the corresponding critical ray H_1 runs at 26^0 to the back-wall rather than 43^0 - see Fig. 7. This might be due to the fact simulations have been carried out for a finite crack. Indeed, the problem is less pronounced in simulations involving larger cracks. To indicate the tentative nature of the ray H_1 it is depicted in Figs. 4 - 6 using the dashed line. Let us proceed with our comparisons. In accordance with Fig. 8, both T_1 and T_2 singularities at about 35^0 and 75^0, respectively, are present in Figs. 10b) and 11b). When the incident beam hits the back-wall before the crack, the creeping diffraction coefficient in Fig. 12b) lacks the T_1 singularity, because in this configuration there can be no rays shown in Fig. 8a). Moving on to the T_0 wave, when the incident beam irradiates the whole crack, we can apply the canonical model of diffraction of plane incident wave. If the angle of incidence is smaller (greater) than $\cos^{-1}(\gamma)$, the configuration is supercritical (subcritical) and we do not have (have) a body L wave incident on the vertical crack face. This explains why there is a singularity at about 18^0 in the standard T GTD diffraction coefficients presented in Fig. 11b) but there is no such singularity in Fig. 10b). When the incident beam hits the back-wall before the crack, the creeping diffraction coefficient in Fig. 12b) has the corresponding singularity at about 15^0. There is a difference between the nature of these singularities: it has been established numerically that in Fig. 10b) we have a single pole at 20^0 and in Fig. 12, the T_0 singularity at 15^0 is a double pole. The latter behaviour is characteristic of the Goodier-Bishop wave. For similar reasons, there a single pole at about 20^0 in the standard L GTD diffraction coefficients presented in Fig. 11a), there is no such singularity in Fig. 10a) and in Fig. 12a) the singularity at 20^0 is a double pole. Note that the corrected diffraction coefficients presented in Figs. 10) - 11 are calculated using the standard recipe described e.g. in [15].

Let us now study the energy variation in the scattered fields as presented in c) and d) parts of Figs. 4 - 6. The comparison is complicated by the difference in amplitude scaling, since in Fig. 5 the incident amplitude appears to be much stronger than in other configurations. It is important to keep in mind that in the ATHENA simulations supercritical incidence was used and thus the situation corresponds to Fig. 10 rather than Fig. 11. To start with, all scattered fields contain the Rayleigh wave travelling along the back-wall. According to Figs. 10a) and 12a), the diffracted L waves are supposed to be very weak. The fact that we see high intensity patches superimposed on the L fronts in Figs. 5c) and d) and Figs. 6c) and d) must be due to the reflection at the vertical crack face of the L component of the creeping wave. Note that there is no high intensity region superimposed on the L front in Figs. 4. This is consistent with the fact that in this supercritical configuration no L wave is incident on the vertical crack face. We now turn attention to the lowest bright spot seen on the T fronts in Figs. 4c) and d). It is natural to assume that it is due to T_1. Indeed, in accordance with Fig. 8b) it appears beneath the shadow boundary running from the crack corner at 35^0 to the back-wall. In Figs. 5c) and d) the corresponding spot is much weaker, which is consistent with the fact that in this configuration the crack is only partially irradiated by the incident beam. However, there are many problems associated with this hypothesis. The ATHENA simulations are more consistent with the idea that the spot is due to interference between T_0 and T_1. This would explain why the spot is not touching the 35^0 ray in

Fig. 4c) and why it is closer to it in Figs. 4d). Furthermore, the new hypothesis would explain why the shift is even stronger in Fig. 5 and why it is accompanied by increase in amplitude. Indeed, in this case the contribution of the T_1 wave is much weaker than when the whole of the crack is irradiated and T_0 has a Goodier-Bishop component whose amplitude increases in a linear fashion in the direction normal to the direction of propagation. The hypothesis also accounts for the weak spot associated with the corresponding ray in Fig. 6c): in this case the T_1 wave is particularly weak, because it is generated by the weak outer beam margin hitting the crack's vertical face. Again, the intensity of the spot grows with time from Fig. 6c) to Fig. 6d), which is hard to explain for any wave other than the Goodier-Bishop type. The fact that this spot is so much brighter in Fig. 5c) than in Fig. 6c) must be partly due to the difference in the incident amplitude.

Moving in the clockwise direction along the circular T front, the other two bright spots in the scattered fields, which sometimes coalesce and which are present in all three figures, must be due to two head waves, H_1 and H_2. The spots lie inside the intersection of both corresponding irradiated zones. The third bright spot present in all figures is easily assigned to the T_2 wave, since it lies just above the 75^0-ray which is, according to Fig. 8b), the T_2 irradiated zone. Other bright regions above this spot are due partly to waves diffracted by the top crack tip and mainly to the single reflection at the back-wall of the incident beam.

We finish this section by the stationary simulation of scatter of a Gaussian beam incident at the critical angle by the traction free interface. It is obtained by applying the inverse Fourier transform to (16) and numerical evaluation of the resulting integrals (Fig. 9). The black arrowed lines in Fig. 9a) indicate the axis of the incident beam and its geometrical optics (GO) reflection. The reflected beam itself demonstrates a well-known lateral shift from the GO path [10]. Fig. 9b) presents the reflected L component. The black line in Fig. 9b) indicates the direction of the main beam propagation, which makes approximately 30^0 with the back-wall.

6. Discussions

There are many controversies surrounding grazing incidence and creeping wave. To start with, because in this limit the incident and reflected L waves coalesce, neither Kirchhoff approximation nor uniform asymptotics of incident and scattered fields taken when in isolation extend to grazing incidence. The solution lies in decomposing the total field into a sum of the grazing waves (incident and reflected plane waves - in the regions where they exist) and the waves diffracted by the crack edge - see in [11], [9] and references therein. When seeking asymptotics, the procedure should be carried out for the incidence angles which are close to grazing and then utilising a limiting procedure. When the limiting incident wave is of the Goodier-Bishop type this should be based on (37). Otherwise, one should just employ continuity considerations [9]. Though arising from mathematical considerations the Goodier-Bishop waves vex many physicists (e.g. [8]). A particular problem arises when the incident wave is plane and transversal, propagating at a critical angle to the surface. In this case the problem is ill-posed. However, no issue surrounds the Goodier-Bishop type waves arising in the problem considered in this paper, because they only figure only in a near boundary approximation. Finally, the term "creeping" wave as used in engineering literature deviates from the accepted usage in applied mathematics. In applied mathematics a creeping wave is a single wave gliding along a curved surface and decaying exponentially (see [12] and references therein). For this reason, some authors - see e.g. [13] - question the existence of a creeping wave in the configuration considered above. In principle, their conclusion is similar to ours: under a critical incidence there arises a disturbance near the boundary which looks like a creeping wave, but is in fact a combination of well known body waves which decays away from its source as $(k_L R)^{-3/2}$. With reference to Fig. 9, the illusion of a surface wave arises at distances from the specular point $(0, 0)$ which are smaller than 10 mm. This corresponds to distances observed in Figs. 4 - 6. Note that when the "creeping" wave hits the crack, apart from exciting diffracted waves each of its components reflects according to the Snell's law. However it is possible to say that the incident "creeping" wave generates reflected and transmitted "creeping" waves [14]. The situation is not identical to the Rayleigh wave incidence, because unlike in the latter case, the interaction between the "creeping" wave and the crack leads to reflected T and L waves as well.

7. Conclusions

In this paper we have studied interaction of a shear Gaussian beam with a back-wall crack. It has been shown that when the incidence is near critical the Gaussian beam excites a strong "creeping" wave which soon attenuates. This "creeping" wave has a T component (the head wave) and L component which - in the near boundary approximation - comprises the plane surface wave and a Goodier-Bishop type wave. We have presented a method for evaluating the diffracted field when the wedge is irradiated by the "creeping" wave and calculated the resulting diffraction coefficients. The full uniform description of the scattered field will be the subject of future investigation.

8. Acknowledgement

This work has been carried out within the framework of the CIVAMONT 2012 collaborative project.

References

[1] Achenbach A, Gautesen A, and McMaken H 1979 *Ray Methods for Waves in Elastic Solids; With Applications to Scattering by Cracks* (New York: Pitman)
[2] Kamotski V, Fradkin L, Babich V M, Borovikov V A, and Samokish B A 2006 *SIAM J. Appl. Math.* **67**(1) 235
[3] Gautesen A and Fradkin L 2011 *SIAM J. Appl. Math.* **70**(8) 3065
[4] Goodier J N and Bishop R E D 1952 *J. Appl. Phys.* **23**(1) 124
[5] Cerveny V and Ravindra R 1971 *Theory of Seismic Head Waves* (Toronto: University of Toronto Press)
[6] Bleistein N and Handelsman R A 1986 *Asymptotic Expansions of Integrals* (New York: Dover)
[7] Borovikov V A 1994 *Uniform Stationary Phase Method* (London: IEE)
[8] Graff K F 1975 *Wave Motion in Elastic Solids* (London: Oxford University Press)
[9] Thomson I 2006 *Proc. R. Soc. A* **462** 1341
[10] Pott J and Harris J G 1984 *J. Acoust. Soc. Am.* **76**(S1) S64
[11] Ufimtsev P Ya 2007 *Fundamentals of the Physical Theory of Diffraction* (Chicester: Wiley-Blackwell)
[12] Tew R H, Chapman S J, King J R, Ockendon J R, Smith B J and I Zafarullah 2000 *Wave Motion* **32** 363
[13] Langerberg K J, Fellinger P and Marklein R 1990 *Res. Nondestr. Eval.* **2** 59
[14] Huet G, Darmon M, Lhemery A and Mahaut S 2009 *Springer Proc. Phys.* **128** 217
[15] Zernov V, Fradkin L and Darmon M 2011 *J. Phys.: Conf. Series* this issue

Ultrasonic non destructive testing based on sparse deconvolution

C Soussen[1], J Idier[2], E Carcreff[2,3], L Simon[3] and C Potel[3,4]

[1] Centre de Recherche en Automatique de Nancy (CRAN, UMR 7039, Nancy-University, CNRS). Campus Sciences, B.P. 70239, F-54506 Vandœuvre-lès-Nancy, France. This work has been performed while the first author was a visiting research fellow at IRCCyN, appointed by the CNRS

[2] Institut de Recherche en Communications et Cybernétique de Nantes (IRCCyN, UMR CNRS 6597), BP 92101, 1 rue de la Noë, F-44321 Nantes Cedex 3, France

[3] Laboratoire d'Acoustique de l'Université du Maine (LAUM, UMR CNRS 6613). Avenue O. Messiaen, F-72085 Le Mans Cedex 9, France

[4] Fédération Acoustique du Nord-Ouest (FANO, FR CNRS 3110), France

E-mail: `Charles.Soussen@cran.uhp-nancy.fr`

Abstract. The acoustic modality yields non destructive testing techniques of choice for in-depth investigation. Given a precise model of acoustic wave propagation in materials of possibly complex structures, acoustical imaging amounts to the so-called acoustic wave inversion. A less ambitious approach consists in processing pulse-echo data (typically, A- or B-scans) to detect localised echoes with the maximum temporal (and lateral) precision. This is a resolution enhancement problem, and more precisely a sparse deconvolution problem which is naturally addressed in the inversion framework. The paper focuses on the main sparse deconvolution methods and algorithms, with a view to apply them to ultrasonic non-destructive testing.

1. Introduction

Non-destructive testing by pulse-echo ultrasound allows the inspection of a complex object and in particular, the detection and localization of inhomogeneities and faults. The acoustic signal measured by the ultrasonic sensor results from the propagation inside the inspected object, of the waves generated by the interaction of the incident wave with the object. Reflections and/or transmissions of the wave induce the presence of echoes of the incident wave in the measured signal. Echoes are closely related to the change of acoustic impedance inside the object. Their detection and localization in terms of time of arrival and relative amplitude can be formulated as a spike deconvolution problem where the impulse response corresponds to the acoustic wave emitted from the input transducer and the spike locations and amplitudes are related to the acoustic reflectivity of the medium. Therefore, their accurate estimation is needed to obtain quantitative information about the inner structure of the object, both in terms of geometry and of structure.

From the information processing viewpoint, the spike deconvolution problem is especially challenging when the material dimension is small in comparison with the wavelength of the incident signal because the echoes overlap. In such a case, the discrimination and precise localisation of the echoes is far from obvious. Another issue occurs when the noise level is high since echoes may be partly or completely drowned in noise due to the signal attenuation while the wave is propagating.

The simplest method to detect and localize echoes is the matched filter [1]. It is based on the maximization of the inter-correlation between the measured signal y and the incident wave h. Let us remark that matched filtering lends itself to a simple interpretation in terms of least-squares principle. It simply consists in considering an isolated echo of unknown delay τ and relative scalar amplitude x, and of minimising a least-squares fidelity-to-data criterion with respect to both x and τ. Since the criterion is a quadratic function of the amplitude x, the minimisation can be carried out explicitly with respect to x, and the remaining minimisation step with respect to τ amounts to maximising the inter-correlation between y and h.

Unfortunately, this technique does not yield satisfying results when the echoes overlap. The approach addressing the deconvolution problem using the assumption that the deconvolved signal is *sparse* (there are a very limited number of spikes) is known to yield better results in that the time resolution of the above mentioned approach can be significantly enhanced [2]. In particular, the approach which models the reflectivity sequence using the Bernoulli-Gaussian model [3–5] is acknowledged to be very efficient. It was successfully applied to the deconvolution of ultrasonic echoes in composite materials [6].

In this paper, we address the spike deconvolution as an inverse problem where the detection and estimation of the spikes are performed jointly. The inversion method explicitly takes into account that the reflectivity signal is sparse. This can be done using either ℓ_1 [7] or ℓ_0 regularisation [8]. In particular, the latter approach has close connection with the Bernoulli-Gaussian deconvolution approach [9]. In both cases, the inversion problem results in a numerical optimisation problem. In the ℓ_1 case, the optimisation problem is continuous and recent efficient algorithms are available [10]. For ℓ_0 regularisation, the optimisation problem turns into a discrete search problem whose goal is to find the spiky configuration (with a limited number of spikes) yielding the best fit to the data [11].

The paper is organised as follows. In Section 2, we formulate the sparse deconvolution problem. In Section 3, we introduce the ℓ_0 and ℓ_1 deconvolution algorithms and we address the estimation of the number of spikes from the algorithm outputs using classical order model selection rules [12]. Section 4 illustrates our approach on two simulated examples: the first one is a classical spike train deconvolution example introduced in the seismic reflection context in [3]. It will enable us to compare ℓ_0 and ℓ_1 deconvolution methods. The second example simulates a more realistic situation of the propagation of an acoustic wave inside a homogeneous material of known dimension and structure. This provides an accurate analysis of the benefit of recent sparse deconvolution for ultrasonic data analysis.

2. Sparse deconvolution as an inverse problem

2.1. Definition of the convolutive model

We use the convolutive observation model

$$y(t) = \sum_{k=1}^{K} x^\star(t_k)\, h(t - t_k) + \varepsilon(t)$$

to describe the propagation of the ultrasonic wave inside a material. h stands for the incident acoustic wave delivered by the input transducer, and t_k and $x^\star(t_k)$ model the time of arrival and the amplitude attenuation corresponding to the k-th echo. The noise $\varepsilon(t)$ includes the measurement noise due to the ultrasonic sensor and the modeling errors. Because the measurements are discrete, we resort to the more realistic formulation

$$y(mT) = \sum_{k=1}^{K} x^\star(t_k)\, h(mT - t_k) + \varepsilon(mT),\ m = 0, \ldots, M-1 \tag{1}$$

where T is the sampling period and M is the number of measurements.

At this point, it is important to notice that t_k are continuous valued. For simplicity reasons, we choose to sample t_k over a grid with sampling period T. This leads to the discrete convolutive model

$$y(mT) = \sum_{n=0}^{N-1} x^\star(nT)\, h((m-n)T) + \varepsilon(mT),\ m = 0, \ldots, M-1 \tag{2}$$

where $x^\star(nT)$ is zero valued for most n, except when nT coincides with the time of arrival of an echo (*i.e.*, a t_k value in (1)). In the following, we use vector notations $\boldsymbol{y} \in \mathbb{R}^M$, $\boldsymbol{h} \in \mathbb{R}^P$ and $\boldsymbol{x}^\star \in \mathbb{R}^N$ to refer to the discrete signals resulting from the sampling of $y(t)$, $h(t)$ and $x(t)$. Because the convolutive model (1) is causal, we set $M = N + P - 1$ so that any echo resulting from $\boldsymbol{x}^\star$ is fully observable in signal $\boldsymbol{y}$. In matrix form, model (2) reads

$$\boldsymbol{y} = \boldsymbol{h} * \boldsymbol{x}^\star + \boldsymbol{\varepsilon} \tag{3}$$

where $*$ stands for the discrete convolution operator.

2.2. *Inversion of the convolutive model*

The sparse deconvolution problem consists in estimating the signal $\boldsymbol{x}^\star$ knowing $\boldsymbol{h}$ and the observation $\boldsymbol{y}$. In practice, $\boldsymbol{h}$ is the emitted ultrasonic wave and $\boldsymbol{y}$ is the recorded signal. For the sake of clarity, $\boldsymbol{x}^\star$ denotes the exact (unknown) sparse signal and we will denote by $\boldsymbol{x}$ an estimated version of $\boldsymbol{x}^\star$ deduced from $\boldsymbol{y}$ and $\boldsymbol{h}$.

As many other inverse problems, our deconvolution problem is *ill-posed* since $\boldsymbol{h}$ plays the role of a filter of limited bandpass. This means that there exist many signals $\boldsymbol{x}$ such that $\boldsymbol{h} * \boldsymbol{x}$ is close to the measured data $\boldsymbol{y}$, some of them being very different from the true solution $\boldsymbol{x}^\star$. The least-squares estimation of $\boldsymbol{x}^\star$ consists in minimising the criterion $\|\boldsymbol{y} - \boldsymbol{h} * \boldsymbol{x}\|^2$, where $\|\,.\|^2$ stands for the usual ℓ_2 norm defined as the sum of squares. However, this basic approach ignores the ill-posed character of the deconvolution problems. It provides solutions that are highly sensitive to the noise embedded in the data [2]. To obtain more stable solutions, it is necessary to introduce appropriate *regularisation* by imposing some constraints on $\boldsymbol{x}$ [13].

In the present context, it is very natural to impose a sparsity constraint on $\boldsymbol{x}$, *i.e.*, that $\boldsymbol{x}$ has only a few non-zero values. This can be done by solving the constrained least-squares problem

$$\min_{\boldsymbol{x}} \|\boldsymbol{y} - \boldsymbol{h} * \boldsymbol{x}\|^2 \quad \text{subject to} \quad \sum_n \varphi(x_n) \leqslant \lambda, \tag{4}$$

where the constraint is designed to favor sparse solutions and the parameter λ controls the compromise between fidelity to data and the level of sparsity. Several choices are available for φ. In the following, we focus on ℓ_0 and ℓ_1 regularisations.

2.3. *ℓ_0 deconvolution*

The ℓ_0 cost function is defined in one dimension as

$$|t|_0 = \begin{cases} 0 \text{ if } t = 0, \\ 1 \text{ otherwise.} \end{cases} \tag{5}$$

By extension, the ℓ_0 pseudo-norm of a vector $\boldsymbol{x}$ is defined as $\|\boldsymbol{x}\|_0 = \sum_m |x_m|_0$. It counts the number of non-zero elements in $\boldsymbol{x}$.

When setting $\varphi(t) = |t|_0$, problem (4) equivalently reads

$$\min_{\boldsymbol{x}} \|\boldsymbol{y} - \boldsymbol{h} * \boldsymbol{x}\|^2 \quad \text{subject to} \quad \|\boldsymbol{x}\|_0 \leqslant K, \tag{6}$$

where $\lambda = K$ is the maximum number of spikes allowed in the signal $\boldsymbol{x}$. (6) is actually a combinatorial optimisation problem over $\{0,1\}^M$ because it consists in finding the support of $\boldsymbol{x}$ of dimension K yielding the least residual error, and there are a finite but large number of configurations with K spikes. Unfortunately, such a problem cannot be solved exactly unless (nearly) all solutions are tested. In practice, ℓ_0 deconvolution methods only explore a subset of all solutions, on the basis of heuristic rules. They are thus *suboptimal*, in the sense that they are not guaranteed to find the true solution of (6).

2.4. ℓ_1 deconvolution

In the sparse signal approximation literature, regularisation using the ℓ_1 norm has become very popular because it leads to convex optimisation problems that can be exactly solved in contrast to their ℓ_0 counterparts, while also yielding sparse solutions.

The ℓ_1 norm is defined as $\|\boldsymbol{x}\|_1 = \sum_m |x_m|$, so the ℓ_1 deconvolution problem can be formulated as

$$\min_{\boldsymbol{x}} \|\boldsymbol{y} - \boldsymbol{h} * \boldsymbol{x}\|^2 \quad \text{subject to} \quad \sum_m |x_m| \leqslant \lambda. \tag{7}$$

The regularisation parameter λ does not stand for the number of spikes anymore. It represents the upper bound on the sum of amplitudes of the signal $\boldsymbol{x}$. However, the lower λ, the larger the number K of spikes in $\boldsymbol{x}$, so that there is an implicit relation between λ and K.

3. Algorithms

This section firstly introduces the algorithms we use to address the ℓ_0 and ℓ_1 deconvolution problems. Their outputs are denoted by $\boldsymbol{x}^0(K)$ and $\boldsymbol{x}^1(\lambda)$, respectively. In a second part, the issue of automatic selection of the regularisation parameter is dealt with, which amounts to determining the model order $K = \|\boldsymbol{x}\|_0$. Our approach can address both ℓ_0 and ℓ_1 cases.

3.1. ℓ_0 deconvolution algorithm

For reasons that will become clear in Subsection 3.3, we preferentially consider *greedy* iterative algorithms to address (6). Such numerical schemes provide a set of solutions $\boldsymbol{x}^0(K)$ for all cardinalities $K = 0, 1, \dots, K_{\max}$ where $K_{\max}$ is a predefined maximal cardinality. *Forward* algorithms start from an empty support and then incrementally introduce new elements (*i.e.*, new spikes) in the support and re-estimate the amplitudes $\boldsymbol{x}$ whenever the support is modified [14, 15].

In deconvolution problems where many echoes frequently overlap, the forward approach is not fully satisfactory: typically, in the presence of two overlapping echoes of similar amplitudes, it is likely to firstly position a false spike in between the two true positions, and then to compensate this wrong initial guess by positioning additional false spikes.

Therefore, we rather favor *forward-backward* algorithms, *i.e.*, strategies that enable both the insertion and the removal of a spike at each iteration. A spike insertion or removal will be termed as a *single replacement*. In the following, we use the Single Best Replacement algorithm (SBR) algorithm to perform ℓ_0 deconvolution. The SBR algorithm is of the *forward-backward* type [9]. It is an extension of the Single Most Likely Replacement (SMLR) algorithm proposed in the context of Bernoulli-Gaussian deconvolution [3–6, 16]. The SMLR algorithm is acknowledged to be very efficient for sparse deconvolution. The difference between SBR and SMLR is that SBR addresses the problem (6) while SMLR is dedicated to a slightly different problem where the regularisation also includes a penalty on the range of non-zero amplitudes x_m.

3.2. ℓ_1 deconvolution algorithm

Although the regularisation parameter λ is continuous valued, ℓ_1 regularisation is very attractive because efficient algorithms are available to compute the solution $\boldsymbol{x}^1(\lambda)$ of (7) for a continuous range of λ values ($\lambda \in [\lambda_{\min}, +\infty)$). The principle of the *homotopy* algorithm is to gradually decrease λ from $+\infty$ to $\lambda_{\min}$ by recursively computing the critical λ values for which the support of $\boldsymbol{x}^1(\lambda)$ is changing. For any critical λ-value, a series of single spike replacements are performed similar to the single replacements occurring in the forward-backward algorithms introduced above [17, 18]. Up to the storage of the intermediate spike replacements occurring during the homotopy iterations, homotopy not only yields a sequence of solutions $\boldsymbol{x}^1(\lambda)$ over the continuous range $\lambda \in [\lambda_{\min}, +\infty)$, but also a series of sparse approximations $\boldsymbol{x}^1(K)$ for all cardinalities $K = 0, 1, \ldots, K_{\max}$ where $K_{\max}$ is a predefined maximum cardinality.

3.3. Model order selection

Model order selection is a well known problem for which many statistical and heuristic rules have been proposed. We refer the reader to [19] for a review of the main information criterion rules and to [12] for their application and adaptation to sparse approximation problems.

In the sparse deconvolution context, model order selection amounts to choose the cardinality K (*i.e.*, the ℓ_0 norm of the solution) which corresponds to selecting a presumed "most acceptable" trade-off between a small least-squares criterion and a high degree of sparsity. Here, we propose to take advantage from the fact that both SBR and homotopy algorithms naturally yield a series of sparse solutions, respectively $\boldsymbol{x}^0(K)$ and $\boldsymbol{x}^1(K)$, for all cardinalities between 0 and $K_{\max}$. Therefore, the best trade-off can be found by direct inspection of $J(\boldsymbol{x}^0(K))$ or $J(\boldsymbol{x}^1(K))$, $K = 0, 1, \ldots, K_{\max}$, as soon as an order selection criterion J has been selected.

Two main categories can be distinguished among the most popular rules for order selection.

The first category proceeds by minimising a compound criterion that directly balances fidelity to data and cardinality. This criterion takes the general form

$$\arg\min_K \left\{ J(K) = M \log(\|\boldsymbol{y} - \boldsymbol{h} * \boldsymbol{x}(K)\|^2) + \alpha K \right\} \tag{8}$$

where M stands for the number of measurements (the size of $\boldsymbol{y}$), $\|\boldsymbol{y} - \boldsymbol{h} * \boldsymbol{x}(K)\|^2$ is the error residual obtained with the solution of cardinality K and α is a constant. α is set to 2, $\log M$ and $2 \log \log M$ for the Akaike Information Criterion (AIC) [20], the Minimum Description Length (MDL) [21] and the Hannan and Quinn Criterion [22], respectively.

The second category is the family of cross validation methods (cross validation is also called PRESS statistic for Predicted REsidual Sums of Squares). Their principle is to partition the data $\boldsymbol{y}$ into two complementary subsets, *e.g.*, by leaving out a single observation y_m leading to a reduced observation signal $\boldsymbol{y}^{[m]}$ of size $M - 1$. Cross validation consists in testing whether the estimation $\boldsymbol{x}(\boldsymbol{y}^{[m]}, K)$ of $\boldsymbol{x}$ knowing the reduced data signal using the sparse approximation algorithm leads to an accurate prediction of observation y_m [23]. Formally, cross validation aims at minimising a criterion of the form

$$J(K) = \frac{1}{M} \sum_m \left(y_m - \hat{y}_m^{[m]}(K) \right)^2 \tag{9}$$

where $\hat{y}_m^{[m]}(K)$ stands for the m-th entry of $\boldsymbol{h} * \boldsymbol{x}(\boldsymbol{y}^{[m]}, K)$. The fact that cross validation effectively penalises high-order models can be explained by the phenomenon of *overfitting*: a solution of excessive order will be more sensitive to noise, so it will not produce a good cross validation score. Generalised cross validation is an improved version where the prediction errors $(y_m - \hat{y}_m^{[m]}(K))^2$ are weighted by a factor depending on K [24]. The special case of sparse approximation algorithms leads to simplified expressions and computation of the cross validation criteria. The reader interested in further details is referred to the book chapters [25, Chapter 5] and [12].

In the simulations below, we compare several automatic order selection rules for the sparse deconvolution problem. We obtain that most of them lead to an over-estimation of the number of spikes and that the MDL criterion yields the best results. For further information regarding MDL, the reader is referred to the comprehensive references [26, 27].

4. Numerical simulations

The goal of the presented simulations is to illustrate that ℓ_0 and ℓ_1 deconvolution algorithms are well suited to acoustic deconvolution problems and to compare their performance. Comparisons are made in terms of ability to detect the presence of spikes and to perform their localisation in a precise manner. The practical efficiency of the order selection criteria introduced above are also compared in order to automatically evaluate the number of spikes. We first consider a classical spike deconvolution problem and then another problem which is typical of acoustic deconvolution.

4.1. Classical spike deconvolution problem

This simulation is close to that introduced in [3]. It is designed in order to analyse in what respect a given algorithm is able to discriminate close echoes and to detect echoes of small amplitudes which are partly drown either in noise of in larger echoes. Figures 1(a,b) display the wavelet $\boldsymbol{h}$ of size $P = 21$ samples, and the simulated data $\boldsymbol{y} = \boldsymbol{h} * \boldsymbol{x}^\star + \boldsymbol{\varepsilon}$ corresponding to $\|\boldsymbol{x}^\star\|_0 = 19$ spikes. The signal-to-noise ratio (SNR) is set to 20 dB. Obviously, there are very close spikes which result in overlapping echoes.

The outputs of the ℓ_0 and ℓ_1 deconvolution algorithms are displayed on figures 1(c,d), respectively, where the number of spikes has been estimated according to the MDL selection criterion. Both algorithms perform accurate detection and localisation of the main spikes. They clearly outperform the algorithms based on the computation of the inter-correlation between $\boldsymbol{y}$ and $\boldsymbol{h}$ such as matched filtering [2]. The zoom in of figures 1(e,f) shows that ℓ_0 deconvolution yields the most accurate results since the spike localisation is obviously better and the signal approximation $\boldsymbol{y} \approx \boldsymbol{h} * \boldsymbol{x}$ is more precise than the approximation obtained with ℓ_1 deconvolution. Figure 2 illustrates the performance of both deconvolution algorithms for all sparsity levels K. Both curves represent the approximation error $\|\boldsymbol{y} - \boldsymbol{h} * \boldsymbol{x}(K)\|^2$ versus the cardinality K. The ℓ_0 curve always lays below the ℓ_1 curve, therefore indicating that ℓ_0 deconvolution leads to better data approximations for any cardinality. Moreover, it is interesting to notice that the ℓ_0 curve lays below the star corresponding to the true solution $\boldsymbol{x}^\star$. This indicates that the ℓ_0 solution of cardinality 19 leads to a residual error which is lower than the noise variance.

In table 1, we test classical order selection criteria. The general trend is that AIC and cross validation tend to over-estimate the number of spikes. For the simulation of figure 1, MDL provides the most realistic order selection since it correctly estimates the true number of spikes (19) for ℓ_0 deconvolution and it yields the number which is the closest to the true number for ℓ_1 deconvolution. For higher noise levels (SNR = 0 dB), MDL under-estimates the number of spikes. This makes sense because the spikes of small amplitudes are totally drowned in noise, justifying that sparse deconvolution algorithms cannot detect them.

Because ℓ_0 deconvolution yields the best results, we set apart ℓ_1 deconvolution and we only consider the ℓ_0 algorithm in the second simulation below.

Journal of Physics: Conference Series **353** (2012) 012018 doi:10.1088/1742-6596/353/1/012018

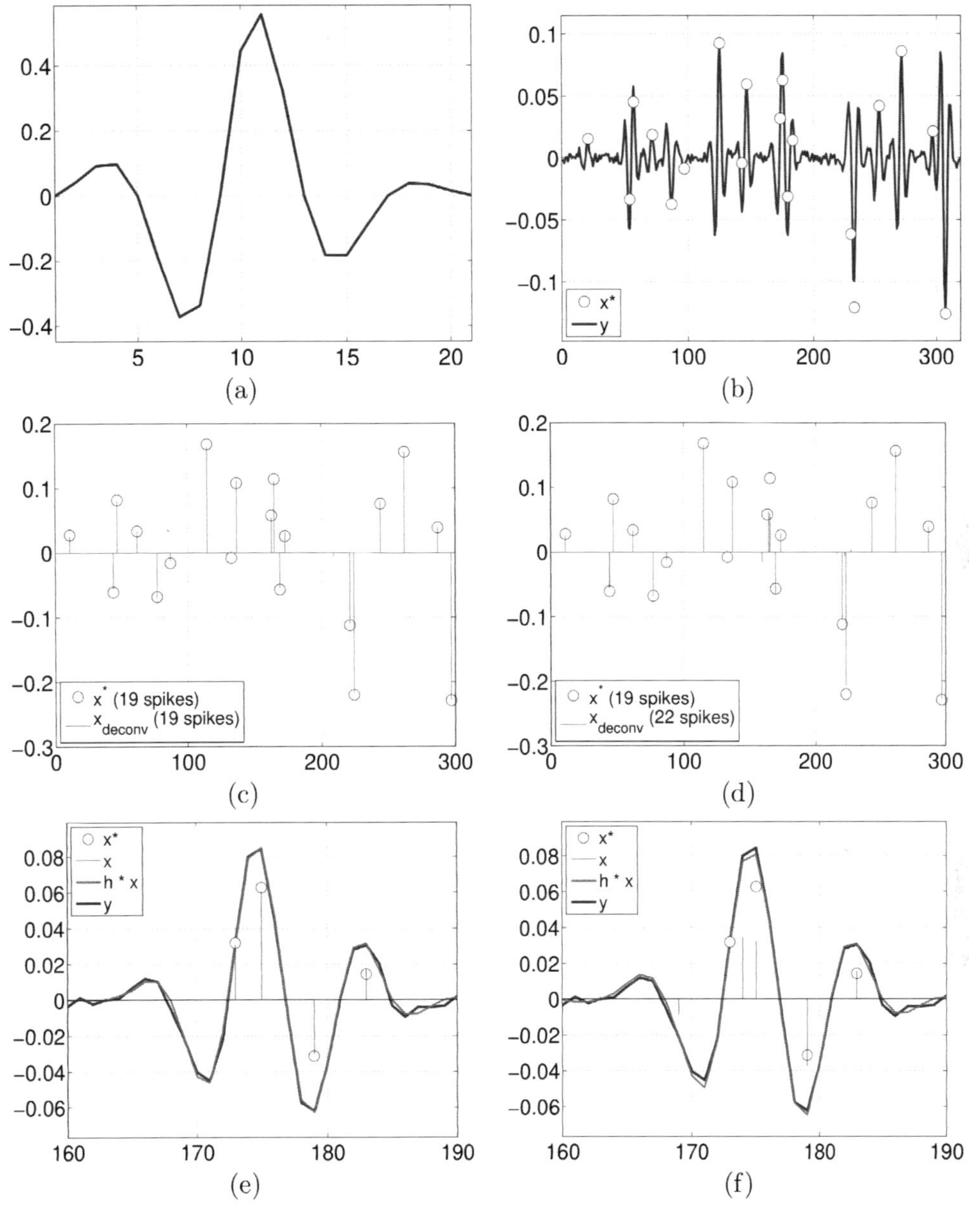

Figure 1. First simulation. (a) Wavelet signal $\boldsymbol{h}$, of size 21. (b) Display of the simulated data $\boldsymbol{y} = \boldsymbol{h} * \boldsymbol{x}^\star + \boldsymbol{\varepsilon}$ together with the true sparse signal $\boldsymbol{x}^\star$ (○). $\boldsymbol{y}$ and $\boldsymbol{x}^\star$ are of size 320 and 300, respectively. $\boldsymbol{x}^\star$ is composed of 19 spikes. The SNR is set to 20 dB. (c) Recovery of $\boldsymbol{x}$ using the ℓ_0 deconvolution method. (d) Recovery of $\boldsymbol{x}$ using the ℓ_1 deconvolution method. (e,f) Data approximations $\boldsymbol{y} \approx \boldsymbol{h} * \boldsymbol{x}$ (zoom in) using the outputs of the ℓ_0 and ℓ_1 deconvolution algorithms, respectively. The number of spikes is automatically estimated using the MDL criterion.

4.2. Processing of simulated acoustic data

This simulation is more specifically dedicated to the acoustic wave propagation problem. We simulate the propagation of the wavelet signal shown on figure 3 through an homogeneous isotropic aluminium material of known geometry according to the propagation model described

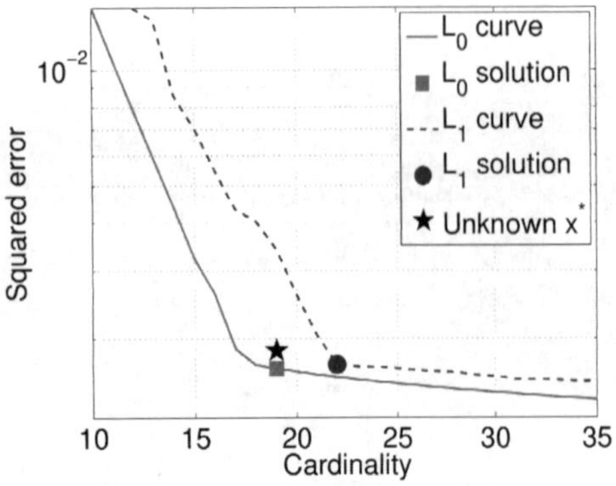

Figure 2. Complement to the simulation of figure 1: curves showing the outputs of the ℓ_0 and ℓ_1 deconvolution algorithms in terms of compromise between approximation error and cardinality. The ℓ_0 and ℓ_1 solutions (labeled ■ and ●) are obtained according to the MDL order selection criterion. The star label represents the cardinality of the true (unknown) solution $\boldsymbol{x}^\star$ and the related approximation error $\|\boldsymbol{h} * \boldsymbol{x}^\star\|^2$ corresponding to the noise power.

Table 1. Comparison of classical order selection criteria: Cross Validation (CV), Generalised Cross Validation (GCV), Akaike Information Criterion (AIC, AICs), Minimum Description Length (MDL, MDLs), Hannan-Quinn information Criterion (HQC). AICs and MDLs are modified versions of AIC and MDL criteria. They were derived to improve AIC and MDL when the number of observations is relatively small [28]. Tests are done on the simulation of figure 1 with moderate noise level (SNR = 20) and on a similar simulation with higher noise level (0 dB).

	CV	GCV	AIC	AICs	**MDL**	MDLs	HQC
SNR = 20 dB							
ℓ_0 deconvolution	41	39	39	28	**19**	18	26
ℓ_1 deconvolution	31	31	31	23	**22**	22	22
SNR = 0 dB							
ℓ_0 deconvolution	45	42	42	21	**11**	9	17
ℓ_1 deconvolution	29	30	30	15	**14**	14	15

in [29]. Obviously, the output signal resulting from various reflections of the wave at the edges of the material, has a content which highly depends on the incident angle and the dimension of the block of aluminium. We consider the case where the incident angle is normal to the material surface and the material thickness (2 mm) is smaller than the longitudinal wavelength (2.92 mm), so that the echoes strongly overlap in the data signal. Figures 4(a,c) display the data simulated for two different noise levels (SNR = 20 and 10 dB). It is noticeable that the echo amplitudes are decreasing with time due to the signal attenuation while the wave is propagating through the material. The ℓ_0 deconvolution method is able to perfectly detect and localise most echoes for SNR = 20 dB without any false detection. For a higher level of noise (SNR = 10 dB), the first main echoes are correctly localised but possible false detections are likely to occur for the smallest echoes which are drown in noise. The MDL selection criterion under-estimates the number of spikes (17 and 19 spikes are found instead of 22) since the smallest spikes cannot be detected. We believe that it is always desirable to under-estimate the number of spikes rather than detecting more spikes together with false alarms.

Finally, let us stress that the presented deconvolution methods are not computationally demanding. For instance, the results of figures 1 and 4 have been obtained in about 0.25

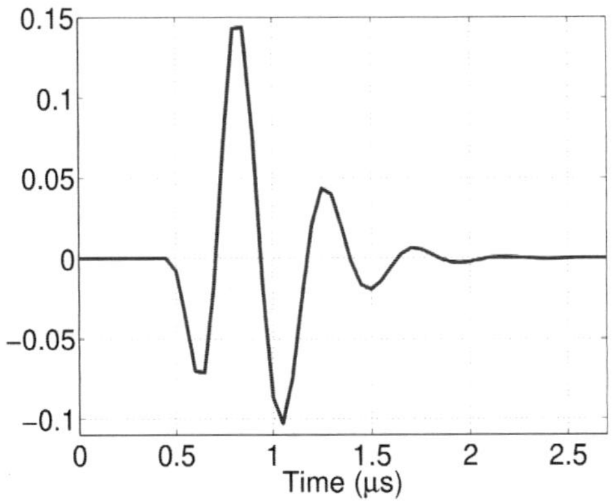

Figure 3. Realistic acoustic wavelet of size 100. It is a typical ultrasound signal propagating through an aluminium object.

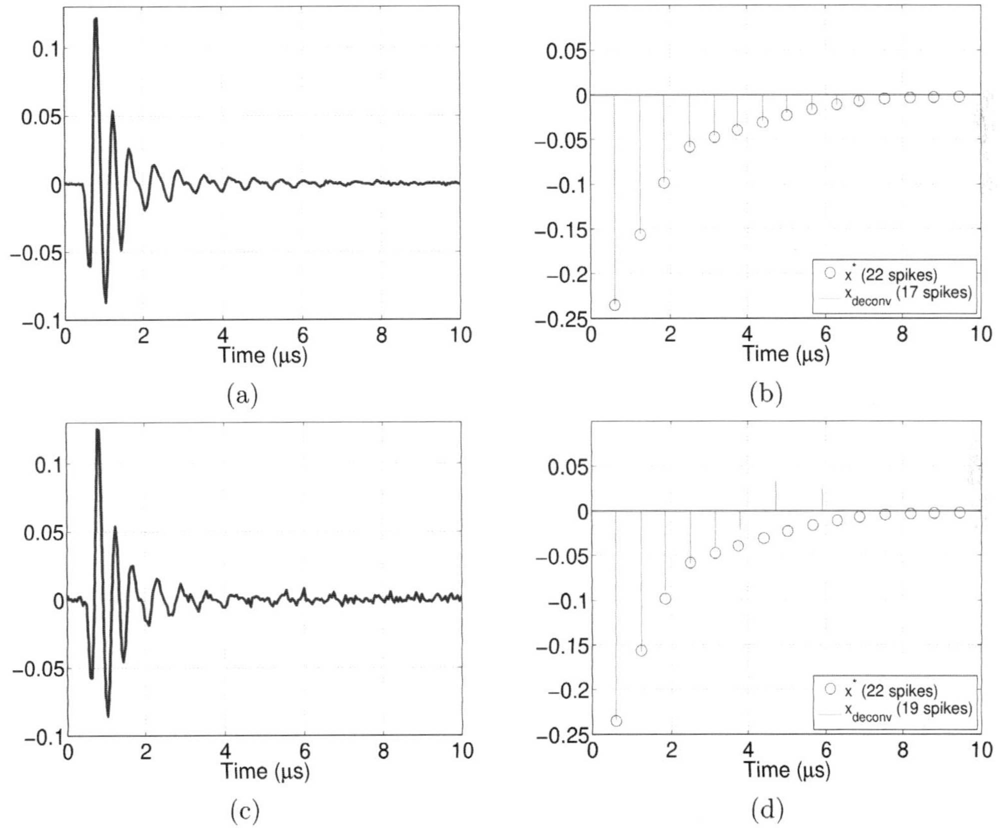

Figure 4. Deconvolution of acoustic simulated data for two noise levels using the ℓ_0 pseudo-norm deconvolution algorithm. The data signal $\boldsymbol{y}$ (of size 2048) corresponds to the propagation of the wavelet displayed in figure 3 through an homogeneous block of aluminium of depth 2 mm. The incident angle is normal to the block surface. When the noise is low (SNR = 20 dB, simulation (a)), the deconvolution algorithm performs accurate estimation of the spike locations (subfigure (b)). For more noisy data (SNR = 10 dB, simulation (c)), the algorithm still locates the main echoes in a precise manner (subfigure (d)). Further echoes cannot be detected since the useful signal is almost completely drowned in noise.

and 3 seconds, respectively, using a code written in Matlab on a laptop PC, Intel Core 2 Duo CPU 1.4 GHz.

5. Conclusions

Sparse deconvolution methods based on ℓ_0 and ℓ_1 regularisations are efficient tools to process acoustic signals resulting from wave propagation inside a material. The accurate detection and localisation of the time of arrival of the acoustic echoes is critical for detecting and localising the inhomogeneities (change of structures, faults) of the insonified object. For the acoustic data we simulated, we considered a simple, homogeneous material of small thickness. However, the method can also be applied to objects of more complex geometry.

In the near future, our goal will be to consider dispersive and/or attenuative cases. In such cases, the shape of the wavefront evolves as the wave propagates, so that proper adaptation of the wavefront must be performed within the deconvolution method, *i.e.*, deconvolution must become *adaptive*. Olofsson and Stepinski's work [30] is a pioneering contribution in this regard. A source of improvement could be to consider a physical attenuation (or dispersion) model, rather than a signal processing-oriented one. Another one would be to address the problem using the tools of sparse deconvolution introduced in the present paper.

References

[1] Van Trees H L 1968 *Detection, estimation and modulation theory* Part 1 (New York, NY: John Wiley)
[2] Champagnat F, Goussard Y, Gautier S and Idier J 2008 *Bayesian Approach to Inverse Problems* ed Idier J (ISTE Ltd and John Wiley & Sons Inc) pp 117–140
[3] Kormylo J J and Mendel J M 1982 *IEEE Trans. Inf. Theory* **28** 482–488
[4] Mendel J M 1983 *Optimal Seismic Deconvolution* (New York, NY: Academic Press)
[5] Champagnat F, Goussard Y and Idier J 1996 *IEEE Trans. Signal Processing* **44** 2988–2998
[6] Benammar A, Drai R and Guessoum A 2008 *Int. J. Simul. Multidisci. Des. Optim.* **2** 107–111
[7] Taylor H, Banks S and McCoy F 1979 *Geophysics* **44** 39–52
[8] Lorenz D A and Trede D 2009 *Signal Processing with Adaptive Sparse Structured Representations (SPARS workshop)* (Saint-Malo, France) pp 1–6
[9] Soussen C, Idier J, Brie D and Duan J 2010 From Bernoulli-Gaussian deconvolution to sparse signal restoration Tech. rep. Centre de Recherche en Automatique de Nancy
[10] Elad M 2010 *Sparse and Redundant Representations: From Theory to Applications in Signal and Image Processing* (Springer)
[11] Duan J, Soussen C, Brie D and Idier J 2009 *Signal Processing with Adaptive Sparse Structured Representations (SPARS workshop)* (Saint-Malo, France) pp 1–6
[12] Wang Y 2004 *Handbook of Computational Statistics* vol 1 ed Gentle J E, Härdle W and Mori Y (Berlin, Germany: Springer-Verlag) pp 437–466
[13] Idier J (ed) 2008 *Bayesian Approach to Inverse Problems* (ISTE Ltd and John Wiley & Sons Inc)
[14] Pati Y C, Rezaiifar R and Krishnaprasad P S 1993 *Proceedings of 27th Asilomar Conference on Signals, Systems and Computers* vol 1 pp 40–44
[15] Chen S S, Donoho D L and Saunders M A 1998 *SIAM J. Sci. Comput.* **20** 33–61
[16] Goussard Y, Demoment G and Idier J 1990 *Proc. IEEE ICASSP* (Albuquerque, NM) pp 1547–1550
[17] Efron B, Hastie T, Johnstone I and Tibshirani R 2004 *Annals Statist.* **32** 407–451
[18] Donoho D L and Tsaig Y 2008 *IEEE Trans. Inf. Theory* **54** 4789–4812
[19] Stoica P and Selén Y 2004 *IEEE Trans. Signal Processing Mag.* **21** 36–47
[20] Akaike H 1974 *IEEE Trans. Automat. Contr.* **AC-19** 716–723
[21] Rissanen J 1983 *Annals Statist.* **11** 416–431
[22] Hannan E J and Quinn B G 1979 *J. R. Statist. Soc. B* **41** 190–195
[23] Wahba G 1977 *SIAM J. Num. Anal.* **14** 651–667
[24] Golub G H, Heath M and Wahba G 1979 *Technometrics* **21** 215–223
[25] Miller A J 2002 *Subset Selection in Regression* 2nd ed (London, UK: Chapman and Hall)
[26] Hansen M H and Yu B 2001 *J. Acoust. Society America* **96** 746–774
[27] Grünwald P D (ed) 2007 *The Minimum Description Length Principle* MIT Press Books (The MIT Press)
[28] de Ridder F, Pintelon R, Schoukens J and Gillikin D P 2005 *IEEE Trans. Instrum. and Meas.* **54** 144–150
[29] Potel C, de Belleval J F and Gargouri Y 1995 *J. Acoust. Society America* **97** 2815–2825
[30] Olofsson T and Stepinski T 2001 *J. Acoust. Society America* **109** 2831–2839

AUTHOR INDEX